Student Solutions Manual

to accompany

College Physics

Second Edition

Alan Giambattista
Cornell University

Betty McCarthy Richardson
Cornell University

Robert C. Richardson
Cornell University

Prepared by
Bill Fellers
Fellers Math & Science

Boston Burr Ridge, IL Dubuque, IA Madison, WI New York San Francisco St. Louis
Bangkok Bogotá Caracas Kuala Lumpur Lisbon London Madrid Mexico City
Milan Montreal New Delhi Santiago Seoul Singapore Sydney Taipei Toronto

The McGraw-Hill Companies

Student Solutions Manual to accompany
COLLEGE PHYSICS, SECOND EDITION
ALAN GIAMBATTISTA, BETTY MCCARTHY RICHARDSON, AND ROBERT C. RICHARDSON

Published by McGraw-Hill Higher Education, an imprint of The McGraw-Hill Companies, Inc., 1221 Avenue of the Americas, New York, NY 10020.

This book is printed on recycled, acid-free paper containing 10% postconsumer waste.

1 2 3 4 5 6 7 8 9 0 QPD / QPD 0 9 8 7 6 5

ISBN-13: 978-0-07-304954-0
ISBN-10: 0-07-304954-9

www.mhhe.com

Student Solutions Manual
to accompany
COLLEGE PHYSICS
Second Edition

Table of Contents

Chapter 1

INTRODUCTION

Problems

1. Strategy Relate the surface area S to the radius r using $S = 4\pi r^2$.

Solution Find the ratio of the new radius to the old.

$S_1 = 4\pi r_1^2$ and $S_2 = 4\pi r_2^2 = 1.160S_1 = 1.160(4\pi r_1^2)$.

$$4\pi r_2^2 = 1.160(4\pi r_1^2)$$
$$r_2^2 = 1.160 r_1^2$$
$$\left(\frac{r_2}{r_1}\right)^2 = 1.160$$
$$\frac{r_2}{r_1} = \sqrt{1.160} = 1.077$$

The radius of the balloon increases by $\boxed{7.7\%}$.

5. Strategy The new fence will be $100\% + 37\% = 137\%$ of the height of the old fence.

Solution Find the height of the new fence.

$1.37 \times 1.8\text{ m} = \boxed{2.5\text{ m}}$

7. Strategy The area of the circular garden is given by $A = \pi r^2$. Let the original and final areas be $A_1 = \pi r_1^2$ and $A_2 = \pi r_2^2$, respectively.

Solution Calculate the percentage increase of the area of the garden plot.

$$\frac{\Delta A}{A} \times 100\% = \frac{\pi r_2^2 - \pi r_1^2}{\pi r_1^2} \times 100\% = \frac{r_2^2 - r_1^2}{r_1^2} \times 100\% = \frac{1.25^2 r_1^2 - r_1^2}{r_1^2} \times 100\% = \frac{1.25^2 - 1}{1} \times 100\% = \boxed{56\%}$$

9. (a) Strategy Rewrite the numbers so that the power of 10 is the same for each. Then add and give the answer with the number of significant figures determined by the less precise of the two numbers.

Solution Perform the operation with the appropriate number of significant figures.

$3.783 \times 10^6\text{ kg} + 1.25 \times 10^8\text{ kg} = 0.03783 \times 10^8\text{ kg} + 1.25 \times 10^8\text{ kg} = \boxed{1.29 \times 10^8\text{ kg}}$

(b) Strategy Find the quotient and give the answer with the number of significant figures determined by the number with the fewest significant figures.

Solution Perform the operation with the appropriate number of significant figures.

$(3.783 \times 10^6\text{ m}) \div (3.0 \times 10^{-2}\text{ s}) = \boxed{1.3 \times 10^8\text{ m/s}}$

13. Strategy Multiply and give the answer in scientific notation with the number of significant figures determined by the number with the fewest significant figures.

Solution Solve the problem.

$$(3.2\text{ m})\times(4.0\times10^{-3}\text{ m})\times(1.3\times10^{-8}\text{ m}) = \boxed{1.7\times10^{-10}\text{ m}^3}$$

17. Strategy There are 3600 seconds in one hour and 1000 m in one kilometer.

Solution Convert 1.00 kilometers per hour to meters per second.

$$\frac{1.00\text{ km}}{1\text{ h}}\times\frac{1\text{ h}}{3600\text{ s}}\times\frac{1000\text{ m}}{1\text{ km}} = \boxed{0.278\text{ m/s}}$$

21. Strategy There are 1000 grams in one kilogram and 100 centimeters in one meter.

Solution Find the density of mercury in units of g/cm^3.

$$\frac{1.36\times10^4\text{ kg}}{1\text{ m}^3}\times\frac{1000\text{ g}}{1\text{ kg}}\times\left(\frac{1\text{ m}}{100\text{ cm}}\right)^3 = \boxed{13.6\text{ g/cm}^3}$$

25. (a) Strategy There are 12 inches in one foot, 2.54 centimeters in one inch, and 60 seconds in one minute.

Solution Express the snail's speed in feet per second.

$$\frac{5.0\text{ cm}}{1\text{ min}}\times\frac{1\text{ min}}{60\text{ s}}\times\frac{1\text{ in}}{2.54\text{ cm}}\times\frac{1\text{ ft}}{12\text{ in}} = \boxed{2.7\times10^{-3}\text{ ft/s}}$$

(b) Strategy There are 5280 feet in one mile, 12 inches in one foot, 2.54 centimeters in one inch, and 60 minutes in one hour.

Solution Express the snail's speed in miles per hour.

$$\frac{5.0\text{ cm}}{1\text{ min}}\times\frac{60\text{ min}}{1\text{ h}}\times\frac{1\text{ in}}{2.54\text{ cm}}\times\frac{1\text{ ft}}{12\text{ in}}\times\frac{1\text{ mi}}{5280\text{ ft}} = \boxed{1.9\times10^{-3}\text{ mi/h}}$$

29. Strategy Replace each quantity in $T^2 = 4\pi^2 r^3/(GM)$ with its dimensions.

Solution Show that the equation is dimensionally correct.

T^2 has dimensions $[\text{T}]^2$ and $\dfrac{4\pi^2 r^3}{GM}$ has dimensions $\dfrac{[\text{L}]^3}{\dfrac{[\text{L}]^3}{[\text{M}][\text{T}]^2}\times[\text{M}]} = \dfrac{[\text{L}]^3}{[\text{M}]}\times\dfrac{[\text{M}][\text{T}]^2}{[\text{L}]^3} = [\text{T}]^2.$

Since $\boxed{[\text{T}]^2 = [\text{T}]^2}$, the equation is dimensionally correct.

33. (a) Strategy and Solution The mass of the lower leg is about 5 kg and that of the upper leg is about 7 kg, so an order of magnitude estimate of the mass of a person's leg is $\boxed{10\text{ kg}}$.

(b) Strategy and Solution The length of a full size school bus is greater than 1 m and less than 100 m, so an order of magnitude estimate of the length of a full size school bus is $\boxed{10\text{ m}}$.

37. Strategy One story is about 3 m high.

Solution Find the order of magnitude of the height in meters of a 40-story building.

$$(3\text{ m})(40) \sim \boxed{100\text{ m}}$$

41. Strategy Use the two temperatures and their corresponding times to find the rate of temperature change with respect to time (the slope of the graph of temperature vs. time). Then, write the linear equation for the temperature with respect to time and find the temperature at 3:35 P.M.

Solution Find the rate of temperature change.

$$m=\frac{\Delta T}{\Delta t}=\frac{101.0°\text{F}-97.0°\text{F}}{4.0\text{ h}}=1.0°\text{F/h}$$

Use the slope-intercept form of a graph of temperature vs. time to find the temperature at 3:35 P.M.

$$T=mt+T_0=(1.0\ °\text{F/h})(3.5\text{ h})+101.0°\text{F}=\boxed{104.5°\text{F}}$$

43. Strategy Put the equation that describes the line in slope-intercept form, $y = mx + b$.

$$at=v-v_0$$
$$v=at+v_0$$

Solution

(a) v is the dependent variable and t is the independent variable, so $\boxed{a}$ is the slope of the line.

(b) The slope-intercept form is $y = mx + b$. Find the vertical-axis intercept.

$v\leftrightarrow y,\ t\leftrightarrow x,\ a\leftrightarrow m,$ so $v_0\leftrightarrow b.$

Thus, $\boxed{+v_0}$ is the vertical-axis intercept of the line.

45. (a) Strategy Plot the decay rate on the vertical axis and the time on the horizontal axis.

Solution The plot is shown.

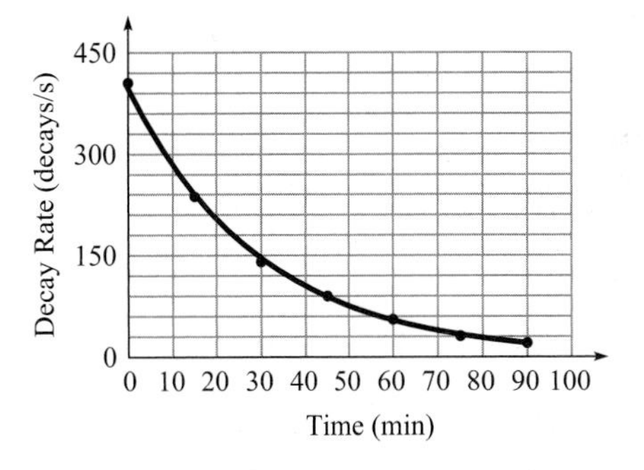

(b) Strategy Plot the natural logarithm of the decay rate on the vertical axis and the time on the horizontal axis.

Solution The plot is shown.

Presentation of the data in this form—as the natural logarithm of the decay rate—might be useful because $\boxed{\text{the graph is linear}}$.

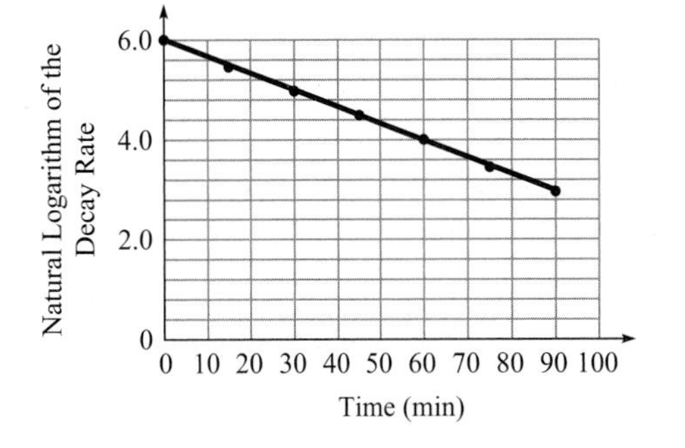

49. Strategy Assuming that the cross section of the artery is a circle, we use the area of a circle, $A = \pi r^2$.

Solution

$A_1 = \pi r_1^2$ and $A_2 = \pi r_2^2 = \pi(2.0r_1)^2 = 4.0\pi r_1^2$.

Form a proportion.

$$\frac{A_2}{A_1} = \frac{4.0\pi r_1^2}{\pi r_1^2} = 4.0$$

The cross-sectional area of the artery increases by a factor of $\boxed{4.0}$.

53. Strategy Use the rules for determining significant figures and for writing numbers in scientific notation.

Solution

(a) 0.00574 kg has three significant figures, 5, 7, and 4. The zeros are not significant, since they are used only to place the decimal point. To write this measurement in scientific notation, we move the decimal point three places to the right and multiply by 10^{-3}.

(b) 2 m has one significant figure, 2. This measurement is already written in scientific notation

(c) 0.450×10^{-2} m has three significant figures, 4, 5, and the 0 to the right of 5. The zero is significant, since it comes after the decimal point and is not used to place the decimal point. To write this measurement in scientific notation, we move the decimal point one place to the right and multiply by 10^{-1}.

(d) 45.0 kg has three significant figures, 4, 5, and 0. The zero is significant, since it comes after the decimal point and is not used to place the decimal point. To write this measurement in scientific notation, we move the decimal point one place to the left and multiply by 10^1.

(e) 10.09×10^4 s has four significant figures, 1, 9, and the two zeros. The zeros are significant, since they are between two significant figures. To write this measurement in scientific notation, we move the decimal point one place to the left and multiply by 10^1.

(f) 0.09500×10^5 mL has four significant figures, 9, 5, and the two zeros to the right of 5. The zeros are significant, since they come after the decimal point and are not used to place the decimal point. To write this measurement in scientific notation, we move the decimal point two places to the right and multiply by 10^{-2}.

The results of parts (a) through (f) are shown in the table below.

	Measurement	**Significant Figures**	**Scientific Notation**
(a)	0.00574 kg	3	5.74×10^{-3} kg
(b)	2 m	1	2 m
(c)	0.450×10^{-2} m	3	4.50×10^{-3} m
(d)	45.0 kg	3	4.50×10^1 kg
(e)	10.09×10^4 s	4	1.009×10^5 s
(f)	0.09500×10^5 mL	4	9.500×10^3 mL

57. Strategy The circumference of a viroid is approximately 300 times 0.35 nm. The diameter is given by $C = \pi d$, or $d = C/\pi$.

Solution Find the diameter of the viroid in the required units.

(a) $d = \dfrac{300(0.35\text{ nm})}{\pi} \times \dfrac{10^{-9}\text{ m}}{1\text{ nm}} = \boxed{3.3\times10^{-8}\text{ m}}$

(b) $d = \dfrac{300(0.35\text{ nm})}{\pi} \times \dfrac{10^{-3}\,\mu\text{m}}{1\text{ nm}} = \boxed{3.3\times10^{-2}\,\mu\text{m}}$

(c) $d = \dfrac{300(0.35\text{ nm})}{\pi} \times \dfrac{10^{-7}\text{ cm}}{1\text{ nm}} \times \dfrac{1\text{ in}}{2.54\text{ cm}} = \boxed{1.3\times10^{-6}\text{ in}}$

61. (a) Strategy a has dimensions $\frac{[\text{L}]}{[\text{T}]^2}$; v has dimensions $\frac{[\text{L}]}{[\text{T}]}$; r has dimension [L].

Solution If we square v and divide by r, we have $\frac{v^2}{r}$, which implies that $\frac{[\text{L}]^2}{[\text{T}]^2}\cdot\frac{1}{[\text{L}]} = \frac{[\text{L}]}{[\text{T}]^2}$, which are the dimensions for a. Therefore, we can write $\boxed{a = K\frac{v^2}{r}\text{, where } K \text{ is a dimensionless constant}}$.

(b) Strategy Divide the new acceleration by the old, and use the fact that the new speed is 1.100 times the old.

Solution Find the percent increase in the radial acceleration.

$$\frac{a_2}{a_1} = \frac{K\frac{v_2^2}{r}}{K\frac{v_1^2}{r}} = \left(\frac{v_2}{v_1}\right)^2 = \left(\frac{1.100v_1}{v_1}\right)^2 = 1.100^2 = 1.210$$

$1.210 - 1 = 0.210$, so the radial acceleration increases by $\boxed{21.0\%}$.

65. Strategy Since there are hundreds of millions of people in the U.S., a reasonable order-of-magnitude estimate of the number of automobiles is 10^8. There are about 365 days per year; that is, about 10^2. A reasonable estimate of the gallons used per day per person is greater than one, but less than one hundred; that is, 10^1.

Solution Calculate the estimate.

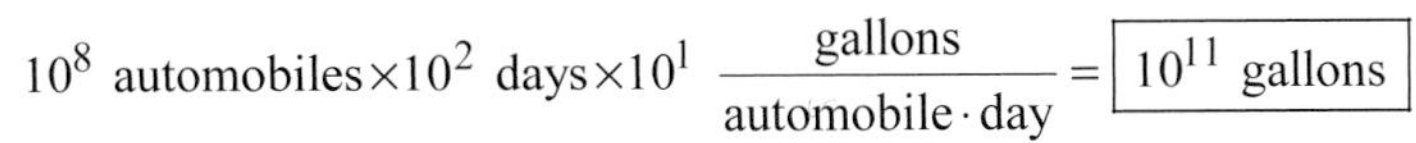

$$10^8\text{ automobiles}\times10^2\text{ days}\times10^1\ \frac{\text{gallons}}{\text{automobile}\cdot\text{day}} = \boxed{10^{11}\text{ gallons}}$$

69. Strategy \$41,000,000,000 has a precision of 1 billion dollars; \$100 has a precision of 100 dollars, so the net worth is the same to one significant figure.

Solution Find the net worth.

$\$41{,}000{,}000{,}000 - \$100 = \boxed{\$41{,}000{,}000{,}000}$

73. (a) Strategy Inspect the units of G, c, and h and use trial-and-error to find the correct combination of these constants.

Solution Through a process of trial and error, we find that the only combination of G, c, and h that has the dimensions of time is $\boxed{\sqrt{\frac{hG}{c^5}}}$.

(b) Strategy Substitute the values of the constants into the formula found in part (a).

Solution Find the time in seconds.

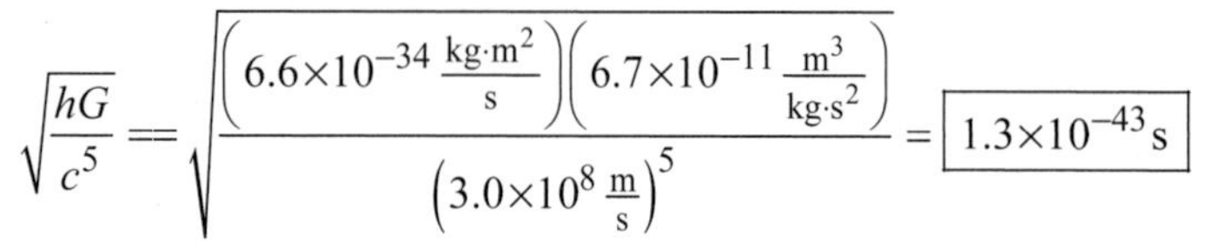

$$\sqrt{\frac{hG}{c^5}} = \sqrt{\frac{\left(6.6\times10^{-34}\ \frac{\text{kg}\cdot\text{m}^2}{\text{s}}\right)\left(6.7\times10^{-11}\ \frac{\text{m}^3}{\text{kg}\cdot\text{s}^2}\right)}{\left(3.0\times10^{8}\ \frac{\text{m}}{\text{s}}\right)^5}} = \boxed{1.3\times10^{-43}\ \text{s}}$$

77. (a) Strategy Plot the data on a graph with mass on the vertical axis and time on the horizontal axis. Then, draw a best-fit smooth curve.

Solution See the graph.

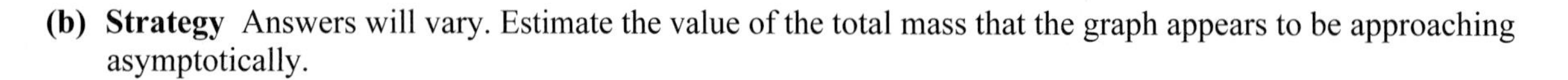

(b) Strategy Answers will vary. Estimate the value of the total mass that the graph appears to be approaching asymptotically.

Solution The graph appears to be approaching asymptotically a maximum value of 100 g, so the carrying capacity is $\boxed{\text{about 100 g}}$.

(c) Strategy Plot the data on a graph with the natural logarithm of m/m_0 on the vertical axis and time on the horizontal axis. Draw a line through the points and find its slope to estimate the intrinsic growth rate.

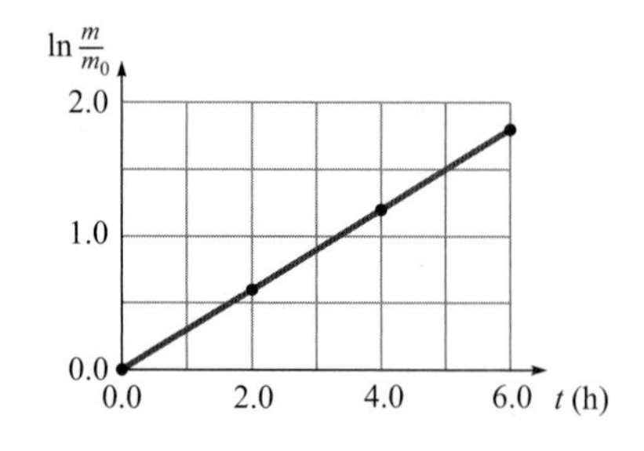

Solution See the graph. From the plot of $\ln\frac{m}{m_0}$ vs. t, the slope r appears to be

$$r = \frac{1.8 - 0.0}{6.0\ \text{s} - 0.0\ \text{s}} = \frac{1.8}{6.0\ \text{s}} = \boxed{0.30\ \text{s}^{-1}}.$$

Chapter 2

FORCE

Problems

1. Strategy Determine the forces *not* acting on the scale.

Solution The scale is in contact with the floor, so a contact force due to the floor is exerted on the scale. The scale is in contact with the person's feet, so a contact force due to the person's feet is exerted on the scale. The scale is in the proximity of a very large mass (Earth), so the weight of the scale is a force exerted on the scale. The weight of the person is a force exerted on the person due to the very large mass, so it is not a force exerted on the scale.

5. Strategy There are 0.2248 pounds per newton.

Solution Find the weight of the astronaut in newtons.

$$175 \text{ lb} \times \frac{1 \text{ N}}{0.2248 \text{ lb}} = \boxed{778 \text{ N}}$$

9. Strategy Use the fact that $|\vec{\mathbf{A}}| = |\vec{\mathbf{B}}|$ and symmetry to determine the direction of $\vec{\mathbf{C}}$; then sketch $\vec{\mathbf{C}}$.

Solution By symmetry, we find that $\vec{\mathbf{C}}$ points downward; the horizontal components cancel when $\vec{\mathbf{A}}$ and $\vec{\mathbf{B}}$ are added. The downward components of each vector have the same magnitude, about 0.7 N. So, the magnitude of $\vec{\mathbf{C}}$ is about 1.4 N. The sketch is shown:

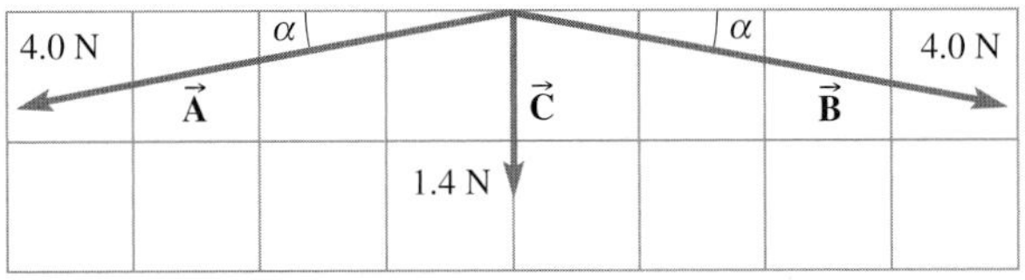

13. Strategy Since the man and mattress are neither moving upward nor downward, the net force must be zero in the vertical direction.

Solution So that the net force is zero, the upward force of the water must be equal to the combined weight of the man and the air mattress, or 806 N.

17. Strategy The force of the lake on the boat must be equal in magnitude and opposite in direction to the weight of the boat. The force of the wind on the boat must be equal in magnitude and opposite in direction to that of the line. Let the subscripts be the following:
s = sailboat e = Earth w = wind l = lake m = mooring line

Solution The free-body diagram is shown.

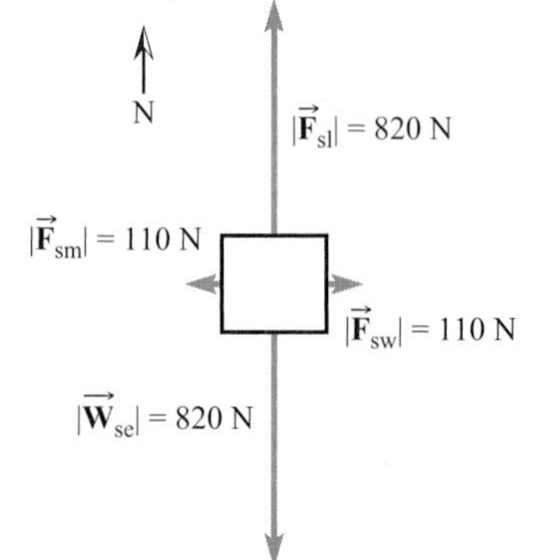

19. Strategy For each object, add the forces to find the net force.

Solution

(a) 10 N left + 40 N right = −10 N right + 40 N right = $\boxed{\text{30 N to the right}}$

(b) The forces balance, so the net force is $\boxed{0}$.

(c) The horizontal forces balance, so the net force is due only to the downward force. The net force is $\boxed{\text{18 N downward}}$.

21. Strategy Since the suitcase is moving at a constant speed, the net force on it must be zero. The force of friction must oppose the force of the pull. So, the force of friction must be equal in magnitude and opposite in direction to the horizontal component of the force of the pull. Draw a free-body diagram to illustrate the situation.

Solution Find the force of friction.

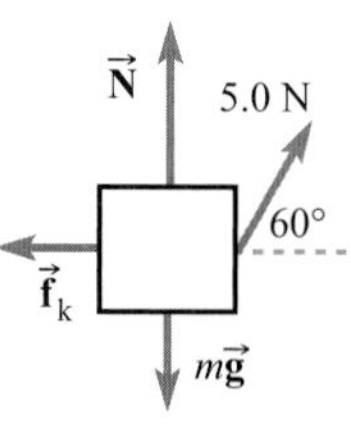

The horizontal component of the pull force is $(5.0\text{ N})\cos 60° = 2.5\text{ N}$. Since the horizontal component of the pull force is equal and opposite to the friction force, the force of friction acting on the suitcase is $\boxed{\text{2.5 N, opposite the direction of motion}}$.

25. Strategy The components of $\vec{\mathbf{a}}$ are given. Since the x-component is negative and the y-component is positive, the vector lies in the second quadrant. Give the angle with respect to the axis to which it lies closest.

Solution Find the magnitude and direction of $\vec{\mathbf{a}}$.

(a) $a = \sqrt{a_x{}^2 + a_y{}^2} = \sqrt{(-3.0\text{ m/s}^2)^2 + (4.0\text{ m/s}^2)^2} = \boxed{5.0\text{ m/s}^2}$

(b) $\theta = \tan^{-1}\dfrac{4.0}{-3.0} = \boxed{37°\text{ CCW from the }+y\text{-axis}}$

29. Strategy Use Newton's laws of motion. Let the y-direction be perpendicular to the canal and the $+x$-direction be parallel to the center line in the direction of motion.

Solution Find the net force on the barge.

$\sum F_y = T\sin 15° - T\sin 15° = 0$ and

$\sum F_x = T\cos 15° + T\cos 15° = 2T\cos 15° = 2(560\text{ N})\cos 15° = 1.1\text{ kN}.$

So, $\vec{\mathbf{F}}_{\text{net}} =$ 1.1 kN forward (along the center line).

Centerline
560 N
560 N
15°
x
y

33. Strategy Consider forces acting on the fish suspended by the line.

Solution

One force acting on the fish is an upward force on the fish by the line; its interaction partner is a downward force on the line by the fish. A second force acting on the fish is the downward gravitational force on the fish; its interaction partner is the upward gravitational force on the Earth by the fish.

35. Strategy Use Newton's first and third laws.

Solution

(a) Margie exerts a downward force on the scale equal to her weight, 543 N. According to the third law, the scale exerts an upward force on Margie equal in magnitude to the magnitude of the force exerted by Margie on it, or 543 N.

(b) Refer to part (a). The interaction partner of the force exerted on Margie by the scale is the contact force of Margie's feet on the scale.

(c) The Earth must hold up both the scale and (indirectly) Margie, since Margie is standing on the scale. So, the Earth must push up on the scale with a force equal to the combined weight of Margie and the scale, or 543 N + 45 N = 588 N.

(d) Refer to part (c). The interaction partner of the force exerted on the scale by the Earth is the contact force on the Earth due to the scale.

37. Strategy Analyze the forces due to and on the three interacting objects: the woman, the chair, and the floor.

Solution

(a) The weight of the woman is directed downward. The forces on the woman due to the seat and armrests are directed upward and total 25 N + 25 N + 500 N = 550 N. The chair and floor must support her entire weight, so the balance of her weight to support is $600\text{ N} - 2(25\text{ N}) - 500\text{ N} = 50\text{ N}.$ Thus, the floor exerts a force on the woman's feet of 50.0 N upward.

(b) The force exerted by the floor on the chair must be equal to the weight of the chair plus the weight of the woman supported by the chair, or 600.0 N + 100.0 N − 50.0 N = 650.0 N. Thus, the floor exerts a force on the chair of 650.0 N upward. (not her feet)

(c) The two forces acting on the woman and chair system are the upward force due to the floor and the downward gravitational force due to the Earth. Let the subscripts be the following: s = woman and chair system, e = Earth, f = floor.

$\vec{\mathbf{F}}_{\text{sf}}$
$\vec{\mathbf{W}}_{\text{se}}$

41. Strategy Use the conversion factor for pounds to newtons, $0.2248 \text{ lb} = 1 \text{ N}$.

Solution

(a) Find the weight of the girl in newtons.
$W = mg = (40.0 \text{ kg})(9.80 \text{ N/kg}) = \boxed{392 \text{ N}}$

(b) Find the weight of the girl in pounds.
$(392 \text{ N})(0.2248 \text{ lb/N}) = \boxed{88.1 \text{ lb}}$

45. Strategy On Earth, $g = 9.80 \text{ N/kg}$.

Solution Find the man's weight on Earth, Mars, Venus, and Earth's moon.
$mg = (65 \text{ kg})(9.80 \text{ N/kg}) = \boxed{640 \text{ N}}$

(a) Find the man's weight on Mars.
$mg = (65 \text{ kg})(3.7 \text{ N/kg}) = \boxed{240 \text{ N}}$

(b) Find the man's weight on Venus.
$mg = (65 \text{ kg})(8.9 \text{ N/kg}) = \boxed{580 \text{ N}}$

(c) Find the man's weight on Earth's moon.
$mg = (65 \text{ kg})(1.6 \text{ N/kg}) = \boxed{100 \text{ N}}$

49. (a) Strategy Use Newton's universal law of gravitation and $r = 3.845\times10^8 \text{ m}$ for the distance between Earth and the Moon.

Solution Find the magnitude of the gravitational force that Earth exerts on the Moon.

$$F = \frac{GM_E M_M}{r^2} = \frac{(6.674\times10^{-11} \text{ N}\cdot\text{m}^2/\text{kg}^2)(5.974\times10^{24} \text{ kg})(7.35\times10^{22} \text{ kg})}{(3.845\times10^8 \text{ m})^2} = \boxed{1.98\times10^{20} \text{ N}}$$

(b) Strategy Use Newton's third law.

Solution According to Newton's third law, the magnitude of the gravitational force that the moon exerts on the Earth is the same as the force that the Earth exerts on the moon.

53. Strategy Draw free-body diagrams for each situation. Let the subscripts be the following:
b = book t = table e = Earth h = hand

Solution The diagrams are shown.

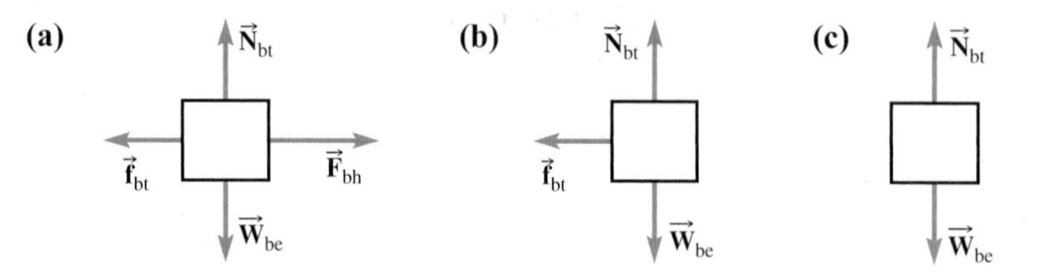

(d) Strategy and Solution In cases (a) and (b), the book is accelerating; so in these cases, the net force is not zero.

(e) Strategy and Solution The normal force on the book is equal to its weight, $(0.50\text{ kg})(9.80\text{ m/s}^2) = 4.9\text{ N}$. The net force acting on the book in part (b) is equal to the force of kinetic friction. The force of kinetic friction is opposite the direction of motion. The magnitude is $\mu_k N = 0.40(4.9\text{ N}) = 2.0\text{ N}$. Thus, the net force on the book is $\boxed{\text{2.0 N opposite the direction of motion}}$.

(f) Strategy and Solution The free-body diagram would look just like the diagram for part (c) and the book would not slow down because there is no net force on the book (friction is zero).

57. (a) Strategy Use Newton's first law of motion.

Solution Since the sleigh is moving with constant speed, the net force acting on the sleigh is $\boxed{\text{zero}}$.

(b) Strategy Since $F_{net} = 0$, the force of magnitude T must be equal to the force of kinetic friction.

Solution Find the coefficient of kinetic friction.

$$T = f_k = \mu_k mg, \text{ so } \mu_k = \boxed{\frac{T}{mg}}.$$

59. Strategy Let the subscripts be the following:

t = table e = Earth 1 = block 1 2 = block 2 3 = block 3 4 = block 4
h = horizontal force 1234 = system of blocks (The blocks are numbered from left to right.)

Solution The diagrams are shown.

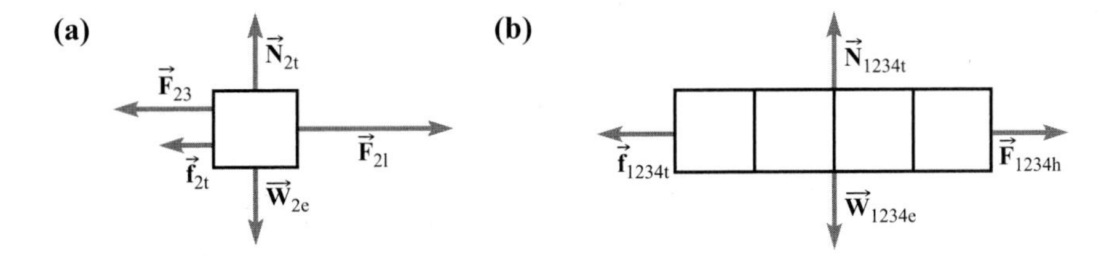

61. Strategy Use Newton's laws of motion.

Solution Without a machine, the force is equal to the weight of the object mg. According to Newton's laws of motion and Fig. 2.32, with a frictionless plane, the force is equal to $mg\sin\phi = mg\frac{h}{d}$. So, $\frac{mg}{mg\frac{h}{d}} = \frac{d}{h}$.

65. Strategy Identify all forces acting on the strut. Decompose the tension into its x- and y-components.

Solution Use Newton's laws of motion. See the diagram.

$\Sigma F_y = T\sin 30.0° - 200.0\text{ N} = 0$, so

$$T = \frac{200.0\text{ N}}{\sin 30.0°} = \boxed{400\text{ N}}.$$

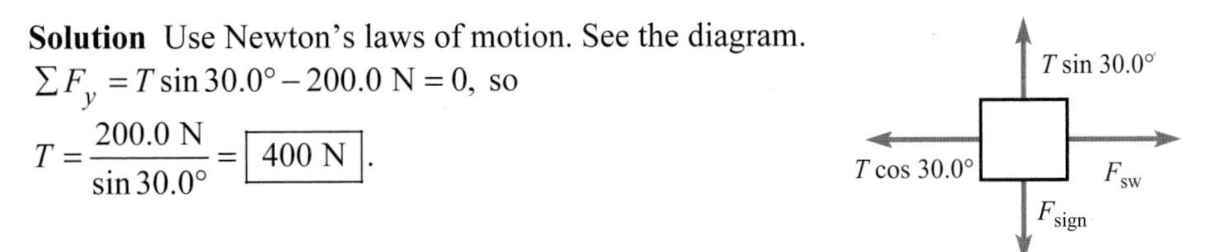

69. Strategy Recall that the tension in the rope is the same along its length.

Solution The tension is equal to the weight at the end of the rope, 120 N. Therefore, $\boxed{\text{both scales read 120 N}}$.

73. Strategy Use Newton's laws of motion. Draw a free-body diagram.

Solution

(a) Find the tension in the rope from which the pulley hangs.
$\sum F_y = T_1 \sin\theta - Mg = 0$ and $\sum F_x = T_1 \cos\theta - T_2 = 0.$
The tension in T_2 is due to the mass M, so $T_2 = Mg$.
Thus, $T_1 \cos\theta = Mg$ and $T_1 \sin\theta = Mg$.
According to these equations, $\cos\theta = \sin\theta$, which is true only if $\theta = 45°$ for $0° \le \theta \le 90°$.
Therefore, $T_1 = \dfrac{Mg}{\cos 45°} = \boxed{\sqrt{2}Mg}$.

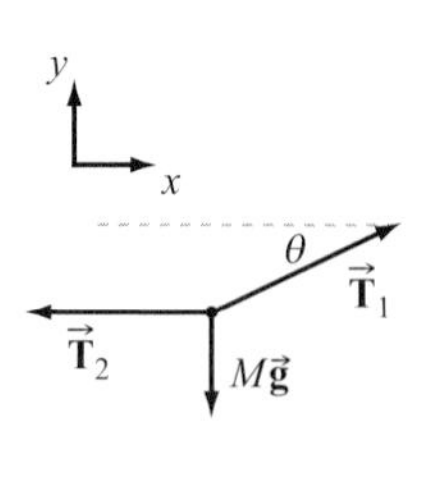

(b) As found in part (a), $\theta = \boxed{45°}$.

77. Strategy Consider the ranges of the forces given.

Solution The range of the strong force is about 10^{-15} m, so it certainly does not have unlimited range. Contact forces are not unlimited, as well, since they are limited to the contact region between objects (and there are no known objects of unlimited size). Both electromagnetic and gravitational forces have unlimited ranges.

81. Strategy Consider each object and its relationships to the others.

Solution

(a) The table must support the weights of both the dictionary and Fernando, so the normal force exerted by the table on the dictionary is $N = m_1 g + m_2 g = (m_1 + m_2)g = (2.0\text{ kg} + 52\text{ kg})(9.80\text{ N/kg}) = \boxed{530\text{ N}}$.

(b) The dictionary must support the weight of Fernando, so the normal force exerted by the dictionary on Fernando is $N = mg = (52\text{ kg})(9.80\text{ N/kg}) = \boxed{510\text{ N}}$.

(c) Fernando and the table are not in contact, so no, there is not a normal force exerted by the table on Fernando.

85. Strategy Use Newton's laws of motion.

Solution

(a) Since the airplane is cruising in a horizontal level flight (straight line) at constant velocity, it is in equilibrium and the net force is zero.

(b) The air pushes upward with a force equal to the weight of the airplane: $\boxed{2.6\times10^4\text{ N}}$.

87. Strategy Set the magnitudes of the forces on the spaceship due to the Earth and the Moon equal. (The forces are along the same line.)

Solution Find the distance from the Earth expressed as a percentage of the distance between the centers of the Earth and the Moon.

$$F_{sE} = \frac{GM_E m}{r_E^2} = F_{sM} = \frac{GM_M m}{r_M^2},\text{ so } r_E = r_M\sqrt{\frac{M_E}{0.0123M_E}} = 9.02r_M.$$

Find the percentage.

$$\frac{r_E}{r_E + r_M} = \frac{9.02r_M}{9.02r_M + r_M} = \frac{9.02}{10.02} = 0.900$$

The distance from the Earth is $\boxed{\text{90.0\% of the Earth-Moon distance}}$.

89. Strategy Use Newton's laws of motion.

Solution

(a) Since the train is moving at constant speed, and air resistance and friction are negligible, the readings on the three scales are $\boxed{\text{all 0}}$.

(b) Air resistance and friction are not considered negligible this time. The engine pulls the cars against these forces. Since the cars are identical, each car contributes about one-third of the total frictional and drag forces. Each spring scale will measure the net force due to the cars behind it, so the relative readings on the three spring scales are $\boxed{A > B > C}$. The free-body diagram is shown.

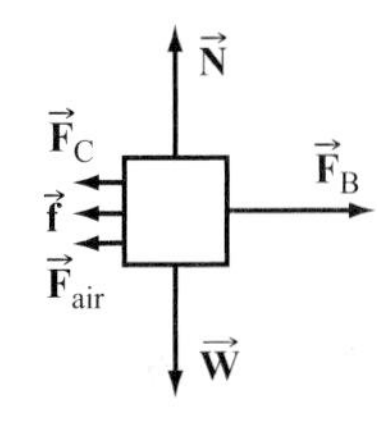

(c) Spring A measures the forces on all 3 cars. Spring B measures the forces on the latter 2 cars. Spring C measures the forces on the final 1 car.
A = 5.5 N + 5.5 N + 5.5 N = $\boxed{16.5\text{ N}}$; B = 5.5 N + 5.5 N = $\boxed{11.0\text{ N}}$; C = $\boxed{5.5\text{ N}}$

93. (a) Strategy The tension due to the weight of the potatoes is divided evenly between the two sets of scales.

Solution Find the tension and, thus, the reading of each scale.

$2T = mg$, so $T = mg/2 = (220.0\text{ N})/2 = \boxed{110.0\text{ N}}$.

(b) Strategy Scales B and D will read 110.0 N as before. Scales A and C will read an additional 5.0 N due to the weights of B and D, respectively.

Solution Find the reading of each scale.

$T_A = 110.0\text{ N} + 5.0\text{ N} = \boxed{115.0\text{ N}} = T_C$ and $T_B = \boxed{110.0\text{ N}} = T_D$.

97. (a) Strategy Let the subscripts be the following:
c = computer d = desk e = Earth

Solution The only forces on the computer are gravity due to the Earth and the normal force due to the desk. The free-body diagram is shown.

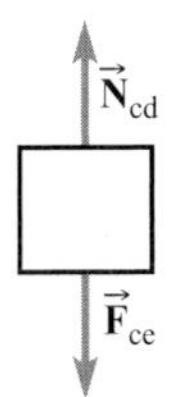

(b) Strategy Consider the nature of friction forces.

Solution Since the only forces acting on the computer are in the vertical direction, the friction force is $\boxed{\text{zero}}$.

(c) Strategy Find the maximum force of static friction on the computer due to the desk; this is the horizontal force necessary to make it begin to slide.

Solution $F = f_s = \mu_s N = \mu_s W = 0.60(87\text{ N}) = \boxed{52\text{ N}}$

101. (a) Strategy Scale A measures the weight of both masses. Scale B only measures the weight of the 4.0-kg mass.

Solution Find the readings of the two scales if the masses of the scales are negligible.

Scale A $= (10.0\text{ kg} + 4.0\text{ kg})(9.80\text{ N/kg}) = \boxed{137\text{ N}}$ and Scale B $= (4.0\text{ kg})(9.80\text{ N/kg}) = \boxed{39\text{ N}}$.

(b) Strategy Scale A measures the weight of both masses and scale B. Scale B only measures the weight of the 4.0-kg mass.

Solution Find the readings if each scale has a mass of 1.0 kg.

Scale A $= (10.0\text{ kg} + 4.0\text{ kg} + 1.0\text{ kg})(9.80\text{ N/kg}) = \boxed{147\text{ N}}$ and Scale B $= \boxed{39\text{ N}}$.

105. Strategy Use Newton's laws of motion and draw a free-body diagram.

Solution Find the tension in the cable.

$$\Sigma F_y = T\cos\theta - mg = 0, \text{ so } T = \boxed{\frac{mg}{\cos\theta}}.$$

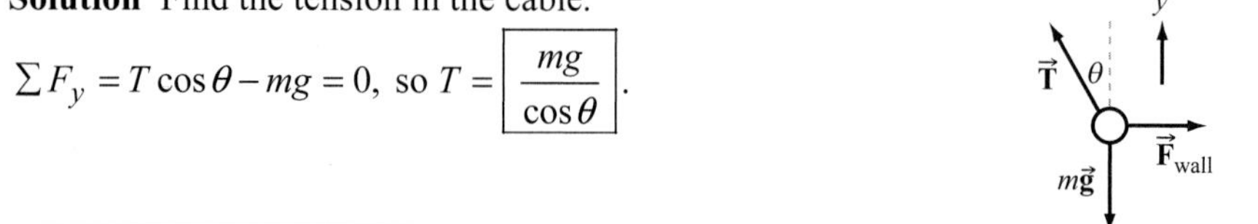

109. (a) Strategy Set the magnitudes of the forces on the spacecraft due to the Earth and the Sun equal.

Solution Find the distance of the spacecraft from Earth.

$$F_{\text{sS}} = \frac{GM_{\text{S}}m}{r_{\text{S}}^2} = F_{\text{sE}} = \frac{GM_{\text{E}}m}{r_{\text{E}}^2}, \text{ so } \frac{r_{\text{E}}}{r_{\text{S}}} = \sqrt{\frac{M_{\text{E}}}{M_{\text{S}}}}.$$

This is the ratio of the Earth-spacecraft distance to the Sun-spacecraft distance. If this is multiplied by the Earth-Sun mean distance, the product is the distance of the spacecraft from the Earth.

$$(1.50\times10^{11}\text{ m})\sqrt{\frac{5.974\times10^{24}\text{ kg}}{1.987\times10^{30}\text{ kg}}} = \boxed{2.60\times10^{8}\text{ m from Earth}}$$

(b) Strategy Imagine the spacecraft is a small distance *d* closer to the Earth and find out which gravitational force is stronger, the Earth's or the Sun's.

Solution At the equilibrium point the net gravitational force is zero. If the spacecraft is closer to the Earth than the equilibrium point distance from the Earth, then the force due to the Earth is greater than that due to the Sun. If the spacecraft is closer to the Sun than the equilibrium point distance from the Sun, then the force due to the Sun is greater than that due to the Earth. So, if the spacecraft is close to, but not at, the equilibrium point, the net force tends to pull it $\boxed{\text{away from}}$ the equilibrium point.

Chapter 3

ACCELERATION AND NEWTON'S SECOND LAW OF MOTION

Problems

1. **Strategy** Use the fact that $|\vec{\mathbf{A}}| = |\vec{\mathbf{B}}|$ and symmetry to determine the direction of $\vec{\mathbf{D}}$.

   Solution The vertical components cancel when $\vec{\mathbf{B}}$ is subtracted from $\vec{\mathbf{A}}$. The direction of the horizontal component of $\vec{\mathbf{B}}$ is reversed due to the subtraction, and so the vector resulting from the subtraction is in the direction of the horizontal component of $\vec{\mathbf{A}}$; that is, to the left. The horizontal components of each vector have the same magnitude, which is about 3.9 cm; so the magnitude of $\vec{\mathbf{D}}$ is $\boxed{\text{about 7.9 cm}}$. The sketch is shown:

5. **Strategy** Add the displacement from Jerry's dorm to the fitness center to the displacement from Cindy's apartment to Jerry's dorm to find the total displacement from Cindy's apartment to the fitness center.

   Solution Add the displacements.
   1.50 mi east + 2.00 mi north + 3.00 mi east = 4.50 mi east + 2.00 mi north
   Let north be along the $+y$-axis and east be along the $+x$-axis. Then, the components of the total displacement are $\Delta x = 4.50$ mi and $\Delta y = 2.00$ mi.
   Find the magnitude.

   $$\Delta r = \sqrt{(\Delta x)^2 + (\Delta y)^2} = \sqrt{(4.50\text{ mi})^2 + (2.00\text{ mi})^2} = \boxed{4.92\text{ mi}}$$

   Find the direction.

   $$\theta = \tan^{-1}\frac{2.00}{4.50} = \boxed{24.0^\circ\text{ north of east}}$$

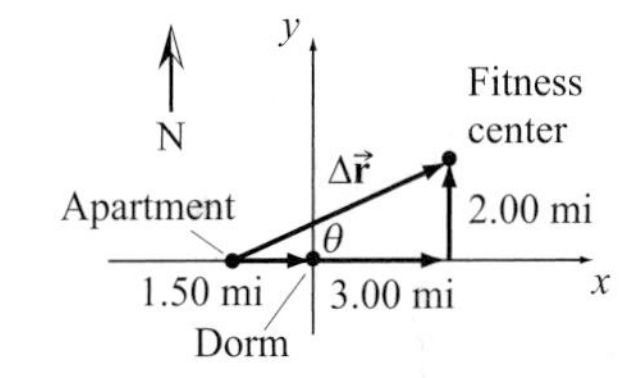

9. **Strategy** Draw the displacement vectors using graph paper, ruler, and protractor. Then use the component method.

   Solution Draw the diagram.

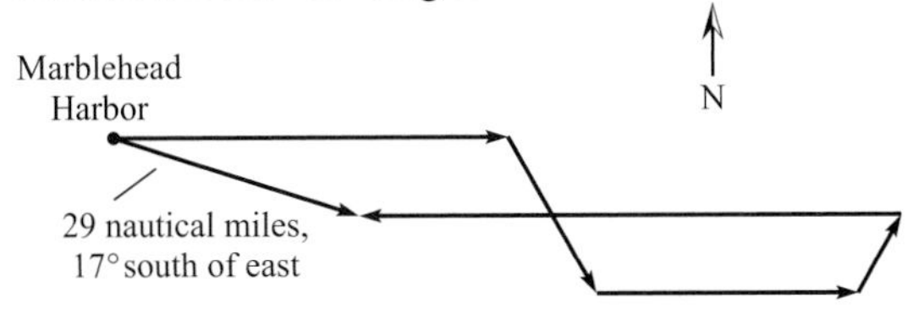

   Use the component method. Let north be $+y$ and east be $+x$.

   $$\Delta x = 45\text{ n.m.} + (20.0\text{ n.m.})\cos 300^\circ + 30.0\text{ n.m.} + (10.0\text{ n.m.})\cos 60^\circ - 62\text{ n.m.} = 28\text{ n.m.}$$

   $$\Delta y = (20.0\text{ n.m.})\sin 300^\circ + (10.0\text{ n.m.})\sin 60^\circ = -8.7\text{ n.m.}$$

   $$\Delta r = \sqrt{(\Delta x)^2 + (\Delta y)^2} = \sqrt{(28\text{ n.m.})^2 + (-8.7\text{ n.m.})^2} = 29\text{ n.m.}$$

   $$\theta = \tan^{-1}\frac{-8.7}{28} = 17^\circ\text{ south of east}$$

   So, $\Delta\vec{\mathbf{r}} = \boxed{29\text{ nautical miles at }17^\circ\text{ south of east}}$.

13. Strategy Set up ratios of speeds to distances.

Solution Find the speed of the baseball v.

$$\frac{v}{60.5\text{ ft}}=\frac{65.0\text{ mph}}{43.0\text{ ft}},\text{ so } v=\frac{65.0\text{ mph}}{43.0\text{ ft}}(60.5\text{ ft})=\boxed{91.5\text{ mph}}.$$

17. (a) Strategy Draw the position vectors with respect to Illium.

Solution

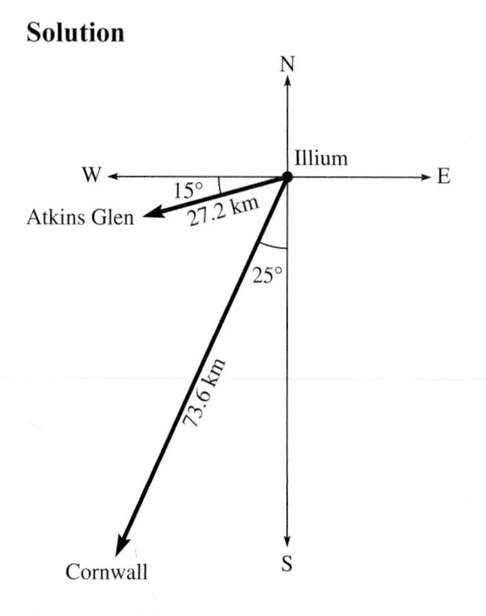

(b) Strategy Use the component method.

Solution

$\Delta\vec{\mathbf{r}}=\vec{\mathbf{r}}_f-\vec{\mathbf{r}}_i;\ \Delta x=x_f-x_i=r_f\cos\theta_f-r_i\cos\theta_i;\ \Delta y=y_f-y_i=r_f\sin\theta_f-r_i\sin\theta_i$

Find the magnitude of the displacement.

$$|\Delta\vec{\mathbf{r}}|=\sqrt{(\Delta x)^2+(\Delta y)^2}$$
$$=\sqrt{[(27.2\text{ km})\cos 195°-(73.6\text{ km})\cos 245°]^2+[(27.2\text{ km})\sin 195°-(73.6\text{ km})\sin 245°]^2}$$
$$=59.9\text{ km}$$

Find the direction of the displacement.

$$\theta=\tan^{-1}\frac{(27.2\text{ km})\sin 195°-(73.6\text{ km})\sin 245°}{(27.2\text{ km})\cos 195°-(73.6\text{ km})\cos 245°}=85°\text{ north of east}$$

So, $\Delta\vec{\mathbf{r}}=\boxed{59.9\text{ km at }85°\text{ north of east}}$.

(c) Strategy Use the definition of average velocity.

Solution

$$\vec{\mathbf{v}}_{av}=\frac{\Delta\vec{\mathbf{r}}}{\Delta t}=\frac{59.9\text{ km at }85°\text{ north of east}}{(45\text{ min})\left(\frac{1\text{ h}}{60\text{ min}}\right)}=\boxed{80\text{ km/h at }85°\text{ north of east}}$$

21. Strategy Use the area under the curve to find the displacement of the skateboard.

Solution The displacement of the skateboard is given by the area under the v vs. t curve. Under the curve for $t=3.00$ s to $t=8.00$ s, there are 16.5 squares and each square represents (1.0 m/s)(1.0 s) = 1.0 m; so the board moves $\boxed{16.5\text{ m}}$.

23. Strategy The slope of the x vs. t curve is equal to v_x. Use the definition of average speed.

Solution Compute the average speed at $t = 2.0$ s.

$$v_x = \frac{6.0\text{ m} - 4.0\text{ m}}{3.0\text{ s} - 1.0\text{ s}} = \boxed{1.0\text{ m/s}}$$

25. Strategy Use vector subtraction to find the change in velocity.

Solution Find the change in velocity of the scooter.

$$\Delta\vec{\mathbf{v}} = \vec{\mathbf{v}}_\text{f} - \vec{\mathbf{v}}_\text{i} = 15\text{ m/s west} - 12\text{ m/s east} = 15\text{ m/s west} - (-12\text{ m/s west}) = \boxed{27\text{ m/s west}}$$

27. Strategy Use the definition of average velocity. Draw a diagram.

Solution Let east be in the $+x$-direction and north be in the $+y$-direction. 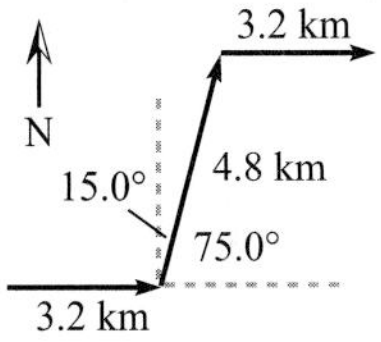
Find the magnitude of $\Delta\vec{\mathbf{r}}$.

N, 3.2 km, 15.0°, 4.8 km, 75.0°, 3.2 km

$$|\Delta\vec{\mathbf{r}}| = \sqrt{[3.2\text{ km} + (4.8\text{ km})\cos 75.0° + 3.2\text{ km}]^2 + [(4.8\text{ km})\sin 75.0°]^2} = 8.9\text{ km}$$

Find the direction of $\Delta\vec{\mathbf{r}}$.

$$\theta = \tan^{-1}\frac{4.6\text{ km}}{7.6\text{ km}} = 31°\text{ north of east}$$

So, $|\vec{\mathbf{v}}_\text{av}| = \dfrac{\Delta r}{\Delta t} = \dfrac{8.94\text{ km}}{0.10\text{ h} + 0.15\text{ h} + 0.10\text{ h}} = 26\text{ km/h}$ and $\vec{\mathbf{v}}_\text{av} = \dfrac{\Delta\vec{\mathbf{r}}}{\Delta t} = \boxed{26\text{ km/h at }31°\text{ north of east}}$.

29. (a) Strategy Find the average speed by dividing the total distance traveled by the total time.

Solution Each distance is given by the product of the speed and time.

$$v_\text{av} = \frac{\text{distance}}{\text{time}} = \frac{(108\text{ km/h})(20.0\text{ min}) + (90.0\text{ km/h})(10.0\text{ min})}{20.0\text{ min} + 10.0\text{ min}} = \boxed{102\text{ km/h}}$$

(b) Strategy Use the definition of average velocity. Draw a diagram.

Solution Compute the distance of each leg of the trip, then draw the diagram.

$$(108\text{ km/h})(20.0\text{ min})\left(\frac{1\text{ h}}{60\text{ min}}\right) = 36.0\text{ km and } (90.0\text{ km/h})(10.0\text{ min})\left(\frac{1\text{ h}}{60\text{ min}}\right) = 15.0\text{ km.}$$

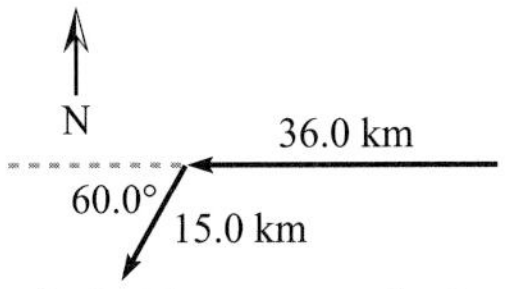

N, 36.0 km, 60.0°, 15.0 km

Find $\Delta\vec{\mathbf{r}}$. Let east be in the $+x$-direction and north be in the $+y$-direction.

$$|\Delta\vec{\mathbf{r}}| = \sqrt{[-36.0\text{ km} + (15.0\text{ km})\cos 240.0°]^2 + [(15.0\text{ km})\sin 240.0°]^2}$$
$$= \sqrt{(-43.5\text{ km})^2 + (-13.0\text{ km})^2} = 45.4\text{ km}$$

$$\theta = \tan^{-1}\frac{-13.0}{-43.5} = 16.6°\text{ south of west}$$

So, $|\vec{\mathbf{v}}_\text{av}| = \dfrac{\Delta r}{\Delta t} = \dfrac{45.4\text{ km}}{(10.0\text{ min} + 20.0\text{ min})\left(\frac{1\text{ h}}{60\text{ min}}\right)} = 90.8\text{ km/h}$ and

$$\vec{\mathbf{v}}_\text{av} = \frac{\Delta\vec{\mathbf{r}}}{\Delta t} = \boxed{90.8\text{ km/h at }16.6°\text{ south of west}}.$$

33. Strategy The acceleration a_x is equal to the slope of the v_x versus t graph. The displacement is equal to the area under the curve.

Solution

(a) a_x is the slope of the graph at $t = 11$ s.

$$a_x = \frac{\Delta v_x}{\Delta t} = \frac{10.0 \text{ m/s} - 30.0 \text{ m/s}}{12.0 \text{ s} - 10.0 \text{ s}} = \boxed{-10 \text{ m/s}^2}$$

(b) Since v_x is constant, $a_x = \boxed{0}$ at $t = 3$ s.

(c) The area under the v_x vs. t curve from $t = 12$ s to $t = 14$ s represents the displacement of the body. Each square represents $(10.0 \text{ m/s})(1.0 \text{ s}) = 1.0\times10^1$ m, and there is $1/2$ square under the curve for $t = 12$ s to $t = 14$ s, so the car travels $\boxed{5.0 \text{ m}}$.

35. Strategy The magnitude of the acceleration is the absolute value of the slope of the graph at $t = 7.0$ s.

Solution

$$a_x = \left|\frac{\Delta v_x}{\Delta t}\right| = \left|\frac{0 - 20.0 \text{ m/s}}{12.0 \text{ s} - 4.0 \text{ s}}\right| = \boxed{2.5 \text{ m/s}^2}$$

37. (a) Strategy Use the definition of average velocity. Draw a diagram.

Solution Let the center of the circle be the origin, then
$\Delta \vec{\mathbf{r}} = \vec{\mathbf{r}}_\text{f} - \vec{\mathbf{r}}_\text{i} = 20.0 \text{ m east} - 20.0 \text{ m south}.$

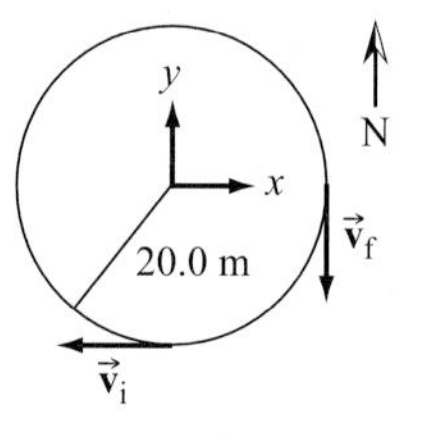

$$|\Delta \vec{\mathbf{r}}| = \sqrt{(20.0 \text{ m})^2 + (-20.0 \text{ m})^2} = 28.3 \text{ m}$$

Let east be the $+x$-direction and north the $+y$-direction.

$$\theta = \tan^{-1}\frac{20.0}{20.0} = 45.0° \text{ north of east}$$

$$\text{So, } |\vec{\mathbf{v}}_\text{av}| = \frac{\Delta r}{\Delta t} = \frac{28.3 \text{ m}}{3.0 \text{ s}} = 9.4 \text{ m/s and } \vec{\mathbf{v}}_\text{av} = \frac{\Delta \vec{\mathbf{r}}}{\Delta t} = \boxed{9.4 \text{ m/s at } 45° \text{ north of east}}.$$

(b) Strategy Use the definition of average acceleration and the fact that $C = 2\pi r$.

Solution Find the average acceleration of the car.

$$|\vec{\mathbf{v}}_\text{f}| = |\vec{\mathbf{v}}_\text{i}| = \frac{2\pi r(3/4)}{\Delta t} = \frac{3\pi r}{2\Delta t}, \text{ so } \vec{\mathbf{a}}_\text{av} = \frac{3\pi r}{2(\Delta t)^2}(\text{south} - \text{west}).$$

$$|\vec{\mathbf{a}}_\text{av}| = \frac{3\pi r}{2(\Delta t)^2}\sqrt{1^2 + (-1)^2} = \frac{3\pi r}{\sqrt{2}(\Delta t)^2}$$

$$\theta = \tan^{-1}\frac{-1}{1} = 45° \text{ south of east}$$

$$\vec{\mathbf{a}}_\text{av} = \frac{\Delta \vec{\mathbf{v}}}{\Delta t} = \frac{3\pi(20.0 \text{ m})}{\sqrt{2}(3.0 \text{ s})^2} \text{ at } 45° \text{ south of east} = \boxed{15 \text{ m/s}^2 \text{ at } 45° \text{ south of east}}$$

(c) Strategy Consider Newton's first law of motion.

Solution Although the magnitude of the velocity is constant, its direction must change continuously for the car to travel in a circle; $\boxed{\text{changing the direction of the velocity requires an acceleration}}$.

41. Strategy Since the particle is moving to the east and is accelerated to the south, its velocity in 8.00 s will be between east and south. Use the component method. Let north be in the $+y$-direction and east be in the $+x$-direction.

Solution

$v_x = 40.0\text{ m/s}$ and $v_y = a_y\Delta t = (-2.50\text{ m/s}^2)(8.00\text{ s}) = -20.0\text{ m/s}.$

$|\vec{\mathbf{v}}| = \sqrt{{v_x}^2 + {v_y}^2} = \sqrt{(40.0\text{ m/s})^2 + (-20.0\text{ m/s})^2} = 44.7\text{ m/s}$

$\theta = \tan^{-1}\frac{v_y}{v_x} = \tan^{-1}\frac{-20.0\text{ m/s}}{40.0\text{ m/s}} = 26.6°\text{ south of east}$

So, $\vec{\mathbf{v}} = \boxed{44.7\text{ m/s at }26.6°\text{ south of east}}$.

45. Strategy Use Newton's second law.

Solution Draw a free-body diagram.

$\Sigma F_y = T\cos\theta - mg = 0,\text{ so } T = \frac{mg}{\cos\theta}.$

$\Sigma F_x = T\sin\theta = ma_x,\text{ so } a_x = \frac{T\sin\theta}{m}.$

$a_x = \frac{mg}{\cos\theta}\left(\frac{\sin\theta}{m}\right) = g\tan\theta = (9.80\text{ m/s}^2)\tan 12° = 2.1\text{ m/s}^2$

The acceleration of the airplane is $\boxed{2.1\text{ m/s}^2\text{ in the direction of motion}}$.

47. Strategy Use Newton's second law for the vertical direction.

Solution Draw a free-body diagram. Find the tension.

$\Sigma F_y = T - mg = ma_y,\text{ so}$

$T = m(a_y + g) = (2010\text{ kg})(1.50\text{ m/s}^2 + 9.80\text{ m/s}^2) = 22.7\text{ kN}.$

The tension in the cable is $\boxed{22.7\text{ kN upward}}$.

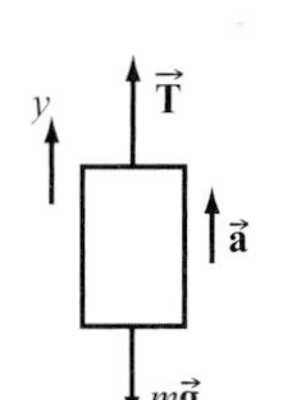

49. (a) Strategy Use Newton's second law for the vertical direction. Let $+y$ be in the upward direction.

Solution Draw a free-body diagram.

$\Sigma F_y = T - mg = ma_y,\text{ so } a_y = \frac{T - mg}{m} = \frac{33.6\text{ kN} - 24.8\text{ kN}}{2530\text{ kg}} = +3.5\text{ m/s}^2.$

So, the acceleration of the elevator is $\boxed{3.5\text{ m/s}^2\text{ up}}$.

(b) Strategy Use the definition of average acceleration and solve for the final speed.

Solution

$a_y = \frac{\Delta v}{\Delta t},\text{ so } \Delta v = v_\text{f} - v_\text{i} = a_y\Delta t,\text{ or } v_\text{f} = v_\text{i} + a_y\Delta t = 1.20\text{ m/s} + (3.5\text{ m/s}^2)(4.00\text{ s}) = 15\text{ m/s}.$

So, the velocity of the elevator 4.00 s later will be $\boxed{15\text{ m/s up}}$.

53. Strategy Use Newton's second law to evaluate the situation.

Solution Since $F = ma$, the maximum acceleration is $a_{\text{max}} = \frac{F}{m} = \frac{T}{m_{\text{car}}} = \frac{2500\text{ N}}{1400\text{ kg}} = \boxed{1.8\text{ m/s}^2}$. With this acceleration, the truck could reach 30 mph in about $\Delta t = \frac{v}{a} = \frac{30\text{ mph}}{1.8\text{ m/s}^2} = 7.5\text{ s}$. This is certainly possible. So, $\boxed{\text{yes}}$, the truck driver should be concerned about the rope breaking, particularly when friction is also considered.

57. Strategy Use Newton's laws of motion.

Solution Find the magnitude of the total force of the stone on the man's hand by first finding the force of the hand on the stone.

$\Sigma F_y = F_{\text{sh}} - mg = ma_y$, so $F_{\text{sh}} = ma_y + mg = m(a_y + g) = (2.0\text{ kg})(1.5\text{ m/s}^2 + 9.80\text{ m/s}^2) = 23\text{ N}$.

Thus, $\vec{\mathbf{F}}_{\text{sh}} = 23$ N upward. According to Newton's third law, $\vec{\mathbf{F}}_{\text{hs}} = -\vec{\mathbf{F}}_{\text{sh}} = -23$ N upward $= \boxed{23\text{ N downward}}$.

59. Strategy Consider the relative motion of the two vehicles. Find $\vec{\mathbf{v}}_{\text{BV}}$ = the velocity of the BMW relative to the VW.

Solution Let north be in the $+x$-direction.

$v_{\text{BR}x}$ = the velocity of the BMW relative to the road = 100.0 km/h

$v_{\text{RV}x}$ = the velocity of the road relative to the VW = $-v_{\text{VR}} = 42$ km/h

$v_{\text{BV}x} = v_{\text{BR}x} + v_{\text{RV}x} = 100.0\text{ km/h} + 42\text{ km/h}$

Find Δt.

$$\Delta t = \frac{\Delta x}{v_{\text{BV}x}} = \frac{10.0\text{ km}}{100.0\text{ km/h} + 42\text{ km/h}}\left(\frac{3600\text{ s}}{1\text{ h}}\right) = \boxed{254\text{ s}}$$

61. Strategy Consider the relative motion of the ship and the water.

Solution The relative speeds are:

$v_{\text{upstream}} = v_{\text{ship}} - v_{\text{water}} = v_{\text{up}} = v_{\text{s}} - v_{\text{w}}$

$v_{\text{downstream}} = v_{\text{ship}} + v_{\text{water}} = v_{\text{d}} = v_{\text{s}} + v_{\text{w}}$

Find the speed of the current, v_{w}.

$\Delta x = v_{\text{up}}\Delta t_{\text{up}} = (v_{\text{s}} - v_{\text{w}})\Delta t_{\text{up}}$, so $\frac{\Delta x}{\Delta t_{\text{up}}} = v_{\text{s}} - v_{\text{w}}$ (1). $\Delta x = v_{\text{d}}\Delta t_{\text{d}} = (v_{\text{s}} + v_{\text{w}})\Delta t_{\text{d}}$, so $\frac{\Delta x}{\Delta t_{\text{d}}} = v_{\text{s}} + v_{\text{w}}$ (2).

Subtract (1) from (2).

$$\frac{\Delta x}{\Delta t_{\text{d}}} - \frac{\Delta x}{\Delta t_{\text{up}}} = 2v_{\text{w}},\text{ so } v_{\text{w}} = \frac{\Delta x}{2}\left(\frac{1}{\Delta t_{\text{d}}} - \frac{1}{\Delta t_{\text{up}}}\right) = \frac{208\text{ km}}{2}\left(\frac{1}{19.2\text{ h}} - \frac{1}{20.8\text{ h}}\right) = \boxed{0.42\text{ km/h}}.$$

65. Strategy The upstream component of the velocity of the boat must be equal in magnitude to that of the current.

Solution Let $+y$-direction be toward the opposite shore.

$(4.0\text{ km/h})\sin\theta = 1.8\text{ km/h}$, so $\theta = \sin^{-1}\frac{1.8}{4.0} = 27°$.

The direction of the velocity of the boat relative to the water is $\boxed{27°\text{ upstream}}$.

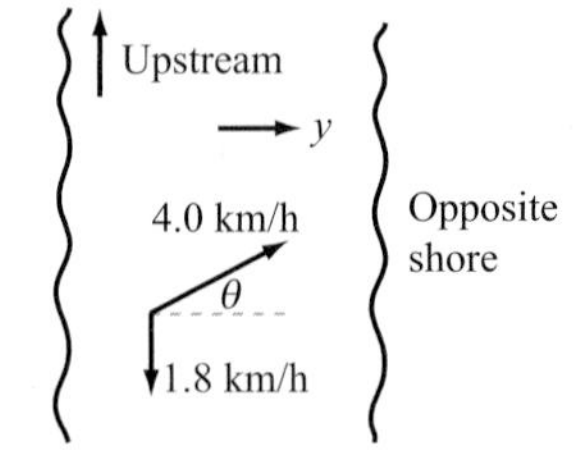

67. (a) Strategy The east-west components of the airplane's and the wind's velocities must be equal in magnitude for the plane to travel north.

Solution Let the $+y$-direction be north and the $+x$-direction be east.

$v_{px} = v_p \cos\theta = v_{air,\,x} = v_{air}\cos 45.00°$, so

$$\theta = \cos^{-1}\frac{v_{air}}{v_p\sqrt{2}} = \cos^{-1}\frac{100.0\ \text{km/h}}{(300.0\ \text{km/h})\sqrt{2}} = \boxed{76.37°\ \text{north of east}}.$$

(b) Strategy The northern or y-component of the plane's velocity relative to the ground is the y-component of its velocity relative to the air minus the y-component of the air's velocity relative to the ground.

Solution Find the time.

$$\Delta t = \frac{\Delta y}{v_y} = \frac{\Delta y}{v_p\sin\theta - v_{air}\sin 45.00°}$$

$$= \frac{600.0\ \text{km}}{(300.0\ \text{km/h})\sin\left[\cos^{-1}\frac{100.0\ \text{km/h}}{(300.0\ \text{km/h})\sqrt{2}}\right] - (100.0\ \text{km/h})\sin 45.00°} = \boxed{2.717\ \text{h}}$$

69. Strategy Consider the relative motion of the water (w) and Sheena (s). Let the $+y$-direction be upstream and the $+x$-direction be toward the opposite bank (b).

Solution

(a) Find the x-component.

$v_x = (3.00\ \text{mi/h})\cos 60.0° = 1.50\ \text{mi/h}$

The y-component is $v_y = v_{Sby} = v_{Swy} + v_{wby}$.

$v_y = (3.00\ \text{mi/h})\sin 60.0° - 1.60\ \text{mi/h} = 1.00\ \text{mi/h}$

Use the Pythagorean theorem.

$v_{Sb} = \sqrt{(1.50\ \text{mi/h})^2 + (1.00\ \text{mi/h})^2} = \boxed{1.80\ \text{mi/h}}$

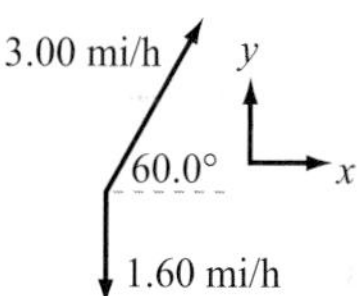

(b) $\Delta t = \dfrac{\Delta x}{v_x} = \dfrac{1.20\ \text{mi}}{1.50\ \text{mi/h}} = (0.800\ \text{h})\left(\dfrac{60\ \text{min}}{\text{h}}\right) = \boxed{48.0\ \text{min}}$

(c) $\Delta y = v_y\Delta t = (1.00\ \text{mi/h})(0.800\ \text{h}) = \boxed{0.800\ \text{mi upstream}}$

(d) The upstream component of her velocity relative to the water must be equal in magnitude to the velocity of the current relative to the bank, or $v_y = 0$.

$(3.00\ \text{mi/h})\sin\theta - 1.60\ \text{mi/h} = 0$, so $\theta = \sin^{-1}\dfrac{1.60}{3.00} = \boxed{32.2°\ \text{upstream}}$.

73. (a) Strategy Draw a diagram and use vector addition.

Solution Find the magnitude of the displacement.

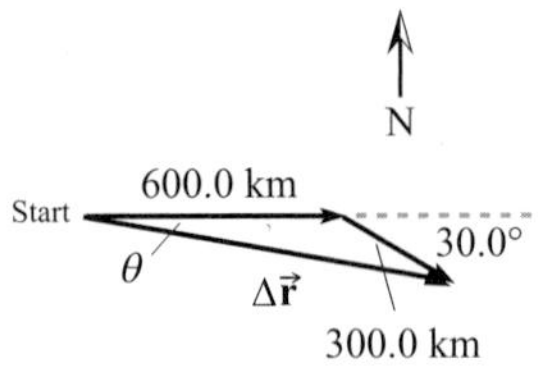

$$|\Delta\vec{\mathbf{r}}| = \sqrt{[600.0\text{ km} + (300.0\text{ km})\cos(-30.0°)]^2 + [(300.0\text{ km})\sin(-30.0°)]^2} = \boxed{873\text{ km}}$$

Start
600.0 km
30.0°
θ
$\Delta\vec{\mathbf{r}}$
300.0 km
N

(b) Strategy Refer to the diagram in part (a). Find the angle between the initial displacement vector and $\Delta\vec{\mathbf{r}}$.

Solution Find the direction of the displacement.

$$\theta = \tan^{-1}\frac{(300.0\text{ km})\sin(-30.0°)}{600.0\text{ km} + (300.0\text{ km})\cos(-30.0°)} = \boxed{9.90°\text{ south of east}}$$

(c) Strategy The flight time is given by the quotient of the distance traveled and the speed of the jetliner.

Solution

$$\Delta t = \frac{d}{v} = \frac{600.0\text{ km} + 300.0\text{ km}}{400.0\text{ km/h}} = \boxed{2.250\text{ h}}$$

(d) Strategy The direct flight time is given by the quotient of the magnitude of the displacement and the speed of the jetliner.

Solution

$$\Delta t = \frac{|\Delta\vec{\mathbf{r}}|}{v} = \frac{873\text{ km}}{400.0\text{ km/h}} = \boxed{2.18\text{ h}}$$

77. Strategy The static and kinetic friction forces always oppose the force applied to the block. Use Newton's laws of motion to analyze the forces on the block and the motion of the block in each case.

Solution

(a) The applied force must overcome the maximum force of static friction. So, the minimum horizontal applied force required to make the block start to slide is $F = f_\text{s} = \mu_\text{s} N = \mu_\text{s} mg = 0.35(4.6\text{ kg})(9.80\text{ m/s}^2) = \boxed{16\text{ N}}$.

(b) Once the block is sliding, the force required to keep it sliding at constant velocity is equal and opposite to the maximum force of kinetic friction. Since the maximum force of kinetic friction is less than the maximum force of static friction, the applied force is greater than necessary for constant velocity; therefore, the block will accelerate.

(c) Use Newton's second law to find the acceleration of the block.

$$\Sigma F_x = F - f_\text{k} = ma, \text{ so } a = \frac{F - f_\text{k}}{m} = \frac{\mu_\text{s} mg - \mu_\text{k} mg}{m} = (\mu_\text{s} - \mu_\text{k})g = (0.35 - 0.22)(9.80\text{ m/s}^2) = \boxed{1.3\text{ m/s}^2}.$$

81. Strategy Let the $+x$-direction be up the incline. Use Newton's second law.

Solution

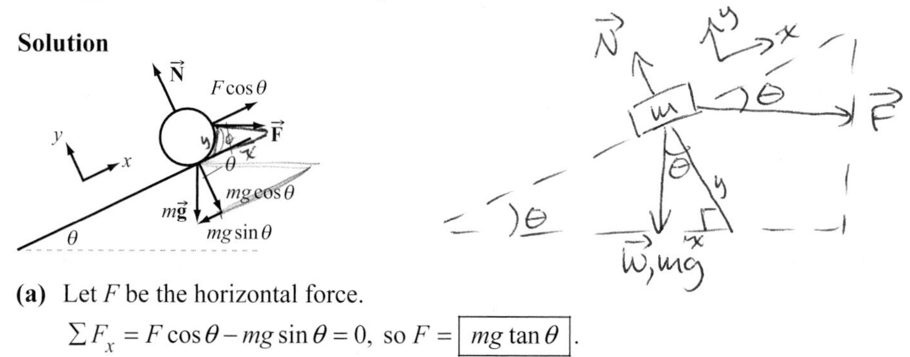

(a) Let F be the horizontal force.

$\Sigma F_x = F\cos\theta - mg\sin\theta = 0$, so $F = \boxed{mg\tan\theta}$.

(b) To roll the crate up at constant speed, the net force is zero, so the force is that from part (a), $\boxed{mg\tan\theta}$.

(c) $\Sigma F_x = F\cos\theta - mg\sin\theta = ma$, so $F = \boxed{mg\tan\theta + \dfrac{ma}{\cos\theta}}$.

85. Strategy If the rocket moves with constant acceleration, its average acceleration is equal to its instantaneous (constant) acceleration.

Solution From Problem 84, $\vec{\mathbf{a}}_{av} = 16\ \text{m/s}^2$ up. Use $\Delta y = v_{iy}\Delta t + \frac{1}{2}a_y(\Delta t)^2$ to find a_y.

$$\Delta y = v_{iy}\Delta t + \tfrac{1}{2}a_y(\Delta t)^2 = (0)\Delta t + \tfrac{1}{2}a_y(\Delta t)^2,\ \text{so}\ a_y = \frac{2\Delta y}{(\Delta t)^2} = \frac{2(160\times10^3\ \text{m})}{(8.0\ \text{min})^2\left(\frac{60\ \text{s}}{1\ \text{min}}\right)^2} = 1.4\ \text{m/s}^2.$$

$\boxed{\vec{\mathbf{a}} \neq \vec{\mathbf{a}}_{av},\ \text{so the acceleration is not constant.}}$

89. (a) Strategy Since $mg = (51\ \text{kg})(9.80\ \text{m/s}^2) = 500\ \text{N} > 408\ \text{N}$, the woman feels less than her normal weight, so the elevator is accelerating downward. Use Newton's second law.

Solution Let the $+y$-direction be up.

$$\Sigma F_y = 408\ \text{N} - mg = ma_y,\ \text{so}\ a_y = \frac{408\ \text{N} - mg}{m} = \frac{408\ \text{N}}{51\ \text{kg}} - 9.80\ \text{m/s}^2 = -1.8\ \text{m/s}^2.$$

Thus, $\vec{\mathbf{a}} = \boxed{1.8\ \text{m/s}^2\ \text{down}}$.

(b) Strategy Find the change in speed of the elevator after 4.0 s at the acceleration found in part (a).

Solution Let down be positive. Find the speed of the elevator.

$\Delta v = v_f - v_i = a_y\Delta t$, so $v_f = v_i + a_y\Delta t = 1.5\ \text{m/s} + (1.8\ \text{m/s}^2)(4.0\ \text{s}) = \boxed{8.7\ \text{m/s}}$.

93. Strategy Use Newton's second law.

Solution

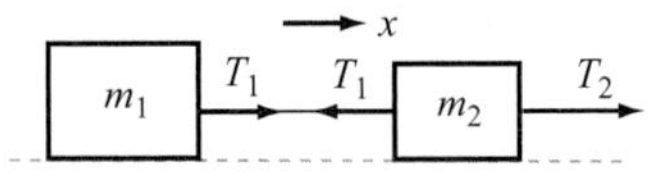

For m_2: $\sum F_x = T_2 - T_1 = m_2 a$

For m_1: $\sum F_x = T_1 = m_1 a$

Find T_2.

$T_2 - T_1 = m_2 a$, so $T_2 = T_1 + m_2 a = m_1 a + m_2 a = (m_1 + m_2)a.$

Find T_1/T_2.

$$\frac{T_1}{T_2} = \frac{m_1 a}{(m_1 + m_2)a} = \boxed{\frac{m_1}{m_1 + m_2}}$$

97. Strategy Analyze the graph to answer each question about the motion of the Engine. The x-component of the engine's velocity is represented by the slope.

Solution

(a) $a_x < 0$ when the engine is moving in the $+x$-direction and slowing down or when it is moving in the $-x$-direction and speeding up. So, at $\boxed{t_3 \text{ and } t_4}$ $a_x < 0$.

(b) $a_x = 0$ when the engine's speed is constant or zero. So, at $\boxed{t_0, t_2, t_5, \text{ and } t_7}$ $a_x = 0$.

(c) $a_x > 0$ when the engine is moving in the $+x$-direction and speeding up and when it is moving in the $-x$-direction and slowing down. So, at $\boxed{t_1 \text{ and } t_6}$ $a_x > 0$.

(d) $v_x = 0$ when the slope of the graph is zero. So, at $\boxed{t_0, t_3, \text{ and } t_7}$ $v_x = 0$.

(e) The speed is decreasing when a_x and v_x have opposite directions. So, at $\boxed{t_6}$ the speed is decreasing.

Chapter 4

MOTION WITH A CHANGING VELOCITY

Problems

1. (a) Strategy Relate the acceleration, speed, and distance using Eq. (4-5).

Solution

$v_{\text{fx}}^2 - v_{\text{ix}}^2 = 2a_x\Delta x$, so $a_x = \dfrac{v_{\text{fx}}^2 - v_{\text{ix}}^2}{2\Delta x} = \dfrac{0-(3.2 \text{ m/s})^2}{2(6.0 \text{ m}-0)} = -0.85 \text{ m/s}^2$.

The magnitude of the average acceleration is $\boxed{0.85 \text{ m/s}^2}$.

(b) Strategy Use Newton's second law. Draw a free-body diagram.

Solution Find μ_k.

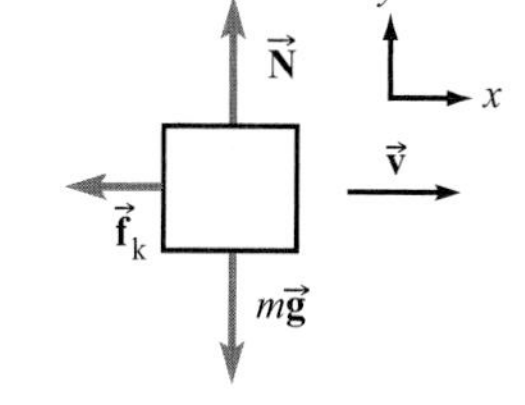

$\sum F_x = -f_\text{k} = ma_x$, so $f_\text{k} = \mu_\text{k} N = -ma_x$ or $\mu_\text{k} = -\dfrac{ma_x}{N}$.

$\sum F_y = N - mg = 0$, so $N = mg$. Thus, we have

$$\mu_\text{k} = -\frac{ma_x}{N} = -\frac{ma_x}{mg} = -\frac{a_x}{g} = -\frac{-0.85 \text{ m/s}^2}{9.80 \text{ m/s}^2} = \boxed{0.087}.$$

5. (a) Strategy Use Eqs. (3-3) and (4-2) since the acceleration is constant.

Solution Find the distance traveled.

$$\Delta x = v_{\text{av},x}\Delta t = \frac{v_{\text{fx}} + v_{\text{ix}}}{2}\Delta t = \frac{27.3 \text{ m/s} + 17.4 \text{ m/s}}{2}(10.0 \text{ s}) = \boxed{224 \text{ m}}.$$

(b) Strategy Use the definition of average acceleration.

Solution Find the magnitude of the acceleration.

$$a = \frac{\Delta v}{\Delta t} = \frac{27.3 \text{ m/s} - 17.4 \text{ m/s}}{10.0 \text{ s}} = \boxed{0.99 \text{ m/s}^2}$$

9. Strategy Use Eq. (4-5) to find the distance the car requires to stop after slamming on the brakes. Then, use Eqs. (3-3) and (4-2) to relate the speeds, distances, time, and acceleration.

Solution Find the distance the car requires to stop.

$\Delta x_\text{c} = \dfrac{v_{\text{fc}}^2 - v_{\text{ic}}^2}{2a} = \dfrac{0-(27.0 \text{ m/s})^2}{2(-7.00 \text{ m/s}^2)} = \boxed{52.1 \text{ m}}$. Since the acceleration of the car is constant, the average speed of the car as it attempts to stop is $v_{\text{c,av}} = (v_{\text{fc}} + v_{\text{ic}})/2 = (0 + 27.0 \text{ m/s})/2 = 13.5 \text{ m/s}$. Thus, the time required for the car to stop is $\Delta t = \Delta x_\text{c}/v_{\text{c,av}} = (52.1 \text{ m})/(13.5 \text{ m/s}) = 3.86 \text{ s}$. The distance the tractor travels in this time is $\Delta x_\text{t} = v_\text{t}\Delta t = (10.0 \text{ m/s})(3.86 \text{ s}) = 38.6 \text{ m}$. Now, $38.6 \text{ m} + 15.0 \text{ m} = 53.6 \text{ m}$, which is $53.6 \text{ m} - 52.1 \text{ m} = \boxed{1.5 \text{ m}}$ beyond the stopping point of the car. Therefore, $\boxed{\text{you won't hit the tractor}}$.

13. (a) Strategy The graph will be a line with a slope of $1.20\ \text{m/s}^2$.

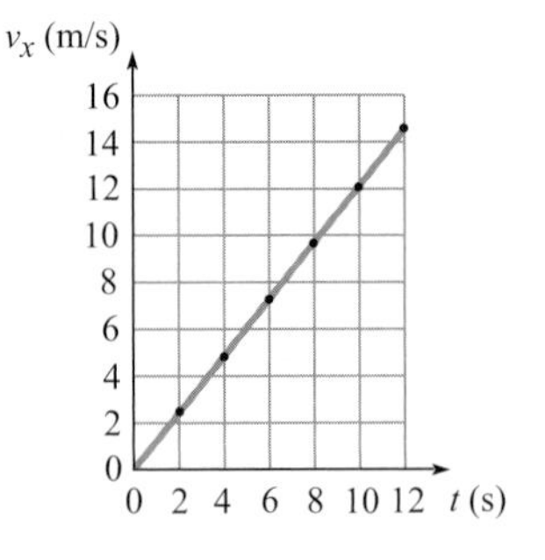

Solution $v_x = 0$ when $t = 0$. The graph is shown.

(b) Strategy Use Eq. (4-4).

Solution Find the distance the train traveled.

$$\Delta x = v_{ix}\Delta t + \frac{1}{2}a_x(\Delta t)^2 = (0)\Delta t + \frac{1}{2}a_x(\Delta t)^2 = \frac{1}{2}a_x(\Delta t)^2$$
$$= \frac{1}{2}(1.20\ \text{m/s}^2)(12.0\ \text{s})^2 = \boxed{86.4\ \text{m}}$$

(c) Strategy Use Eq. (4-1).

Solution Find the final speed of the train.

$v_{fx} - v_{ix} = v_{fx} - 0 = a_x\Delta t$, so $v_{fx} = a_x\Delta t = (1.20\ \text{m/s}^2)(12.0\ \text{s}) = \boxed{14.4\ \text{m/s}}$.

17. Strategy Refer to the figure. Analyze graphically and algebraically.

Solution Graphical analysis: Find the slope of the graph.

$$a_{\text{av},x} = \frac{40\ \text{m/s} - 20\ \text{m/s}}{9.0\ \text{s} - 5.0\ \text{s}} = 5.0\ \text{m/s}^2$$

Algebraic solution: $v_{fx} - v_{ix} = a_{\text{av},x}\Delta t$, so $a_{\text{av},x} = \dfrac{v_{fx} - v_{ix}}{\Delta t} = \dfrac{40\ \text{m/s} - 20\ \text{m/s}}{9.0\ \text{s} - 5.0\ \text{s}} = 5.0\ \text{m/s}^2$.

The average acceleration is $\boxed{5.0\ \text{m/s}^2 \text{ in the } +x\text{-direction}}$.

19. Strategy Let the $+x$-direction be down the incline. Use Newton's second law and Eq. (4-5).

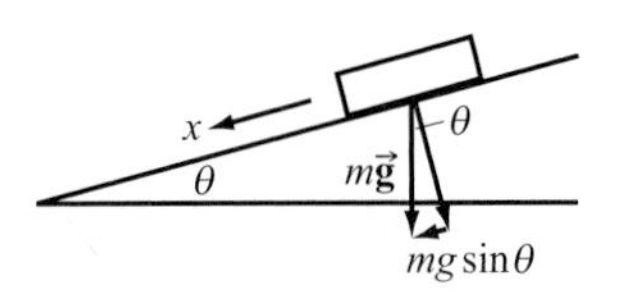

Solution Find the acceleration of the glider.

$\sum F_x = mg\sin\theta = ma_x$, so $a_x = g\sin\theta$.

Find the angle of inclination.

$v_{fx}^2 - v_{ix}^2 = v_{fx}^2 - 0 = 2a_x\Delta x = 2g\sin\theta\Delta x$, so

$$\theta = \sin^{-1}\frac{v_{fx}^2}{2g\Delta x} = \sin^{-1}\frac{(0.250\ \text{m/s})^2}{2(9.80\ \text{m/s}^2)(0.500\ \text{m})} = \boxed{0.365°}.$$

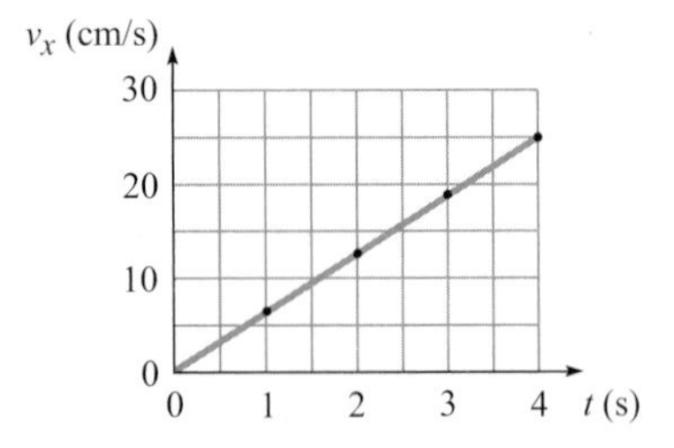

The slope is $a_x = g\sin\theta = 0.0625\ \text{m/s}^2 = 6.25\ \text{cm/s}^2$.

The positions and times are shown in the graph.

21. Strategy Use Eq. (4-5) to find the distance of the block from the top of the incline.

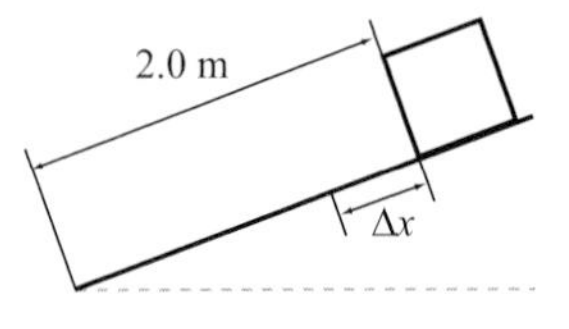

Solution

$v_f^2 - v_i^2 = v_f^2 - 0 = v_f^2 = 2ad$ and $(0.50v_f)^2 = 2a\Delta x = 0.25v_f^2$.

Form a proportion.

$$\frac{v_f^2}{0.25v_f^2} = \frac{2ad}{2a\Delta x}, \text{ so } \Delta x = \frac{d}{4.0} = \frac{2.0\ \text{m}}{4.0} = \boxed{0.50\ \text{m}}.$$

23. Strategy Ignoring air resistance, the golf ball is in free fall. Use Eq. (4-9).

Solution

(a) Find the time it takes the golf ball to fall 12.0 m.

$$v_{iy}=0, \text{ so } \Delta y=-\frac{1}{2}g(\Delta t)^2 \text{ and } \Delta t=\sqrt{-\frac{2\Delta y}{g}}=\sqrt{\frac{-2(0-12.0\text{ m})}{9.80\text{ m/s}^2}}=\boxed{1.6\text{ s}}.$$

(b) Find how far the golf ball would fall in $2\sqrt{\frac{-2(0-12.0\text{ m})}{9.80\text{ m/s}^2}}=3.13\text{ s}$.

$$\Delta y=-\frac{1}{2}g(\Delta t)^2=-\frac{1}{2}(9.80\text{ m/s}^2)(3.13\text{ s})^2=-48\text{ m}$$

The golf ball would fall $\boxed{48\text{ m}}$.

25. Strategy The final speed is zero. Use Eq. (4-10).

Solution Find the initial speed.

$$v_{fy}{}^2-v_{iy}{}^2=0-v_{iy}{}^2=-2g\Delta y, \text{ so } v_{iy}=\sqrt{2g\Delta y}=\sqrt{2(9.80\text{ m/s}^2)(1.3\text{ m}-0)}=\boxed{5.0\text{ m/s}}.$$

27. Strategy Use Eq. (4-10).

Solution Find the sandbag's speed when it hits the ground.

$$v_{fy}{}^2-v_{iy}{}^2=-2g\Delta y, \text{ so } v_{fy}=\sqrt{v_{iy}{}^2-2g\Delta y}=\sqrt{(10.0\text{ m/s})^2-2(9.80\text{ m/s}^2)(-40.8\text{ m})}=\boxed{30.0\text{ m/s}}.$$

29. Strategy The acceleration of the camera is given by $v_{y1}/\Delta t_1$, where $v_{y1}=3.3\text{ m/s}$ and $\Delta t_1=2.0\text{ s}$. Use Eq. (4-9).

Solution After 4.0 s, the camera has fallen

$$\Delta y=\frac{1}{2}a_y(\Delta t)^2=\frac{1}{2}\left(\frac{v_{y1}}{\Delta t_1}\right)(\Delta t)^2=\frac{3.3\text{ m/s}}{2(2.0\text{ s})}(4.0\text{ s})^2=\boxed{13\text{ m}}.$$

33. (a) Strategy Use Eq. (4-9) to find the height and Eq. (4-7) to find the speed of the rocket when it runs out of fuel. Then, use Eq. (4-10) to find the maximum height of the rocket, where the height and speed of the rocket when it runs out of fuel are the initial conditions and its final speed is zero.

Solution Find the height h_1 and speed v_1 of the rocket when it runs out of fuel.

$$\Delta y=v_{iy}\Delta t+\frac{1}{2}a_y(\Delta t)^2=0+\frac{1}{2}a_y(\Delta t)^2=\frac{1}{2}a_y(\Delta t)^2=h_1 \text{ and } \Delta v_y=a_y\Delta t=v_1.$$

Find the maximum height.

$$v_{fy}^2-v_{iy}^2=-2g\Delta y=0-v_{iy}^2=-2g(y_f-y_i), \text{ so}$$

$$y_f=y_i+\frac{v_{iy}^2}{2g}=h_1+\frac{v_1^2}{2g}=\frac{1}{2}a_y(\Delta t)^2+\frac{(a_y\Delta t)^2}{2g}=\frac{1}{2}(17.5\text{ m/s}^2)(1.5\text{ s})^2+\frac{[(17.5\text{ m/s}^2)(1.5\text{ s})]^2}{2(9.80\text{ m/s}^2)}=\boxed{55\text{ m}}.$$

(b) Strategy The rocket runs out of fuel after $\Delta t_1 = 1.5$ s. Then it travels for a time Δt_2 before it reaches its maximum height. Finally, it falls freely (from rest) for a time Δt_3 until it reaches the ground.

Solution Find Δt_2 using Eq. (4-7).

$$v_{fy} - v_{iy} = a_y \Delta t = -g\Delta t_2 = 0 - v_1, \text{ so } \Delta t_2 = \frac{v_1}{g} = \frac{a_y \Delta t_1}{g}.$$

Find Δt_3 using Eq. (4-9).

$$y_f = \frac{1}{2} g(\Delta t_3)^2, \text{ so } \Delta t_3 = \sqrt{\frac{2y_f}{g}}.$$

Find the total time of flight.

$$\Delta t_1 + \Delta t_2 + \Delta t_3 = \Delta t_1 + \frac{a_y \Delta t_1}{g} + \sqrt{\frac{2y_f}{g}} = 1.5 \text{ s} + \frac{(17.5 \text{ m/s}^2)(1.5 \text{ s})}{9.80 \text{ m/s}^2} + \sqrt{\frac{2(55 \text{ m})}{9.80 \text{ m/s}^2}} = \boxed{7.5 \text{ s}}$$

37. (a) Strategy Ignoring air resistance, the bomb will travel a distance equal to the speed of the plane times the time is takes the bomb to reach ground level.

Solution Use Eq. (4-9) to find the time it takes for the bomb to reach ground level.

$$\Delta y = v_{iy}\Delta t + \frac{1}{2} a_y (\Delta t)^2 = 0 - \frac{1}{2} g(\Delta t)^2, \text{ so } \Delta t = \sqrt{\frac{-2\Delta y}{g}}.$$ So, the bomb should be released

$$\Delta x = v_x \Delta t = v_x \sqrt{\frac{-2\Delta y}{g}} = (40.0 \text{ m/s})\sqrt{\frac{-2(-125 \text{ m})}{9.80 \text{ m/s}^2}} = \boxed{202 \text{ m}}$$ horizontally from the target.

(b) Strategy The horizontal component of the velocity of the bomb is 40.0 m/s. Find the vertical component of the velocity and use the components to find the direction the bomb is traveling just before it hits the target.

Solution Use Eq. (4-10) to find the vertical component of the bomb's velocity.

$$v_{fy}^2 - v_{iy}^2 = 2a_y \Delta y = -2g\Delta y = v_{fy}^2 - 0, \text{ so } v_{fy} = \sqrt{-2g\Delta y}.$$ Find the angle.

$$\theta = \tan^{-1}\frac{v_{fy}}{v_{fx}} = \tan^{-1}\frac{\sqrt{-2g\Delta y}}{v_{fx}} = \tan^{-1}\frac{\sqrt{-2(9.80 \text{ m/s}^2)(-125 \text{ m})}}{40.0 \text{ m/s}}$$

$$= \boxed{51.1° \text{ below the horizontal}}$$

41. (a) Strategy Consider each quantity's dependence on time.

Solution

$\Delta x = v_x \Delta t$, so x increases linearly with time.

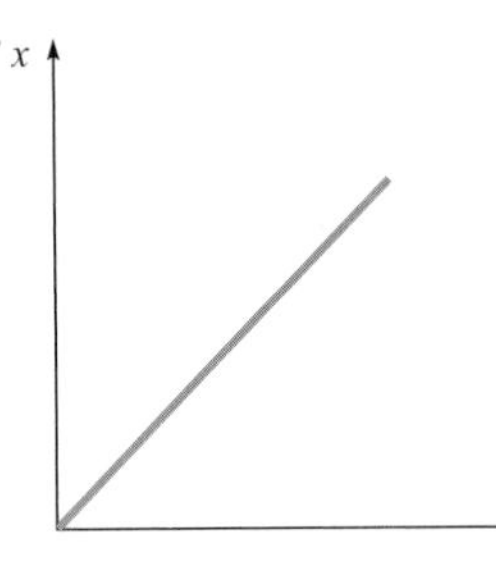

According to Eq. (4-9), y is parabolic.

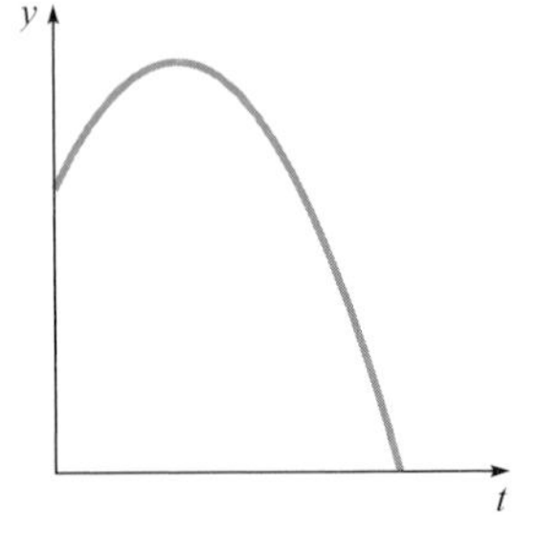

Since the net acceleration of the stone in the horizontal direction is zero, v_x is constant.

v_y starts positive and decreases linearly.

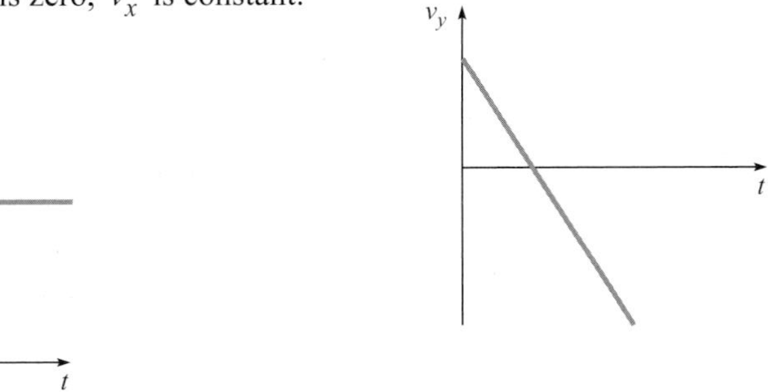

(b) Strategy Find v_i in terms of Δx, Δt, and θ.

Solution Solve for the initial speed.

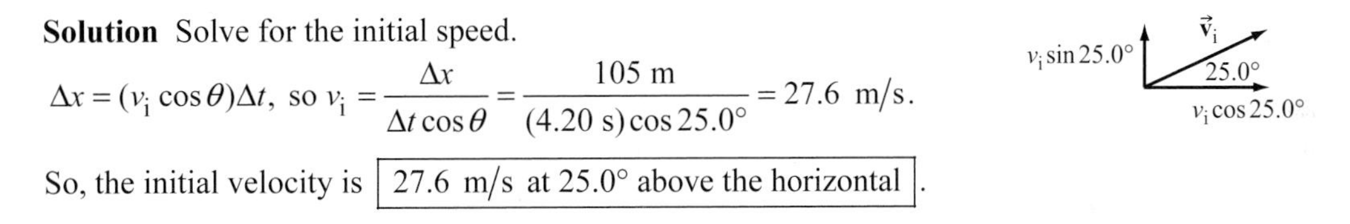

$$\Delta x = (v_i \cos\theta)\Delta t, \text{ so } v_i = \frac{\Delta x}{\Delta t\cos\theta} = \frac{105 \text{ m}}{(4.20 \text{ s})\cos 25.0°} = 27.6 \text{ m/s}.$$

So, the initial velocity is $\boxed{27.6 \text{ m/s at } 25.0° \text{ above the horizontal}}$.

(c) Strategy Find h using the result for v_i found in part (b).

Solution Use Eq. (4-9).

$$\Delta y = v_{iy}\Delta t + \frac{1}{2}a_y(\Delta t)^2 = (v_i \sin\theta)\Delta t - \frac{1}{2}g(\Delta t)^2 = \left(\frac{\Delta x}{\Delta t\cos\theta}\sin\theta\right)\Delta t - \frac{1}{2}g(\Delta t)^2 = \Delta x\tan\theta - \frac{1}{2}g(\Delta t)^2$$

$$= (105 \text{ m})\tan 25.0° - \frac{1}{2}(9.80 \text{ m/s}^2)(4.20 \text{ s})^2 = -37.5 \text{ m}$$

So, $h = \boxed{37.5 \text{ m}}$.

(d) Strategy Set $v_y = 0$ to find the time when the stone reaches its maximum height.

Solution Use Eq. (4-7) to find the time.

$$v_{fy} - v_{iy} = a_y\Delta t = -g\Delta t, \text{ so } v_{fy} = v_i \sin\theta - g\Delta t = 0, \text{ or } \Delta t = \frac{v_i \sin\theta}{g}.$$

Find H.

$$H = h + (v_i \sin\theta)\Delta t - \frac{1}{2}g(\Delta t)^2 = h + v_i \sin\theta\left(\frac{v_i \sin\theta}{g}\right) - \frac{1}{2}g\left(\frac{v_i \sin\theta}{g}\right)^2 = h + \frac{v_i^2 \sin^2\theta}{g} - \frac{v_i^2 \sin^2\theta}{2g}$$

$$= h + \frac{v_i^2 \sin^2\theta}{2g} = 37.5 \text{ m} + \frac{(27.6 \text{ m/s})^2 \sin^2 25.0°}{2(9.80 \text{ m/s}^2)} = \boxed{44.4 \text{ m above the ground}}$$

45. (a) Strategy At the maximum height of the cannonball's trajectory, $v_{fy} = 0$. Use Eq. (4-10).

Solution Find the maximum height reached by the cannonball.

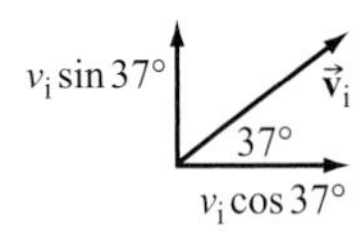

$$v_{fy}^2 - v_{iy}^2 = 0 - (v_i \sin\theta)^2 = 2a_y\Delta y = -2g(y_f - y_i), \text{ so}$$

$$y_f = y_i + \frac{v_i^2 \sin^2\theta}{2g} = 7.0 \text{ m} + \frac{(40 \text{ m/s})^2 \sin^2 37°}{2(9.80 \text{ m/s}^2)} = \boxed{37 \text{ m}}.$$

(b) Strategy Solve $\Delta x = v_x \Delta t$ for the time and substitute the result into Eq. (4-9). Then, solve for Δx to find the horizontal distance from the release point.

Solution When the cannonball hits the ground, $\Delta y = -7.0$ m.

$$\Delta x = v_x \Delta t = (v_i \cos\theta)\Delta t, \text{ so } \Delta t = \frac{\Delta x}{v_i \cos\theta}. \text{ Substitute.}$$

$$\Delta y = y_f - y_i = v_{iy}\Delta t + \frac{1}{2}a_y(\Delta t)^2 = (v_i \sin\theta)\frac{\Delta x}{v_i\cos\theta} - \frac{1}{2}g\frac{(\Delta x)^2}{v_i^2\cos^2\theta} = \Delta x\tan\theta - \frac{g(\Delta x)^2}{2v_i^2\cos^2\theta}, \text{ so}$$

$$0 = \frac{g}{2v_i^2\cos^2\theta}(\Delta x)^2 - (\tan\theta)\Delta x + \Delta y. \text{ Use the quadratic formula.}$$

$$\Delta x = \frac{\tan\theta \pm \sqrt{\tan^2\theta - \frac{4g\Delta y}{2v_i^2\cos^2\theta}}}{\frac{2g}{2v_i^2\cos^2\theta}} = \frac{\tan 37° \pm \sqrt{\tan^2 37° - \frac{2(9.80\text{ m/s}^2)(-7.0\text{ m})}{(40\text{ m/s})^2\cos^2 37°}}}{\frac{9.80\text{ m/s}^2}{(40\text{ m/s})^2\cos^2 37°}} = 170\text{ m or } -9\text{ m}$$

Since the catapult doesn't fire backward, -9 m is extraneous. So, the cannonball lands $\boxed{170\text{ m}}$ from its release point.

(c) Strategy The x-component is the same as the initial value. Find the y-component using Eq. (4-10).

Solution The x-component of the velocity is $v_{fx} = v_{ix} = v_i\cos\theta = (40\text{ m/s})\cos 37° = \boxed{32\text{ m/s}}$.
Find the y-component of the velocity.
$v_{fy}^2 - v_{iy}^2 = v_{fy}^2 - v_i^2\sin^2\theta = 2a_y\Delta y = -2g\Delta y$, so
$v_{fy} = \pm\sqrt{v_i^2\sin^2\theta - 2g\Delta y} = \pm\sqrt{(40\text{ m/s})^2\sin^2 37° - 2(9.80\text{ m/s}^2)(-7.0\text{ m})} = \boxed{-27\text{ m/s}}$,
where the negative sign was chosen because the cannonball is on its way down.

49. Strategy Use the result for the range given in Problem 48, part (b).

Solution Compute the ranges.

(a) $R = \dfrac{v_i^2\sin 2\theta}{g} = \dfrac{(36.2\text{ m/s})^2\sin 2(36.0°)}{9.80\text{ m/s}^2} = \boxed{127\text{ m}}$

$R = \dfrac{v_i^2\sin 2\theta}{g} = \dfrac{(36.2\text{ m/s})^2\sin 2(54.0°)}{9.80\text{ m/s}^2} = \boxed{127\text{ m}}$

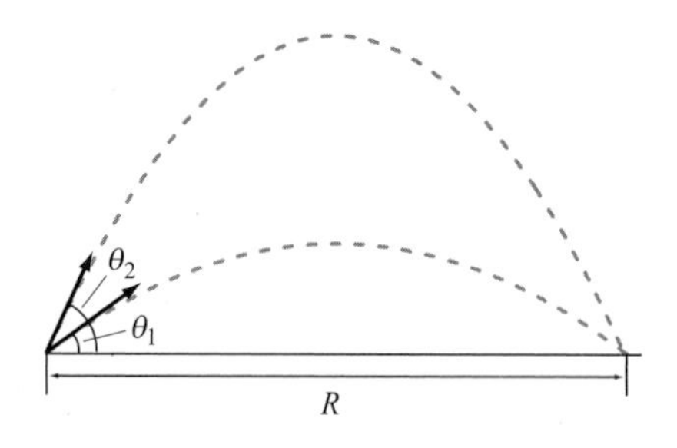

(b) $R = \dfrac{v_i^2\sin 2\theta}{g} = \dfrac{(36.2\text{ m/s})^2\sin 2(23.0°)}{9.80\text{ m/s}^2} = \boxed{96.2\text{ m}}$

$R = \dfrac{v_i^2\sin 2\theta}{g} = \dfrac{(36.2\text{ m/s})^2\sin 2(67.0°)}{9.80\text{ m/s}^2} = \boxed{96.2\text{ m}}$

(c) $R = \dfrac{v_i^2\sin 2\theta}{g} = \dfrac{(36.2\text{ m/s})^2\sin 2(45.0°)}{9.80\text{ m/s}^2} = \boxed{134\text{ m}}$

(d) The ranges are the same for each pair of complementary angles. The largest range occurred for an angle of 45.0° above the horizontal.

53. Strategy When Jaden is on the ground, his weight is equal to $mg = 600$ N. While on the accelerating elevator, his apparent weight is 550 N. Since 550 N < 600 N, the acceleration must be downward.

Solution According to Newton's second law, $\Sigma F_y = N - W = ma_y$, so

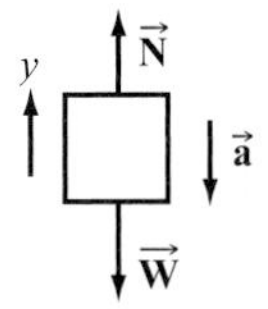

$$a_y = \frac{N-W}{m} = \frac{W'-W}{W/g} = g\left(\frac{W'}{W} - 1\right) = (9.80\ \text{m/s}^2)\left(\frac{550\ \text{N}}{600\ \text{N}} - 1\right) = -0.8\ \text{m/s}^2,\ \text{or}$$

$\vec{\mathbf{a}} = \boxed{0.8\ \text{m/s}^2\ \text{downward}}$.

55. Strategy Refer to Example 4.12.

Solution

(a) The elevator is accelerating downward, so $a_y = -0.50\ \text{m/s}^2$.

$$W' = \frac{W}{g}(g + a_y) = W\left(1 + \frac{a_y}{g}\right) = (598\ \text{N})\left(1 + \frac{-0.50\ \text{m/s}^2}{9.80\ \text{m/s}^2}\right) = \boxed{567\ \text{N}}$$

(b) Since the elevator is moving downward and slowing down, it is accelerating upward, so $a_y = 0.50\ \text{m/s}^2$.

$$W' = W\left(1 + \frac{a_y}{g}\right) = (598\ \text{N})\left(1 + \frac{0.50\ \text{m/s}^2}{9.80\ \text{m/s}^2}\right) = \boxed{629\ \text{N}}$$

57. Strategy The apparent weight is given by $W' = W(1 + a_y/g)$.

Solution Up is the positive direction. Find Felipe's actual weight.

$$W' = W\left(1 + \frac{a_y}{g}\right),\ \text{so}\ W = \frac{W'}{1 + \frac{a_y}{g}} = \frac{750\ \text{N}}{1 + \frac{2.0\ \text{m/s}^2}{9.80\ \text{m/s}^2}} = \boxed{620\ \text{N}}.$$

61. (a) Strategy Use Eq. (4-13). The force of air resistance is directed opposite the paratrooper's motion.

Solution Find the magnitude of the force of air resistance.

$F_\text{d} = bv^2 = (0.14\ \text{N}\cdot\text{s}^2/\text{m}^2)(64\ \text{m/s})^2 = 570\ \text{N}$, so $\vec{\mathbf{F}}_\text{d} = \boxed{570\ \text{N up}}$.

(b) Strategy Use Newton's second law.

Solution Up is the positive direction.

$\Sigma F_y = F_\text{d} - W = ma_y$, so

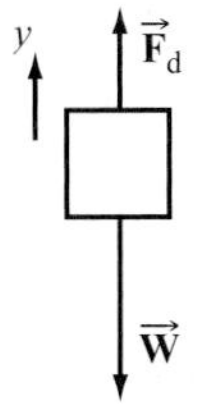

$$a_y = \frac{F_\text{d} - W}{m} = \frac{F_\text{d} - mg}{m} = \frac{F_\text{d}}{m} - g = \frac{570\ \text{N}}{120\ \text{kg}} - 9.80\ \text{m/s}^2 = -5.0\ \text{m/s}^2.$$

The acceleration is $\boxed{5.0\ \text{m/s}^2\ \text{downward}}$.

(c) Strategy Set $W = F_\text{d}$, or $mg = bv_\text{t}^2$, and solve for v_t.

Solution Find the terminal speed.

$$v_\text{t} = \sqrt{\frac{mg}{b}} = \sqrt{\frac{(120\ \text{kg})(9.80\ \text{m/s}^2)}{0.14\ \text{N}\cdot\text{s}^2/\text{m}^2}} = \boxed{92\ \text{m/s}}$$

65. (a) Strategy Find the time it takes the coconut to strike the ground using Eq. (4-9).

Solution Solve for Δt. The initial velocity in the vertical direction is zero.

$$\Delta y = v_{iy}\Delta t + \frac{1}{2}a_y(\Delta t)^2 = (0)\Delta t - \frac{1}{2}g(\Delta t)^2, \text{ so}$$

$$\Delta t = \sqrt{-\frac{2\Delta y}{g}} = \sqrt{-\frac{2(-100 \text{ m})}{9.80 \text{ m/s}^2}} = \boxed{4.5 \text{ s}}.$$

(b) Strategy Use the result of part (a) and $\Delta x = v_x \Delta t$ to find the horizontal distance the coconut travels, Δx.

Solution The horizontal distance the coconut travels from the release point is

$$\Delta x = v_x \Delta t = (18 \text{ m/s})(4.5 \text{ s}) = \boxed{81 \text{ m}}.$$

67. Strategy Let $+y$ be downward. $v_{ix} = v_x = v_i$ and $v_{iy} = 0$, so $\Delta x = v_i \Delta t$ and $\Delta y = \frac{1}{2}g(\Delta t)^2$.

Let the step number $n = \frac{x}{0.30 \text{ m}} = \frac{y}{0.18 \text{ m}}$, such that $n = 0$ to 1 represents step 1, $n = 1$ to 2 represents step 2, etc.

Solution Find the step the marble strikes first.

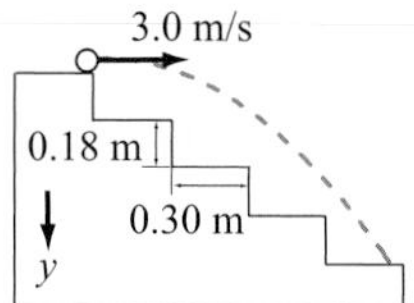

$$n = \frac{v_i \Delta t}{0.30 \text{ m}} = \frac{g(\Delta t)^2}{2(0.18 \text{ m})}, \text{ so } \Delta t = \frac{2(0.18 \text{ m})v_i}{(0.30 \text{ m})g} = 1.2\frac{v_i}{g}.$$

Therefore,

$$n = \frac{v_i}{0.30 \text{ m}}\left(1.2\frac{v_i}{g}\right) = (4.0 \text{ m}^{-1})\frac{v_i^2}{g} = (4.0 \text{ m}^{-1})\frac{(3.0 \text{ m/s})^2}{9.80 \text{ m/s}^2} = 3.7.$$

The value of n is between 3 and 4, so the marble first strikes $\boxed{\text{step 4}}$.

69. Strategy $v_t v/g$ has dimensions of meters. Let $v \approx v_t$.

Solution

(a) $$\frac{v_t v}{g} \approx \frac{v_t^2}{g} = \frac{(7 \text{ m/s})^2}{9.80 \text{ m/s}^2} = \boxed{5 \text{ m}}$$

(b) $$\frac{(100 \text{ m/s})^2}{9.80 \text{ m/s}^2} = \boxed{1 \text{ km}}$$

73. Strategy Use Eqs. (4-4) and (4-9). Let $+x$ be east and $+y$ be north.

Solution Find the components of the position vector $\vec{\mathbf{r}}$.

$$x_f = x_i + v_{ix}\Delta t + \frac{1}{2}a_x(\Delta t)^2 = 2.0 \text{ m} + (0)(2.0 \text{ s}) + \frac{1}{2}(5.0 \text{ m/s}^2)(2.0 \text{ s})^2 = 12 \text{ m}$$

and

$$y_f = y_i + v_{iy}\Delta t + \frac{1}{2}a_y(\Delta t)^2 = 0 + (20 \text{ m/s})(2.0 \text{ s}) + \frac{1}{2}(0)(2.0 \text{ s})^2 = 40 \text{ m}.$$

So, $\vec{\mathbf{r}} = \boxed{\text{12 m east and 40 m north}}$.

77. Strategy Each car has traveled the same distance Δx in the same time Δt when they meet.

Solution Using Eq. (4-4), we have

$\Delta x = v_i\Delta t + \frac{1}{2}a(\Delta t)^2 = 0 + \frac{1}{2}a(\Delta t)^2 = v\Delta t$, so $\Delta t = \frac{2v}{a}$. The speed of the police car is $v_p = a\Delta t = a(2v/a) = \boxed{2v}$.

81. Strategy $v_{1fx}{}^2 - v_{1ix}{}^2 = 2a_1d_1$, where $a_1 = 10.0\ \text{ft/s}^2$ and d_1 is the distance to the point of no return. $v_{2fx}{}^2 - v_{2ix}{}^2 = 2a_2d_2$, where $a_2 = -7.00\ \text{ft/s}^2$ and d_2 is the distance from the point of no return to the end of the runway. The initial speed v_{1ix} and the final speed v_{2fx} are zero. The speed at the point of no return is $v_{1fx} = v_{2ix}$. Let $v_{1fx} = v_{2ix} = v$ for simplicity. Also, $d = d_1 + d_2$ is the length of the runway.

Solution From the setup, we have $v^2 = 2a_1d_1$ and $-v^2 = 2a_2d_2 = 2a_2(d - d_1)$.

d_1 d_2
$d = 1.50$ mi

Eliminate v^2.

$$\begin{aligned} 2a_1d_1 &= 2a_2(d_1 - d) \\ a_1d_1 &= a_2d_1 - a_2d \\ (a_1 - a_2)d_1 &= -a_2d \\ d_1 &= \frac{a_2}{a_2 - a_1}d = \frac{-7.00\ \text{ft/s}^2}{-7.00\ \text{ft/s}^2 - 10.0\ \text{ft/s}^2}(1.50\ \text{mi})\left(\frac{5280\ \text{ft}}{1\ \text{mi}}\right) = \boxed{3260\ \text{ft}} \end{aligned}$$

Find the time to d_1 using Eq. (4-4).

$$d_1 = \frac{1}{2}a_1(\Delta t)^2, \text{ so } \Delta t = \sqrt{\frac{2d_1}{a_1}} = \sqrt{\frac{2(3260\ \text{ft})}{10.0\ \text{ft/s}^2}} = \boxed{25.5\ \text{s}}.$$

85. (a) Strategy The time required for the round-trip is equal to the round-trip distance divided by the cruising speed.

Solution

$$\Delta t = \frac{\text{round-trip distance}}{\text{cruising speed}} = \frac{2(5.80\times10^3\ \text{km})}{350.0\ \text{km/h}} = \boxed{33.1\ \text{h}}$$

(b) Strategy Consider the relative motion of the plane, air, and the ground.

Solution Subscripts: airplane = p; air = a; ground = g.

$$\begin{aligned} \Delta t &= \Delta t_{\text{tailwind}} + \Delta t_{\text{headwind}} \\ &= \frac{d}{v_{pa} + v_{ag}} + \frac{d}{v_{pa} - v_{ag}} = \frac{5.80\times10^3\ \text{km}}{350.0\ \text{km/h} + 60.0\ \text{km/h}} + \frac{5.80\times10^3\ \text{km}}{350.0\ \text{km/h} - 60.0\ \text{km/h}} = \boxed{34.1\ \text{h}} \end{aligned}$$

(c) Strategy Let $+y$ be antiparallel to the crosswind. Set $v_{pgy} = 0$, so the plane can travel in a straight line. $v_{pax} = v_{pa}\cos\theta$ is the speed of the plane along the straight line between the cities.

Solution

y x v_{pay} $\vec{\mathbf{v}}_{pa}$ θ New York v_{pax} Paris $\vec{\mathbf{v}}_{ag}$

$$v_{pgy} = v_{pay} + v_{agy} = v_{pa}\sin\theta - v_{ag} = 0, \text{ so } \theta = \sin^{-1}\frac{v_{ag}}{v_{pa}} = \sin^{-1}\frac{60.0}{350.0} = 9.87°.$$

$$\text{Thus, } \Delta t = \frac{2\Delta x}{v_{pa}\cos\theta} = \frac{2(5.80\times10^3\ \text{km})}{(350.0\ \text{km/h})\cos 9.87°} = \boxed{33.6\ \text{h}}.$$

89. Strategy Use $\Delta x = v_x \Delta t$ and Eq. (4-9).

Solution Find Δt in terms of Δx and v_{ix}.

$\Delta x = v_x \Delta t = v_{ix} \Delta t$, so $\Delta t = \dfrac{\Delta x}{v_{ix}}$.

Substitute the expression for Δt into Eq. (4-9), where $y_i = 0$, since the projectile is launched from the origin.

$$y_f = y_i + v_{iy}\Delta t + \frac{1}{2}a_y(\Delta t)^2 = 0 + v_{iy}\left(\frac{\Delta x}{v_{ix}}\right) - \frac{1}{2}g\left(\frac{\Delta x}{v_{ix}}\right)^2 = \left(\frac{v_{iy}}{v_{ix}}\right)\Delta x + \left(\frac{-g}{2v_{ix}^2}\right)(\Delta x)^2$$

93. Strategy Use the second witness's information to determine the average speed of the flowerpot as it passed the 18th-story window. Then, use this speed to determine the height above the window from which the flowerpot fell.

Solution The average speed of the flowerpot as it passed the window was

$$v_{av} = \frac{\Delta y}{\Delta t} = \frac{1.5 \text{ m}}{0.044 \text{ s}} = 34 \text{ m/s}.$$

This is the approximate speed of the flowerpot as it passed the middle of the window. Assuming the flowerpot started at rest, use Eq. (4-10) to find the distance above the window from which the flowerpot fell.

$$v_{fy}^2 - v_{iy}^2 = v_{fy}^2 - 0 = 2a_y\Delta y = -2g\Delta y, \text{ so } \Delta y = -\frac{v_{fy}^2}{2g} = -\frac{(-34 \text{ m/s})^2}{2(9.80 \text{ m/s}^2)} = -59 \text{ m}.$$

Thus, the flowerpot fell from about 59 m above the middle of the 18th-story window; that is, it fell from $59 \text{ m} + 75 \text{ m} - (1.5 \text{ m})/2 = 133 \text{ m}$ above the ground. If the flowerpot had fallen from the 24th story, it would have fallen $94 \text{ m} - 75 \text{ m} = 19 \text{ m}$ before it passed the second witness's window. This is not far enough for the speed to be as measured; so no, the flowerpot could not have fallen with zero initial velocity from the 24th-story window. It either fell from 133 m high or, if it came from a lower location (such as the 24th floor), it was thrown downward. The first witness is not credible.

Chapter 5

CIRCULAR MOTION

Problems

1. Strategy Find the arc length swept out by the carnival swing.

Solution Use Eq. (5-4).

$$s = r\theta = (8.0\text{ m})(120°)\left(\frac{2\pi\text{ rad}}{360°}\right) = \boxed{17\text{ m}}$$

5. Strategy Use Eq. (5-9) to find the angular speed of the bicycle's tires.

Solution

$$|\omega| = \frac{v}{r} = \frac{9.0\text{ m/s}}{0.35\text{ m}} = \boxed{26\text{ rad/s}}$$

7. (a) Strategy and Solution There are 2π radians per revolution and 60 seconds per minute, so

$$\left(\frac{33.3\text{ rev}}{\text{min}}\right)\left(\frac{2\pi\text{ rad}}{\text{rev}}\right)\left(\frac{1\text{ min}}{60\text{ s}}\right) = \boxed{3.49\text{ rad/s}}.$$

(b) Strategy Use the relationship between linear speed and angular speed.

Solution Find the speed of the doll.

$$v = r|\omega| = (0.13\text{ m})(3.49\text{ rad/s}) = \boxed{0.45\text{ m/s}}$$

9. Strategy Use the conversion factor between degrees and radians and $s = r\theta$, where $s = 100.0$ ft, $\theta = 1.5°$, and r is the radius of curvature.

Solution Find the radius of curvature of a "1.5° curve".

$$r = \frac{s}{\theta} = \frac{100.0\text{ ft}}{1.5°}\left(\frac{360°}{2\pi\text{ rad}}\right) = \boxed{3800\text{ ft}}$$

13. (a) Strategy Use the relationship between linear speed and radial acceleration.

Solution The radius r is half the length of the rod.

$$a_\text{r} = \frac{v^2}{r},\text{ so } v = \sqrt{ra_\text{r}} = \sqrt{(1.0\text{ m})(980\text{ m/s}^2)} = \boxed{31\text{ m/s}}.$$

(b) Strategy Use the relationship between angular speed and radial acceleration.

Solution

$$a_\text{r} = \omega^2 r,\text{ so } |\omega| = \sqrt{\frac{a_\text{r}}{r}} = \sqrt{\frac{980\text{ m/s}^2}{1.0\text{ m}}} = \boxed{31\text{ rad/s}}.$$

17. (a) Strategy Use Newton's second law and the relationship between linear speed and radial acceleration.

Solution According to Newton's second law,

$$\Sigma F_{\mathrm{r}} = T = ma_{\mathrm{r}} = m\frac{v^2}{r} = m\frac{v^2}{L}, \text{ thus, } T = \boxed{\frac{mv^2}{L}}.$$

(b) Strategy Draw a free-body diagram for the rock. Use Newton's second law.

Solution Decompose the force into vertical (y) and radial (r) components.

$$\Sigma F_{\mathrm{r}} = T_{\mathrm{r}} = \frac{mv^2}{L} \text{ and } \Sigma F_y = T_y - mg = 0.$$

Find the magnitude of the tension.

$$T = \sqrt{T_{\mathrm{r}}^2 + T_y^2} = \boxed{\sqrt{\left(\frac{mv^2}{L}\right)^2 + (mg)^2}}$$

Find the direction of the tension with respect to the horizontal.

$$\theta = \tan^{-1}\frac{T_y}{T_{\mathrm{r}}} = \tan^{-1}\frac{mg}{\frac{mv^2}{L}} = \boxed{\tan^{-1}\frac{gL}{v^2}}$$

21. Strategy Let the x-axis point toward the center of curvature and the y-axis point upward. Draw a free-body diagram. Use Newton's second law and the relationship between radial acceleration and linear speed.

Solution

$$\Sigma F_y = N\cos\theta - mg = 0, \text{ so } N\cos\theta = mg, \text{ and } \Sigma F_x = N\sin\theta = ma_{\mathrm{r}} = m\frac{v^2}{r}.$$

Solve for v.

$$\frac{N\sin\theta}{N\cos\theta} = \frac{m\frac{v^2}{r}}{mg}, \text{ so } v^2 = rg\tan\theta, \text{ or}$$

$$v = \sqrt{rg\tan\theta} = \sqrt{(120\text{ m})(9.80\text{ m/s}^2)\tan 3.0°} = \boxed{7.9\text{ m/s}}.$$

25. Strategy Let the x-axis point toward the center of curvature and the y-axis point upward. Draw a free-body diagram. Use Newton's second law and the relationship between radial acceleration and linear speed.

Solution

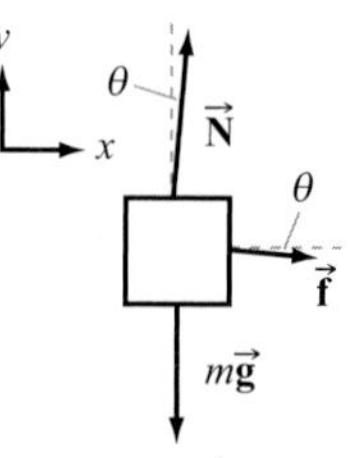

(a) $\Sigma F_y = N\cos\theta - mg - f\sin\theta = 0$ and $\Sigma F_x = N\sin\theta + f\cos\theta = mv^2/r$.

Solve for N in the first equation and substitute into the second.

$$N = \frac{f\sin\theta + mg}{\cos\theta}, \text{ so}$$

$$\frac{f\sin\theta + mg}{\cos\theta}\sin\theta + f\cos\theta = m\frac{v^2}{r}$$

$$f\sin^2\theta + mg\sin\theta + f\cos^2\theta = m\frac{v^2}{r}\cos\theta$$

$$f(\sin^2\theta+\cos^2\theta)=m\frac{v^2}{r}\cos\theta-mg\sin\theta$$

$$f(1)=m\left(\frac{v^2}{r}\cos\theta-g\sin\theta\right)$$

$$f=(1400\text{ kg})\left[\frac{(32\text{ m/s})^2}{410\text{ m}}\cos5.0°-(9.80\text{ m/s}^2)\sin5.0°\right]=\boxed{2300\text{ N}}$$

(2288 N)

(b) Set the expression found for the force of friction equal to zero.

$$f=0=m\left(\frac{v^2}{r}\cos\theta-g\sin\theta\right),\text{ so }\frac{v^2}{r}\cos\theta=g\sin\theta,\text{ or}$$

$$v=\sqrt{gr\tan\theta}=\sqrt{(9.80\text{ m/s}^2)(410\text{ m})\tan5.0°}=\boxed{19\text{ m/s}}.$$

(18.8 m/s)

29. Strategy Use $v=r\omega$ and $\omega=2\pi/T$, where the radius is the average Earth-Sun distance and the period is one year.

Solution

$$v=r\omega=r\left(\frac{2\pi}{T}\right)=\frac{2\pi r}{T}=\frac{2\pi(1.50\times10^{11}\text{ m})}{1\text{ y}}\left(\frac{1\text{ y}}{3.156\times10^7\text{ s}}\right)=\boxed{2.99\times10^4\text{ m/s}}$$

31. Strategy According to Kepler's third law, $r^3\propto T^2$. Form a proportion.

Solution Find the orbital period of the second satellite.

$$\left(\frac{4.0r}{r}\right)^3=64=\left(\frac{T_{4.0}}{T}\right)^2=\frac{T_{4.0}{}^2}{T^2},\text{ so }T_{4.0}{}^2=64T^2,\text{ or }T_{4.0}=8.0T=8.0(16\text{ h})=\boxed{130\text{ h}}.$$

33. Strategy Use Eq. (5-14) with the mass of Jupiter in place of the mass of the Sun.

Solution Solve for r.

$$\frac{4\pi^2}{GM_\text{J}}r^3=T^2,\text{ so }r^3=\frac{GM_\text{J}}{4\pi^2}T^2,\text{ or }r=\sqrt[3]{\frac{GM_\text{J}}{4\pi^2}T^2}.$$

Compute the distance from the center of Jupiter for each satellite.

$$r_\text{Io}=\sqrt[3]{\frac{GM_\text{J}}{4\pi^2}T_\text{Io}{}^2}=\sqrt[3]{\frac{(6.674\times10^{-11}\text{ N}\cdot\text{m}^2/\text{kg}^2)(1.9\times10^{27}\text{ kg})}{4\pi^2}(1.77\text{ d})^2\left(\frac{86,400\text{ s}}{1\text{ d}}\right)^2}=\boxed{420,000\text{ km}}$$

$$r_\text{Europa}=\sqrt[3]{\frac{GM_\text{J}}{4\pi^2}T_\text{Europa}{}^2}=\sqrt[3]{\frac{(6.674\times10^{-11}\text{ N}\cdot\text{m}^2/\text{kg}^2)(1.9\times10^{27}\text{ kg})}{4\pi^2}(3.54\text{ d})^2\left(\frac{86,400\text{ s}}{1\text{ d}}\right)^2}=\boxed{670,000\text{ km}}$$

37. Strategy Use Newton's second law and law of universal gravitation, as well as the relationship between radial acceleration and linear speed.

Solution

$$\Sigma F_\mathrm{r} = \frac{GmM_\mathrm{J}}{(3.0\ R_\mathrm{J})^2} = \frac{mv^2}{3.0R_\mathrm{J}}$$

Now, the gravitational field strength of Jupiter is given by $g_\mathrm{J} = \dfrac{GM_\mathrm{J}}{R_\mathrm{J}^2}$. Find the period of the spacecraft's orbit.

$\dfrac{mv^2}{3.0R_\mathrm{J}} = \dfrac{mg_\mathrm{J}}{9.0}$, so $v^2 = \dfrac{g_\mathrm{J}R_\mathrm{J}}{3.0} = \left(\dfrac{2\pi r}{T}\right)^2 = \dfrac{4\pi^2(3.0R_\mathrm{J})^2}{T^2}$. Solving for T, we have

$$T = 2\pi\sqrt{\frac{27R_\mathrm{J}}{g_\mathrm{J}}} = 2\pi\sqrt{\frac{27(71{,}500\times10^3\ \mathrm{m})}{23\ \mathrm{N/kg}}}\left(\frac{1\ \mathrm{h}}{3600\ \mathrm{s}}\right) = \boxed{16\ \mathrm{h}}.$$

41. Strategy Use Newton's second law and the relationship between radial acceleration and linear speed.

Solution The only forces acting on the car are gravity and the normal force of the ground pushing on the car. The radial acceleration is downward, or toward the center of the radius of curvature. Let up be positive.

$$\Sigma F_\mathrm{r} = N - mg = ma_\mathrm{r} = m\left(-\frac{v^2}{r}\right)$$

When the car is just in contact with the ground, the normal force must be zero. If the car goes any faster it will lose contact with the road. Solve for the speed.

$$-\frac{mv^2}{r} = N - mg = 0 - mg,\ \text{so}\ v^2 = gr,\ \text{or}\ v = \sqrt{gr} = \sqrt{(9.80\ \mathrm{m/s^2})(55.0\ \mathrm{m})} = \boxed{23.2\ \mathrm{m/s}}.$$

43. Strategy and Solution Refer to the figure. If the $+y$-direction is radial and the $+x$-direction is tangential to the left, according to Newton's second law, $\Sigma F_x = mg\sin\theta = ma_\mathrm{t}$, or $a_\mathrm{t} = \boxed{g\sin\theta}$.

45. Strategy Equations (5-18) and (5-19) are $\Delta\omega = \omega_\mathrm{f} - \omega_\mathrm{i} = \alpha\Delta t$ and $\Delta\theta = \frac{1}{2}(\omega_\mathrm{f} + \omega_\mathrm{i})\Delta t$, respectively.

Solution Solve for Δt in Eq. (5-18), then substitute the result into Eq. (5-19) and simplify.

$\Delta\omega = \omega_\mathrm{f} - \omega_\mathrm{i} = \alpha\Delta t$, so $\Delta t = \dfrac{\omega_\mathrm{f} - \omega_\mathrm{i}}{\alpha}$. Substitute.

$$\Delta\theta = \frac{1}{2}(\omega_\mathrm{f} + \omega_\mathrm{i})\Delta t = \frac{1}{2}(\omega_\mathrm{f} + \omega_\mathrm{i})\left(\frac{\omega_\mathrm{f} - \omega_\mathrm{i}}{\alpha}\right) = \frac{\omega_\mathrm{f}^2 - \omega_\mathrm{i}^2}{2\alpha},\ \text{so}\ \omega_\mathrm{f}^2 - \omega_\mathrm{i}^2 = 2\alpha\Delta\theta,\ \text{which is Eq. (5-21).}$$

47. Strategy Use Eq. (5-20) to find the constant angular acceleration.

Solution Since the cyclist starts from rest, the initial angular velocity is zero.

$$\Delta\theta = \omega_\mathrm{i}\Delta t + \frac{1}{2}\alpha(\Delta t)^2 = (0)\Delta t + \frac{1}{2}\alpha(\Delta t)^2 = \frac{1}{2}\alpha(\Delta t)^2,\ \text{so}\ \alpha = \frac{2\Delta\theta}{(\Delta t)^2} = \frac{2(8.0\ \mathrm{rev})\left(\frac{2\pi\ \mathrm{rad}}{\mathrm{rev}}\right)}{(5.0\ \mathrm{s})^2} = \boxed{4.0\ \mathrm{rad/s^2}}.$$

49. Strategy Use Eq. (5-21) to find the angular acceleration.

Solution

$$\omega_\text{f}^2 - \omega_\text{i}^2 = 2\alpha\Delta\theta, \text{ so } \alpha = \frac{\omega_\text{f}^2 - \omega_\text{i}^2}{2\Delta\theta} = \frac{(0.50 \text{ rev/s})^2 - 0}{2(2.0 \text{ rev})}\left(\frac{2\pi \text{ rad}}{\text{rev}}\right) = \boxed{0.39 \text{ rad/s}^2}.$$

53. (a) Strategy Use Eq. (5-18) to find the time it takes for the rotor to come to rest.

Solution The acceleration is opposite the rotation of the rotor, so it is negative.

$$\Delta t = \frac{\omega_\text{f} - \omega_\text{i}}{\alpha} = \frac{0 - 5.0\times10^5 \text{ rad/s}}{-0.40 \text{ rad/s}^2} = \boxed{1.3\times10^6 \text{ s}}$$

(b) Strategy Use Eq. (5-21) to find the number of revolutions the rotor spun before it stopped.

Solution

$$\Delta\theta = \frac{\omega_\text{f}^2 - \omega_\text{i}^2}{2\alpha}$$

$$\left(\frac{1 \text{ rev}}{2\pi \text{ rad}}\right)\Delta\theta = \frac{0 - (5.0\times10^5 \text{ rad/s})^2}{2(-0.40 \text{ rad/s}^2)}\left(\frac{1 \text{ rev}}{2\pi \text{ rad}}\right) = \boxed{5.0\times10^{10} \text{ rev}}$$

57. Strategy Use $a_\text{r} = \omega^2 r$ to find the angular speed required.

Solution The magnitude of the radial acceleration must be the same as the magnitude of the gravitational field strength.

$$a_\text{r} = \omega^2 r = g, \text{ so } \omega = \sqrt{\frac{g}{r}} = \sqrt{\frac{9.80 \text{ m/s}^2}{0.20 \text{ m}}} = \boxed{7.0 \text{ rad/s}}.$$

59. (a) Strategy At the top, $\vec{\mathbf{g}}$ and $\vec{\mathbf{a}}$ are both directed downward. Draw a free-body diagram. Use Newton's second law and the relationship between radial acceleration and angular speed.

Solution

$$\Sigma F_\text{r} = N - mg = ma_\text{r} = m(-\omega^2 R), \text{ so } W' = N = \boxed{m(g - \omega^2 R)}.$$

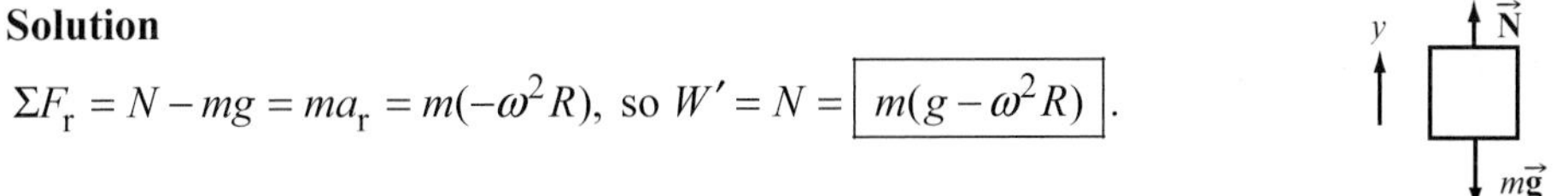

(b) Strategy At the bottom, $\vec{\mathbf{g}}$ is directed downward and $\vec{\mathbf{a}}$ is directed upward. Draw a free-body diagram.

Solution

$$\Sigma F_\text{r} = N - mg = ma_\text{r} = m(\omega^2 R), \text{ so } W' = N = \boxed{m(g + \omega^2 R)}.$$

61. Strategy The strength of the artificial gravity is equal to the radial acceleration.

Solution Compute the magnitude of the radial acceleration.

$$a_\text{r} = \omega^2 r = \left(\frac{\omega^2 r}{g}\right)g = \frac{(4.0 \text{ rev/s})^2\left(\frac{2\pi \text{ rad}}{\text{rev}}\right)^2(0.25 \text{ m})}{9.80 \text{ m/s}^2}g = \boxed{16g}$$

65. Strategy Use the relationship between linear speed and angular speed.

Solution Compute the linear speed of the tip of the nylon cord.

$v = \omega r = (660 \text{ rad/s})(0.23 \text{ m}) = \boxed{150 \text{ m/s}}$

69. Strategy Use Eq. (5-20) to find the number of rotations each gear goes through in 2.0 s. Refer to Problem 68.

Solution

$$\Delta\theta_A = \omega_i \Delta t + \frac{1}{2}\alpha(\Delta t)^2$$

$$\left(\frac{1 \text{ rotation}}{2\pi \text{ rad}}\right)\Delta\theta_A = \left(\frac{1 \text{ rotation}}{2\pi \text{ rad}}\right)\left[2\pi(0.955 \text{ Hz})(2.0 \text{ s}) + \frac{1}{2}(3.0 \text{ rad/s}^2)(2.0 \text{ s})^2\right] = \boxed{2.9 \text{ rotations}}$$

From Problem 68, we know that gear *B* has an angular speed that is twice that of gear *A*. Thus, gear *B* rotates twice for each rotation of gear *A*. Therefore,

$$\left(\frac{1 \text{ rotation}}{2\pi \text{ rad}}\right)\Delta\theta_B = 2\left(\frac{1 \text{ rotation}}{2\pi \text{ rad}}\right)\left[2\pi(0.955 \text{ Hz})(2.0 \text{ s}) + \frac{1}{2}(3.0 \text{ rad/s}^2)(2.0 \text{ s})^2\right] = \boxed{5.7 \text{ rotations}}.$$

73. (a) Strategy Use the relationship between radial acceleration and linear speed.

Solution

$$a_r = \frac{v^2}{r} = \frac{(2.0\pi \text{ m/s})^2}{0.50 \text{ m}} = \boxed{8.0\pi^2 \text{ m/s}^2 = 79 \text{ m/s}^2}$$

(b) Strategy Use Newton's second law.

Solution

$\Sigma F_r = T = ma_r$, so $T = (0.50 \text{ kg})(8.0\pi^2 \text{ m/s}^2) = \boxed{4.0\pi^2 \text{ N} = 39 \text{ N}}$.

77. Strategy Use Eq. (5-18).

Solution After one minute:

$\omega_f - \omega_i = 0.80\omega - \omega = -0.20\omega = \alpha\Delta t$

After three minutes:

$\omega_f - \omega_i = \omega_f - \omega = \alpha(3\Delta t) = 3\alpha\Delta t = 3(-0.20\omega)$, so $\omega_f = -0.60\omega + \omega = \boxed{0.40\omega}$.

81. Strategy Use the relationship between linear speed and angular speed.

Solution

$$v = r|\omega| = \left(\frac{0.65 \text{ m}}{2}\right)(101 \text{ rad/s})\left(\frac{1 \text{ km}}{1000 \text{ m}}\right)\left(\frac{3600 \text{ s}}{\text{h}}\right) = \boxed{120 \text{ km/h}}$$

85. Strategy For each revolution of the flagellum, the bacterium moves the distance of the pitch.

Solution Compute the speed of the bacterium.

$v = (1.0 \text{ μm/rev})(110 \text{ rev/s}) = \boxed{110 \text{ μm/s}}$

REVIEW AND SYNTHESIS: CHAPTERS 1–5

Review Exercises

1. **Strategy** Replace the quantities with their units.

   Solution Find the units of the spring constant k.

   $F = kx$, so $k = \frac{F}{x}$, and the units of k are $\boxed{\text{N/m}} = \frac{\text{kg}\cdot\text{m/s}^2}{\text{m}} = \boxed{\text{kg/s}^2}$.

5. **Strategy** To reach the other side of the river in as short a time as possible, Paula must swim in the direction perpendicular to the river's flow. Find the time it takes for Paula to cross. Then use this time and the speed of the river to find how far downstream she travels while crossing.

   Solution It takes Paula a time $\Delta t = \Delta x/v_{\text{swim}}$ to cross the river. During this time, she travels

   $\Delta y = v_{\text{river}}\Delta t = v_{\text{river}}\left(\frac{\Delta x}{v_{\text{swim}}}\right) = \frac{(1.43\ \text{m/s})(10.2\ \text{m})}{0.833\ \text{m/s}} = \boxed{17.5\ \text{m}}$ downstream.

   10.2 m
   0.833 m/s
   1.43 m/s

9. **(a) Strategy and Solution** Answers will vary, but a reasonable magnitude of the force required to pull out a single hair is $\boxed{1\ \text{N}}$.

   (b) Strategy The total force exerted on all of the hairs is the weight of the prince, $W = mg$. Dividing his weight by the number of hairs gives the average force pulling on each strand of hair.

   Solution Will Rapunzel be bald?

   $F_{\text{per hair}} = \frac{W}{100{,}000} = \frac{(60\ \text{kg})(9.80\ \text{N/kg})}{100{,}000} = \boxed{6\ \text{mN}}$

   Since 6 mN << 1 N, $\boxed{\text{Rapunzel will most certainly not be made bald}}$ by the prince climbing up her hair.

13. **Strategy** The point at which the gravitational field is zero is somewhere along the line between the centers of the two stars. Use Newton's law of universal gravitation.

    Solution The distance between the stars is $d = d_1 + d_2$, where d_1 is the distance to the $F = 0$ point from the star with mass M_1 and d_2 is the distance to the $F = 0$ point from the other star. The forces are equal in magnitude and opposite in direction at the $F = 0$ point. Let m be a test mass at the $F = 0$ point.

    $\frac{GM_1m}{d_1^2} = \frac{G(4.0M_1)m}{d_2^2}$, so $d_2 = 2.0d_1$.

    $d = d_1 + d_2 = d_1 + 2.0d_1 = 3.0d_1$, so $d_1 = \frac{d}{3.0} = 0.33d$.

    $\boxed{\text{The gravitational field is zero approximately one third (0.33) of the distance between the stars as measured from the star with mass } M_1.}$

17. (a) Strategy Use Eq. (4-10). Let the $+y$-direction be down and neglect air resistance.

Solution Let the initial speed for all three rocks be v_i and the vertical distance from the cliff to the ground be h. For the first rock (thrown straight down):

$v_{fy}^2 - v_{iy}^2 = v_f^2 - v_i^2 = 2gh$, so $v_f = \sqrt{v_i^2 + 2gh}$.

For the second rock (thrown straight up):

$v_{fy}^2 - v_{iy}^2 = v_f^2 - v_i^2 = 2gh$, so $v_f = \sqrt{v_i^2 + 2gh}$.

For the third rock (thrown horizontally):

$v_{fy}^2 - v_{iy}^2 = v_{fy}^2 - 0 = v_{fy}^2 = 2gh$ and $v_{fx} = v_{ix} = v_i$, so $v_f = \sqrt{v_{fx}^2 + v_{fy}^2} = \sqrt{v_i^2 + 2gh}$.

Therefore, just before the rocks hit the ground at the bottom of the cliff, all three have the same final speed.

(b) Strategy Use the result obtained for the final speed in part (a).

Solution Compute the final speed.

$$v_f = \sqrt{v_i^2 + 2gh} = \sqrt{(10.0\text{ m/s})^2 + 2(9.80\text{ m/s}^2)(15.00\text{ m})} = \boxed{19.8\text{ m/s}}$$

21. Strategy Draw a diagram. Use Newton's second law and the relationship between radial acceleration and linear speed. Refer to Example 5.7.

Solution From Example 5-7, the banking angle is given by $\theta = \tan^{-1}[v_b^2/(rg)]$, where v_b is the speed that a car can navigate the curve without any friction. Find the relationship for the *slowest* speed the car can go around the curve without sliding *down* the bank.

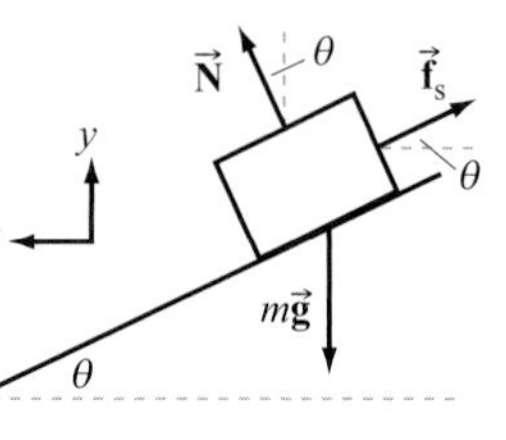

$$\Sigma F_y = N_y - mg + f_y = 0 \text{ and } \Sigma F_x = N_x - f_x = ma_r = m\frac{v^2}{r}, \text{ so}$$

$$m\frac{v^2}{r} = N_x - f_x = N\sin\theta - \mu_s N\cos\theta, \text{ or } v = \sqrt{\frac{rN}{m}(\sin\theta - \mu_s\cos\theta)}.$$

Find N.

$$N_y - mg + f_y = N\cos\theta - mg + \mu_s N\sin\theta = 0, \text{ so } N(\cos\theta + \mu_s\sin\theta) = mg, \text{ or } N = \frac{mg}{\cos\theta + \mu_s\sin\theta}.$$

Substitute for N in v and substitute for θ.

$$v = \sqrt{\frac{r}{m}\left(\frac{mg}{\cos\theta + \mu_s\sin\theta}\right)(\sin\theta - \mu_s\cos\theta)} = \sqrt{\frac{gr(\sin\theta - \mu_s\cos\theta)}{\cos\theta + \mu_s\sin\theta}} = \sqrt{\frac{gr(\tan\theta - \mu_s)}{1 + \mu_s\tan\theta}}$$

$$= \sqrt{\frac{gr[\tan\tan^{-1} v_b^2/(rg) - \mu_s]}{1 + \mu_s\tan\tan^{-1} v_b^2/(rg)}} = \sqrt{\frac{gr[v_b^2/(rg) - \mu_s]}{1 + \mu_s[v_b^2/(rg)]}} = \sqrt{\frac{v_b^2 - \mu_s gr}{1 + \mu_s v_b^2/(rg)}}$$

Thus, the slowest allowed speed is $v = \sqrt{\frac{(15.0\text{ m/s})^2 - 0.120(9.80\text{ m/s}^2)(75.0\text{ m})}{1 + 0.120(15.0\text{ m/s})^2/[(9.80\text{ m/s}^2)(75.0\text{ m})]}} = \boxed{11.5\text{ m/s}}$.

25. Strategy Use the equations of motion for changing velocity.

Solution $\Delta t = \Delta t_1 + \Delta t_2 = 0.10\text{ s} + 0.15\text{ s} = 0.25\text{ s}$ has elapsed since the first pellet was fired from the toy gun. The components of its velocity are given by $v_{1x} = v_i \cos\theta$ and $v_{1y} = v_i \sin\theta - g\Delta t$. The components of the second pellet's velocity are given by $v_{2x} = v_i \cos\theta$ and $v_{2y} = v_i \sin\theta - g\Delta t_2$. The horizontal components of the two velocities are the same, but the vertical components differ. Compute this difference.

$$v_{1y} - v_{2y} = v_i \sin\theta - g\Delta t - (v_i \sin\theta - g\Delta t_2) = -g(\Delta t - \Delta t_2) = -(9.80\text{ m/s}^2)(0.25\text{ s} - 0.15\text{ s}) = -0.98\text{ m/s}$$

So, the velocity of the first pellet with respect to the second after the additional 0.15 s have passed is $\boxed{0.98\text{ m/s directed downward}}$.

29. (a) Strategy The circumference of the track is given by $C = 2\pi r$. The distance traveled is three-fourths of this.

Solution Compute the distance traveled by the runner before the collision.

$$\text{distance traveled} = \frac{3}{4}(2\pi r) = \frac{3}{2}\pi(60.0\text{ m}) = \boxed{283\text{ m}}$$

(b) Strategy Let the center of the circle be the origin, and let the runner begin at $\theta = 0°$ and collide at $\theta = 270°$.

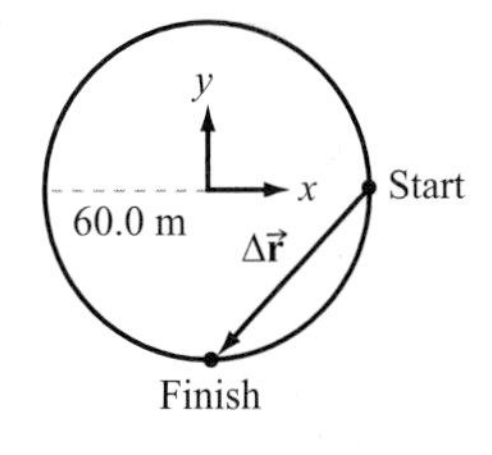

Solution Find the components of the runner's displacement.

$r_{ix} = 60.0\text{ m}$, $r_{iy} = 0$, $r_{fx} = 0$ and $r_{fy} = -60.0\text{ m}$.

Find the magnitude of the displacement.

$$|\Delta\vec{\mathbf{r}}| = \sqrt{(r_{fx} - r_{ix})^2 + (r_{fy} - r_{iy})^2} = \sqrt{(-60.0\text{ m})^2 + (-60.0\text{ m})^2} = \boxed{84.9\text{ m}}$$

MCAT Review

1. Strategy and Solution Gravity contributes an acceleration of $-g$. Air resistance is always opposite an object's direction of motion, so the vertical component of the acceleration contributed by air resistance is negative as well. According to Newton's second law, $F = ma$, so the magnitude of the acceleration due to air resistance is $a_R = F_R/m = bv^2/m$. Since we want the vertical component of acceleration, the correct answer is $\boxed{D}$, $-g - (bvv_y)/(0.5\text{ kg})$.

2. Strategy Use the result for the range derived in Problem 4.48b.

Solution Assuming air resistance is negligible, the horizontal distance the projectile travels before returning to the elevation from which it was launched is $R = \dfrac{v_i^2 \sin 2\theta}{g} = \dfrac{(30\text{ m/s})^2 \sin[2(40°)]}{9.80\text{ m/s}^2} = 90\text{ m}$. Thus, the correct

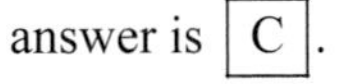

answer is $\boxed{C}$.

3. Strategy and Solution The magnitude of the horizontal component of air resistance is $F_R \cos\theta = bv^2 \cos\theta = bv(v\cos\theta) = bvv_x$. Thus, the correct answer is $\boxed{D}$.

4. **Strategy** Use Newton's second law to analyze each case. For simplicity, consider only vertical motion.

   Solution Let the positive *y*-direction be up.
   On the way up:

   $$\Sigma F_y = -mg - bv^2 = ma_y, \text{ so } a_y = -g - \frac{bv^2}{m}.$$

   On the way down:

   $$\Sigma F_y = -mg + bv^2 = ma_y, \text{ so } a_y = -g + \frac{bv^2}{m}.$$

   The magnitude of the acceleration is greater on the way up than on the way down. On the way up, the magnitude of the acceleration is never less than *g*. On the way down, it may be as small as zero. The projectile must travel the same distance in each case. So, when a projectile is rising, it begins with an initial speed which is reduced to zero relatively quickly due to the relatively large negative acceleration it experiences. When a projectile is falling, it begins with zero speed and is accelerated toward the ground by a smaller acceleration relative to when it is rising. Thus, it must take the projectile longer to reach the ground than to reach its maximum height; therefore, the correct answer is $\boxed{\text{C}}$.

5. **Strategy** Find the time it takes to cross the river. Use this time and the speed of the river to find how far downstream the raft travels while crossing. Then use the Pythagorean theorem to find the total distance traveled.

   Solution Let x be the width of the river and y be the distance traveled down the river during the crossing. The raft takes the time $\Delta t = x/v_{\text{raft}}$ to cross the river. During this time, the raft travels the distance $y = v_{\text{river}}\Delta t = v_{\text{river}}(x/v_{\text{raft}})$ down the river. Compute the distance traveled.

   $$\sqrt{x^2 + y^2} = \sqrt{x^2 + \left(\frac{v_{\text{river}}}{v_{\text{raft}}}x\right)^2} = x\sqrt{1 + \left(\frac{v_{\text{river}}}{v_{\text{raft}}}\right)^2} = (200\text{ m})\sqrt{1 + \left(\frac{2\text{ m/s}}{2\text{ m/s}}\right)^2} = 283\text{ m}$$

   The correct answer is $\boxed{\text{C}}$.

6. **Strategy** To row directly across the river, the component of the raft's velocity that is antiparallel to the current of the river must equal the speed of the current, 2 m/s.

   Solution Since the angle is relative to the shore, the antiparallel component of the raft's velocity is $(3\text{ m/s})\cos\theta$. Set this equal to the speed of the current and solve for θ.

   $(3\text{ m/s})\cos\theta = 2\text{ m/s}$, so $\cos\theta = \frac{2}{3}$, or $\theta = \cos^{-1}\frac{2}{3}$. The correct answer is $\boxed{\text{D}}$.

7. **Strategy** Use Eq. (4-9).

   Solution Find the time it takes the rock to reach the ground.

   $$\Delta y = v_{iy}\Delta t + \frac{1}{2}a_y(\Delta t)^2 = (0)\Delta t - \frac{1}{2}g(\Delta t)^2, \text{ so } \Delta t = \sqrt{-\frac{2\Delta y}{g}} = \sqrt{-\frac{2(0 - 100\text{ m})}{10\text{ m/s}^2}} = 4.5\text{ s}.$$

   The correct answer is $\boxed{\text{A}}$.

Chapter 6

CONSERVATION OF ENERGY

Problems

1. Strategy Use Eq. (6-1).

Solution Find the work done by Denise dragging her basket of laundry.

$W = F\Delta r\cos\theta = (30.0\text{ N})(5.0\text{ m})\cos 60.0° = \boxed{75\text{ J}}$

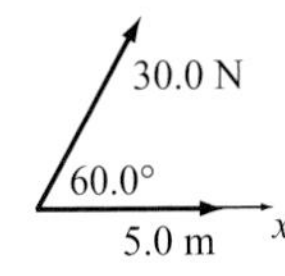

3. Strategy and Solution Since the book undergoes no displacement, $\boxed{\text{no work is done}}$ *on the book* by Hilda.

5. Strategy Use Newton's second law and Eq. (6-2).

Solution Find the net force on the barge.

$\Sigma F_y = T\sin\theta - T\sin\theta = 0$ and $\Sigma F_x = T\cos\theta + T\cos\theta = F_x$.

Find the work done on the barge.

$W = F_x\Delta x = (2T\cos\theta)\Delta x = 2(1.0\text{ kN})\cos 45°(150\text{ m}) = \boxed{210\text{ kJ}}$

9. Strategy Use Eq. (6-2). Let the x-axis point in the direction of motion.

Solution The force of friction is opposite the motion of the box, and according to Newton's second law, it is equal to $f_k = \mu_k N = \mu_k mg$. Juana's horizontal force is in the direction of motion. Solve for the displacement.

$W = F_x\Delta x$, so

$$\Delta x = \frac{W}{F_x} = \frac{W}{F - f_k} = \frac{W}{F - \mu_k mg} = \frac{74.4\text{ J}}{124\text{ N} - 0.120(56.8\text{ kg})(9.80\text{ m/s}^2)} = \boxed{1.3\text{ m}}.$$

11. Strategy The work done on the briefcase by the executive is equal to the change in kinetic energy of the briefcase. Use Eqs. (6-6) and (6-7).

Solution Find the work done by the executive on the briefcase.

$$W = \Delta K = \frac{1}{2}m(v_f^2 - v_i^2) = \frac{1}{2}(5.00\text{ kg})\left[(2.50\text{ m/s})^2 - 0\right] = \boxed{15.6\text{ J}}$$

13. Strategy The kinetic energy of the sack is equal to the work done on it by Sam. Use Eqs. (6-2), (6-6), and (6-7).

Solution

(a) Compute the kinetic energy of the sack.

$$\Delta K = \frac{1}{2}m(v_f^2 - v_i^2) = \frac{1}{2}mv^2 - 0 = K = W = F_x\Delta x,\text{ so } K = (2.0\text{ N})(0.35\text{ m}) = \boxed{0.70\text{ J}}.$$

(b) Solve for the speed of the sack.

$$K=\frac{1}{2}mv^2,\ \text{so}\ v=\sqrt{\frac{2K}{m}}=\sqrt{\frac{2(0.70\ \text{J})}{10.0\ \text{kg}}}=\boxed{0.37\ \text{m/s}}.$$

17. Strategy Use Eqs. (6-6) and (6-7).

Solution Compute the work done by the wall on the skater.

$$W_{\text{total}}=\Delta K=K_\text{f}-K_\text{i}=0-\frac{1}{2}mv^2=-\frac{1}{2}(69.0\ \text{kg})(11.0\ \text{m/s})^2=\boxed{-4.17\ \text{kJ}}$$

21. (a) Strategy and Solution Since the floor is level, the motion of the desk is perpendicular to the force due to gravity; therefore, the change in the desk's gravitational potential energy is $\boxed{\text{zero}}$.

(b) Strategy The motion of the desk is in the direction of the applied constant force. Use Eq. (6-2).

Solution Compute the work done by Justin.

$$W=F_x\Delta x=(340\ \text{N})(10.0\ \text{m})=\boxed{3.4\ \text{kJ}}$$

(c) Strategy and Solution Justin did work against friction, not gravity, so the energy has been $\boxed{\text{dissipated as heat}}$ by friction between the bottom of the desk and the floor.

23. Strategy Use Eq. (6-9).

Solution

(a) Since the orange returns to its original position $(\Delta y=0)$ and air resistance is ignored, the change in its potential energy is $\boxed{0}$.

(b) Let the y-axis point upward and the initial position be $y=0$.

$$\Delta U_{\text{grav}}=mg\Delta y=(0.30\ \text{kg})(9.80\ \text{m/s}^2)(-1.0\ \text{m}-0)=\boxed{-2.9\ \text{J}}$$

25. Strategy Use Newton's second law and Eq. (6-10).

Solution The total work is given by $W_{\text{total}}=W_{\text{vs. friction}}+W_{\text{vs. gravity}}$.

Find the work done against friction.

$\Sigma F_y=N-mg\cos\phi=0,\ \text{so}\ f=\mu N=\mu mg\cos\phi.$

$$W_{\text{vs. friction}}=f\Delta x=\mu mg\cos\phi L=\mu mg\frac{\sqrt{L^2-h^2}}{L}L=\mu mgh\sqrt{\left(\frac{L}{h}\right)^2-1}$$ and the work done against gravity is

$$W_{\text{vs. gravity}}=mgh.\ \text{So,}\ W_{\text{total}}=mgh\left(\mu\sqrt{\left(\frac{L}{h}\right)^2-1}+1\right)=(1400\ \text{N})(1.0\ \text{m})\left(0.20\sqrt{\left(\frac{4.0\ \text{m}}{1.0\ \text{m}}\right)^2-1}+1\right)=\boxed{2.5\ \text{kJ}}.$$

27. Strategy Use conservation of energy.

Solution Find a general expression for the speed of the cart.

$E_4 = K_4 + U_4 = \frac{1}{2}mv_4^2 + mgy_4 = E_i$, so $E_n = K_n + U_n = \frac{1}{2}mv_n^2 + mgy_n$, where $n = 1$, 2, or 3.

Solve for v_n.

$E_f = \frac{1}{2}mv_n^2 + mgy_n = E_i = \frac{1}{2}mv_4^2 + mgy_4$, so $v_n = \sqrt{v_4^2 + 2g(y_4 - y_n)}$.

Compute the speed at each position.

$v_1 = \sqrt{(15\text{ m/s})^2 + 2(9.80\text{ m/s}^2)(20.0\text{ m} - 0)} = \boxed{25\text{ m/s}}$

$v_2 = \sqrt{(15\text{ m/s})^2 + 2(9.80\text{ m/s}^2)(20.0\text{ m} - 15.0\text{ m})} = \boxed{18\text{ m/s}}$

$v_3 = \sqrt{(15\text{ m/s})^2 + 2(9.80\text{ m/s}^2)(20.0\text{ m} - 10.0\text{ m})} = \boxed{21\text{ m/s}}$

29. Strategy Use conservation of energy.

Solution Find the maximum height achieved by the swinging child.

$K_f - K_i = \frac{1}{2}mv^2 - 0 = U_i - U_f = mgy_{top} - mgy_{bottom}$, so

$$y_{top} = \frac{v^2}{2g} + y_{bottom} = \frac{(4.9\text{ m/s})^2}{2(9.80\text{ m/s}^2)} + 0.70\text{ m} = \boxed{1.9\text{ m}}.$$

33. Strategy Since energy is conserved and nonconservative forces do no work, $\Delta K = -\Delta U$. Let the y-axis point upward.

Solution The initial speeds are zero and the final speeds are the same (due to the rope). Since $m_1 < m_2$, block 1 moves up the incline and block 2 falls. Let d be the distance block 1 moves along the incline, then $\Delta r_1 = d\sin\theta$ and $\Delta r_2 = -d$.

$\Delta K = \Delta K_1 + \Delta K_2 = \frac{1}{2}m_1v^2 + \frac{1}{2}m_2v^2 = -\Delta U = -\Delta U_1 - \Delta U_2 = -m_1g\Delta r_1 - m_2g\Delta r_2 = -m_1gd\sin\theta - m_2g(-d)$, so

$$v = \sqrt{\frac{2gd(-m_1\sin\theta + m_2)}{m_1 + m_2}} = \sqrt{\frac{2(9.80\text{ m/s}^2)(1.4\text{ m})[-(12.4\text{ kg})\sin 36.9° + 16.3\text{ kg}]}{12.4\text{ kg} + 16.3\text{ kg}}} = \boxed{2.9\text{ m/s}}.$$

37. Strategy In the equation for the escape speed found in Example 6.8, replace the values for Earth with appropriate values for the fictional planet. Use proportional reasoning and the relationship between the volume of a sphere and its radius to relate the mass and radius of the planet with those of Earth.

Solution Find the escape speed.

Earth:

$$v_{esc} = \sqrt{\frac{2GM_E}{R_E}} \text{ and } \rho_E = \text{density} = \frac{M_E}{\frac{4}{3}\pi R_E^3}.$$

Planet:

$$v_{esc} = \sqrt{\frac{2GM}{R}} = \sqrt{\frac{2G}{R}\rho_E V} = \sqrt{\frac{2G}{2R_E}\left(\frac{M_E}{\frac{4}{3}\pi R_E^3}\right)\left[\frac{4}{3}\pi(2R_E)^3\right]} = \sqrt{\frac{8GM_E}{R_E}}$$

$$= \sqrt{\frac{8(6.674\times10^{-11}\text{ N}\cdot\text{m}^2/\text{kg}^2)(5.974\times10^{24}\text{ kg})}{6.37\times10^6\text{ m}}} = \boxed{22.4\text{ km/s}}$$

41. Strategy Use the result for escape speed found in Example 6.8.

Solution The magnitude of the gravitational field is given by $GM/R^2 = 30.0\ \text{m/s}^2$. Find the escape speed.

$$v_{\text{esc}} = \sqrt{\frac{2GM}{R}} = \sqrt{2\left(\frac{GM}{R^2}\right)R} = \sqrt{2(30.0\ \text{m/s}^2)(6.00\times10^7\ \text{m})} = \boxed{60.0\ \text{km/s}}$$

45. Strategy Use proportional reasoning.

Solution The force is linear with respect to the displacement of the string. If the string is pulled back half as far (20.0 cm) as in Example 6.9 (40.0 cm), the average force is only half that as in Example 6.9. Therefore, the work done is $\left(\frac{1}{2}\right)\left(\frac{1}{2}\right)(32\ \text{J}) = \boxed{8\ \text{J}}$.

47. (a) Strategy The increase in the force is F and the length that the tendon increases is x if the tendon is modeled as a spring and Hooke's law is used.

Solution Compute the spring constant.

$$k = \frac{F}{x} = \frac{4800\ \text{N} - 3200\ \text{N}}{0.50\ \text{cm}} = \frac{1600\ \text{N}}{0.50\ \text{cm}} = \boxed{3200\ \text{N/cm}}$$

(b) Strategy The triangular area under a force of the muscle vs. the stretch of the tendon graph is equal to the work done by the muscle.

Solution Compute the work done by the muscle in stretching the tendon.

$$W = \frac{1}{2}Fx = \frac{1}{2}(4800\ \text{N} - 3200\ \text{N})(0.0050\ \text{m}) = \boxed{4.0\ \text{J}}$$

49. (a) Strategy Solve for k in Hooke's law.

Solution Compute the spring constant.

$$k = \frac{F}{x} = \frac{120\ \text{N}}{2.0\ \text{nm}}\left(\frac{10^9\ \text{nm}}{1\ \text{m}}\right) = \boxed{6.0\times10^{10}\ \text{N/m}}$$

(b) Strategy Solve for x in Hooke's law and use the value for k found in part (a).

Solution Compute the compression of the block.

$$x = \frac{F}{k} = \frac{480\ \text{N}}{6.0\times10^{10}\ \text{N/m}} = \boxed{8.0\ \text{nm}}$$

(c) Strategy Since the forces due to the block are opposite to the directions of compression, the block does negative work during the compression. The work done by the applied forces is positive and equal to the negative of the work done by the block.

Solution Compute the work done.

$$W = -W_{\text{block}} = \frac{1}{2}kx^2 = \frac{1}{2}(6.0\times10^{10}\ \text{N/m})(8.0\times10^{-9}\ \text{m})^2 = \boxed{1.9\ \mu\text{J}}$$

53. Strategy The work done by the force on the object is represented by the area between the curve and the x-axis. The area under the axis represents negative work done.

Solution Compute the work done by the force.

$$W = \frac{1}{2}(2.0\text{ N})(1.0\text{ m}) + \frac{1}{2}(1.0\text{ N})(1.0\text{ m}) + (-1.0\text{ N})(1.0\text{ m}) = \boxed{0.5\text{ J}}$$

57. Strategy The elastic potential energy of the catapult, $(1/2)kx^2$, is converted into gravitational potential energy of the pebble, mgh.

Solution Find the maximum height achieved by the pebble.

$$mgh = \frac{1}{2}kx^2, \text{ so } h = \frac{kx^2}{2mg} = \frac{(320\text{ N/m})(0.20\text{ m})^2}{2(0.051\text{ kg})(9.80\text{ m/s}^2)} = \boxed{13\text{ m}}.$$

61. (a) Strategy The increase in kinetic energy of the block is equal to the decrease in its potential energy. Let the potential energy be zero at $y = 0.25$ m.

Solution To find the speed of the block, set Eqs. (6-6) and (6-13) equal and solve for v.

$$\frac{1}{2}mv^2 = mgy, \text{ so } v = \sqrt{2gy} = \sqrt{2(9.80\text{ m/s}^2)(0.25\text{ m})} = \boxed{2.2\text{ m/s}}.$$

(b) Strategy The elastic potential energy increase of the spring is equal to the decrease in gravitational potential energy of the block. Let the potential energy be zero at $y = 0$ m.

Solution To find the compression of the spring, set Eqs. (6-24) and (6-13) equal and solve for x.

$$\frac{1}{2}kx^2 = mgy, \text{ so } x = \sqrt{\frac{2mgy}{k}} = \sqrt{\frac{2(2.0\text{ kg})(9.80\text{ m/s}^2)(0.50\text{ m})}{450\text{ N/m}}} = \boxed{0.21\text{ m}}.$$

(c) Strategy and Solution Since the surface is frictionless, no nonconservative forces do work on the block. So, the block will return to its previous height, or $\boxed{0.50\text{ m}}$.

65. Strategy Assume that friction is negligible. Use Eq. (6-27).

Solution The rate at which gravity does work on the bicycle and rider is $P = mgv\cos\theta$, where θ is the angle between $\vec{\mathbf{v}}$ and $\vec{\mathbf{g}}$, or $\theta = 90° + \phi$. Find ϕ.

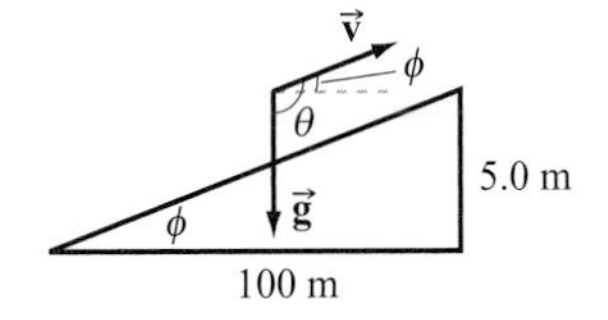

$$\tan\phi = \frac{5.0\text{ m}}{100\text{ m}}, \text{ so } \phi = \tan^{-1} 0.050.$$

The power output of the rider is equal to the rate of change of potential energy, which equals $-P$. Therefore,

$$P_{\text{rider}} = -P = -(75\text{ kg})(9.80\text{ m/s}^2)(4.0\text{ m/s})\cos(90° + \tan^{-1} 0.050) = \boxed{150\text{ W}}.$$

69. Strategy Use the definition of average power. The change in energy is equal to the change in kinetic energy.

Solution Determine the average mechanical power the engine must supply.

$$P_{\text{av}} = \frac{\Delta E}{\Delta t} = \frac{\frac{1}{2}m(v_{\text{f}}^2 - v_{\text{i}}^2)}{\Delta t} = \frac{(1200\text{ kg})\left[(30.0\text{ m/s})^2 - (20.0\text{ m/s})^2\right]}{2(5.0\text{ s})} = \boxed{60\text{ kW}}$$

73. Strategy Use the definition of average power. The change in energy is equal to the change in kinetic energy.

Solution Find the engine's average power output.

$$P_{\text{av}} = \frac{\Delta E}{\Delta t} = \frac{\Delta K}{\Delta t} = \frac{\frac{1}{2}m(v_{\text{f}}^2 - v_{\text{i}}^2)}{\Delta t} = \frac{(500.0\text{ kg})\left[(125\text{ m/s})^2 - 0\right]}{2(4.2\text{ s})} = \boxed{930\text{ kW}}$$

75. Strategy Use conservation of energy.

Solution Compute the required speed of the high jumper.

$$K_{\text{i}} + U_{\text{i}} = \frac{1}{2}mv^2 + 0 = \frac{1}{2}mv^2 = K_{\text{f}} + U_{\text{f}} = 0 + mgh = mgh, \text{ so}$$
$$v = \sqrt{2gh} = \sqrt{2(9.80\text{ m/s}^2)(1.2\text{ m})} = \boxed{4.8\text{ m/s}}.$$

77. Strategy Use conservation of energy. Neglect drag.

Solution Find the speed of the hang glider.

$$\Delta K = \frac{1}{2}mv_{\text{f}}^2 - \frac{1}{2}mv_{\text{i}}^2 = -\Delta U = mgh, \text{ so } v_{\text{f}} = \sqrt{2gh + v_{\text{i}}^2} = \sqrt{2(9.80\text{ m/s}^2)(8.2\text{ m}) + (9.5\text{ m/s})^2} = \boxed{16\text{ m/s}}.$$

81. Strategy Use Hooke's law and Newton's laws.

Solution

(a) The mass connected to the lower spring exerts a force on the lower spring equal to its weight, W. The spring stretches an amount $x = F/k = W/k$. The lower spring exerts a force on the upper spring equal to $F = W$, and causes it to stretch by $x = F/k = W/k$. Thinking of the two springs as a single spring:

$$2x = \frac{F}{k} + \frac{F}{k} = \frac{2F}{k} = x', \text{ so } F = \frac{k}{2}x' = k'x'. \text{ Therefore, } \boxed{\frac{k}{2}} = k', \text{ the effective spring constant.}$$

(b) Sum the forces on the mass.

$$F + F - W = kx + kx - W = 2kx - W = 0, \text{ so } W = 2kx = k'x.$$

Therefore, $\boxed{2k} = k'$, the effective spring constant.

85. Strategy Use conservation of energy.

Solution The elastic potential energy of the spring is converted to gravitational potential energy, so $\frac{1}{2}kx^2 = mgh = mg(l\sin\theta)$ where l is the distance the object travels up the incline.

$$\text{Thus, } l = \frac{kx^2}{2mg\sin\theta} = \frac{(40.0\text{ N/m})(0.20\text{ m})^2}{2(0.50\text{ kg})(9.80\text{ N/kg})\sin 30.0°} = \boxed{0.33\text{ m}}.$$

89. (a) Strategy Let $+x$ be up the incline. Use Newton's second law and Eq. (6-27).

Solution Compute the power the engine must deliver.

$\Sigma F_x = F_{\text{air}} - mg\sin\phi = 0$ at terminal speed.

The rate at which air resistance dissipates energy is $P_{\text{air}} = F_{\text{air}}v\cos 180° = -F_{\text{air}}v = -mg\sin\phi v.$

(We use cos 180° since the force of air resistance is opposite the car's velocity.)

The power the engine must deliver to drive the car on level ground is

$$P_{\text{engine}} = -P_{\text{air}} = mg\sin\phi v = (1500\text{ kg})(9.80\text{ m/s}^2)(20.0\text{ m/s})\sin 2.0° = \boxed{10\text{ kW}}.$$

(b) Strategy The power available to climb the hill is the power delivered by the engine minus the dissipating power of air resistance.

Solution From part (a), for a slope of ϕ:

$$P = mgv\sin\phi,\ \text{so}\ \phi = \sin^{-1}\frac{P}{mgv} = \sin^{-1}\frac{40.0\times10^3\ \text{W} - 10.26\times10^3\ \text{W}}{(1500\ \text{kg})(9.80\ \text{m/s}^2)(20.0\ \text{m/s})} = \boxed{5.8^\circ}.$$

93. Strategy The basal metabolic rate is equal to the number of kilocalories per day required by a person resting under standard conditions.

Solution

(a) Compute Jermaine's basal metabolic rate.

$$\text{BMR} = \left(\frac{1\ \text{kcal}}{0.010\ \text{mol}}\right)\left(\frac{0.015\ \text{mol}}{\text{min}}\right)\left(\frac{1440\ \text{min}}{\text{day}}\right) = \boxed{2200\ \text{kcal/day}}$$

(b) Find the mass of fat lost.

$$\frac{2160\ \text{kcal/day}}{9.3\ \text{kcal/g}}\left(\frac{2.2\ \text{lb}}{10^3\ \text{g}}\right) = 0.51\ \text{lb/day}$$

Since Jermaine is not resting the entire time, he loses $\boxed{\text{more than 0.51 lb}}$.

97. Strategy Draw a diagram. Then, use Newton's second law and conservation of energy. Let the positive direction be away from the slope.

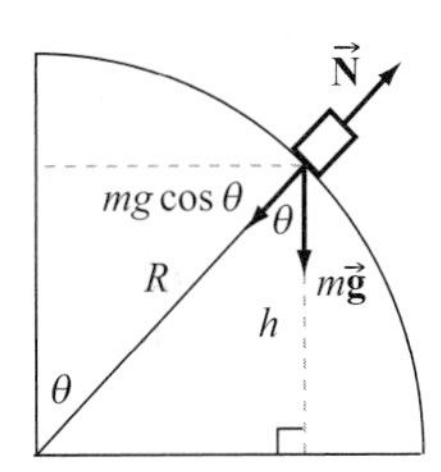

Solution According to Newton's second law, $\Sigma F_r = N - mg\cos\theta = ma_r = -mv^2/R$.

When the normal force becomes zero, we have $mg\cos\theta = mv^2/R$, or $mgR\cos\theta = mv^2$. When this condition is true, the skier leaves the surface of the ice. Note from the figure that $h = R\cos\theta$, thus, the condition becomes $mgh = mv^2$. Now, mgh is the final gravitational potential energy ($U_i = mgR$) and mv^2 is twice the final kinetic energy ($K_i = 0$).

So, the condition becomes $U_f = 2K_f$, or $K_f = U_f/2$. Use conservation of energy to find h in terms of R.

$$0 = \Delta K + \Delta U = K_f - K_i + U_f - U_i = \frac{1}{2}U_f - 0 + U_f - U_i = \frac{3}{2}U_f - U_i,\ \text{so}\ \frac{3}{2}mgh - mgR = \frac{3}{2}h - R = 0,\ \text{or}$$

$$h = \boxed{\frac{2}{3}R}.$$

99. (a) Strategy Use Hooke's law and Newton's laws of motion.

Solution According to Hooke's law, $F_1 = k_1x_1$ and $F_2 = k_2x_2$.

Imagine that the springs are suspended from a ceiling such that the bottom of each is at the same height. Then a mass m is attached to the bottom of both, the springs stretch, and the system comes to equilibrium. Assume that the masses of the springs are negligible. Sum the vertical forces.

$F_1 + F_2 - W = 0$, so $W = F_1 + F_2 = k_1x_1 + k_2x_2$.

Assuming the springs are attached to the same point on the top of the mass, $x_1 = x_2 = x$.

$W = k_1x_1 + k_2x_2 = k_1x + k_2x = (k_1 + k_2)x = kx = W$

So, in response to a force that stretches the springs (W, in this case), the springs act like one spring with a spring constant $\boxed{k = k_1 + k_2}$.

(b) Strategy Use the result from part (a) and Eq. (6-24).

Solution Compute the potential energy stored in the spring.

$$U=\frac{1}{2}kx^2=\frac{1}{2}(k_1+k_2)x^2=\frac{1}{2}(500\ \text{N/m}+300\ \text{N/m})(0.020\ \text{m})^2=\boxed{0.16\ \text{J}}$$

101. Strategy Use Newton's second law, Hooke's law, and Eq. (6-24).

Solution Find the distance that the tendon stretches.

$$\sum F=T-kx=0,\ \text{so}\ x=\frac{T}{k}=\frac{4.7\ \text{kN}}{350\ \text{kN/m}}=\boxed{1.3\ \text{cm}}.$$

Find the stored elastic energy.

$$U=\frac{1}{2}kx^2=\frac{1}{2}k\frac{T^2}{k^2}=\frac{T^2}{2k}=\frac{(4.7\times10^3\ \text{N})^2}{2(350\times10^3\ \text{N/m})}=\boxed{32\ \text{J}}$$

105. Strategy and Solution The kinetic energy of a volume of wind passing through the circular area swept out by the rotor blades in time Δt is $\frac{1}{2}mv^2$, where $m=\rho V=\rho Ad=\rho(\pi L^2)(v\Delta t)$ and v is the speed of the wind; therefore, the kinetic energy is given by $K=\frac{1}{2}\rho\pi L^2\Delta t v^3$. The average power generated is $P_{\text{av}}=\frac{\Delta E}{\Delta t}=\frac{\varepsilon\Delta K}{\Delta t}$, where ε is the efficiency of the energy conversion from kinetic energy to electrical energy.

Therefore, $P_{\text{av}}=\frac{\varepsilon\Delta K}{\Delta t}=\frac{\varepsilon\rho\pi L^2}{2}v^3\propto v^3$.

Chapter 7

LINEAR MOMENTUM

Problems

1. Strategy Use conservation of momentum.

Solution Find the total momentum of the two blocks after the collision.

$$\begin{aligned}\Delta p_2 &= -\Delta p_1\\ p_{2f} - p_{2i} &= p_{1i} - p_{1f}\\ p_{1f} + p_{2f} &= p_{1i} + p_{2i}\\ (m_1 + m_2)v_f &= m_1 v_{1i} + m_2 v_{2i}\\ p_f &= (2.0\text{ kg})(1.0\text{ m/s}) + (1.0\text{ kg})(0) = 2.0\text{ kg}\cdot\text{m/s} = p_{1i}\end{aligned}$$

Since p_{1i} was directed to the right, and $p_f = p_{1i}$, the total momentum of the two blocks after the collision is $\boxed{2.0\text{ kg}\cdot\text{m/s to the right}}$.

5. Strategy Add the momenta of the three particles.

Solution Find the total momentum of the system.

$$\begin{aligned}\vec{\mathbf{p}}_{\text{tot}} &= \vec{\mathbf{p}}_1 + \vec{\mathbf{p}}_2 + \vec{\mathbf{p}}_3 = m_1\vec{\mathbf{v}}_1 + m_2\vec{\mathbf{v}}_2 + m_3\vec{\mathbf{v}}_3 = m_1 v_1\text{ north} + m_2 v_2\text{ south} + m_3 v_3\text{ north}\\ &= (m_1 v_1 - m_2 v_2 + m_3 v_3)\text{ north} = [(3.0\text{ kg})(3.0\text{ m/s}) - (4.0\text{ kg})(5.0\text{ m/s}) + (7.0\text{ kg})(2.0\text{ m/s})]\text{ north}\\ &= \boxed{3\text{ kg}\cdot\text{m/s north}}\end{aligned}$$

7. Strategy The initial momentum is toward the wall and the final momentum is away from the wall.

Solution Find the change in momentum.

$\Delta p = p_f - p_i = mv_f - mv_i = m(v_f - v_i) = (5.0\text{ kg})(-2.0\text{ m/s} - 2.0\text{ m/s}) = -20\text{ kg}\cdot\text{m/s}$, so

$\Delta\vec{\mathbf{p}} = \boxed{20\text{ kg}\cdot\text{m/s in the }-x\text{-direction}}$.

9. Strategy Use the definition of linear momentum and Eq. (4-7). Let up be the positive direction.

Solution $v_f = v_{fy} = v_{iy} - g\Delta t = -g\Delta t$, since the object starts from rest.

Find Δp.

$\Delta p = p_f - p_i = m(v_f - v_i) = m(-g\Delta t - 0) = -mg\Delta t = -(3.0\text{ kg})(9.80\text{ m/s}^2)(3.4\text{ s}) = -1.0\times10^2\text{ kg}\cdot\text{m/s}$, so

$\Delta\vec{\mathbf{p}} = \boxed{1.0\times10^2\text{ kg}\cdot\text{m/s downward}}$.

13. Strategy Use the impulse-momentum theorem. Let the positive direction be in the direction of motion.

Solution Find the average horizontal force exerted on the automobile during breaking.

$$F_{\text{av}} = \frac{\Delta p}{\Delta t} = \frac{m(v_f - v_i)}{\Delta t} = \frac{(1.0\times10^3\text{ kg})(0 - 30.0\text{ m/s})}{5.0\text{ s}} = -6.0\times10^3\text{ N}$$

So, $\vec{\mathbf{F}}_{\text{av}} = \boxed{6.0\times10^3\text{ N opposite the car's direction of motion}}$.

17. Strategy and Solution Impulse $= F\Delta t$, so the SI unit is $\text{N}\cdot\text{s} = \text{kg}\cdot\text{m/s}^2\cdot\text{s} = \text{kg}\cdot\text{m/s}$. $p = mv$, so the SI unit is $\text{kg}\cdot\text{m/s}$. Therefore, the SI unit of impulse is the same as the SI unit of momentum.

21. Strategy Use conservation of momentum.

Solution Find the recoil speed of the thorium nucleus.
$\vec{\mathbf{p}}_\text{i} = 0 = -\vec{\mathbf{p}}_\text{f}$, so if n = nucleus and p = particle,

$$\vec{\mathbf{p}}_\text{n} + \vec{\mathbf{p}}_\text{p} = m_\text{n}\vec{\mathbf{v}}_\text{n} + m_\text{p}\vec{\mathbf{v}}_\text{p} = 0, \text{ so } |\vec{\mathbf{v}}_\text{n}| = \frac{m_\text{p}}{m_\text{n}}|-\vec{\mathbf{v}}_\text{p}| = \frac{4.0\text{ u}}{234\text{ u}}\left[0.050(2.998\times10^8\text{ m/s})\right] = \boxed{2.6\times10^5\text{ m/s}}.$$

23. Strategy Use conservation of momentum.

Solution Find the recoil speed of the railroad car.
$\vec{\mathbf{p}}_\text{i} = 0 = -\vec{\mathbf{p}}_\text{f}$, and since we are only concerned with the horizontal direction, we have:

$$m_\text{c}v_{\text{c}x} = m_\text{s}v_{\text{s}x}, \text{ so } v_{\text{c}x} = \frac{m_\text{s}}{m_\text{c}}v_{\text{s}x} = \frac{98\text{ kg}}{5.0\times10^4\text{ kg}}(105\text{ m/s})\cos 60.0° = \boxed{0.10\text{ m/s}}.$$

25. Strategy Use the law of conservation of linear momentum to determine the astronaut's speed.

Solution According to the law of conservation of linear momentum, the astronaut will move toward the ship with linear momentum equal in magnitude to the magnitude of the combined momentum of the objects thrown. Find the speed of the astronaut after he throws the mallet.
$\Sigma p_\text{objects} = p_\text{w} + p_\text{s} + p_\text{m} = m_\text{w}v_\text{w} + m_\text{s}v_\text{s} + m_\text{m}v_\text{m} = p_\text{A} = m_\text{A}v_\text{A}$, so

$$v_\text{A} = \frac{m_\text{w}v_\text{w} + m_\text{s}v_\text{s} + m_\text{m}v_\text{m}}{m_\text{A}}$$

$$= \frac{(0.72\text{ kg})(5.0\text{ m/s}) + (0.80\text{ kg})(8.0\text{ m/s}) + (1.2\text{ kg})(6.0\text{ m/s})}{58\text{ kg}} = \boxed{0.30\text{ m/s}}.$$

27. Strategy Use the component form of the definition of center of mass.

Solution Find the location of particle B.
Find x_CM.

$$x_\text{CM} = \frac{m_\text{A}x_\text{A} + m_\text{B}x_\text{B}}{m_\text{A} + m_\text{B}} = \frac{0 + m_\text{B}x_\text{B}}{m_\text{A} + m_\text{B}}, \text{ so}$$

$$x_\text{B} = \frac{m_\text{A} + m_\text{B}}{m_\text{B}}x_\text{CM} = \frac{30.0\text{ g} + 10.0\text{ g}}{10.0\text{ g}}(2.0\text{ cm}) = 8.0\text{ cm}.$$

Similarly,

$$y_\text{B} = \frac{30.0\text{ g} + 10.0\text{ g}}{10.0\text{ g}}(5.0\text{ cm}) = 20\text{ cm}.$$

The coordinates of particle B are $(x_\text{B}, y_\text{B}) = \boxed{(8.0\text{ cm, }20\text{ cm})}$.

y (cm)
CM
B?
(2.0, 5.0)
A
x (cm)

29. Strategy Since no y-components of the positions have changed, the center of mass moves only in the x-direction. Use the component form of the definition of center of mass.

Solution Find the displacement of the center of mass of the three bodies.

$$x_i = \frac{mx_{1i} + mx_{2i} + mx_{3i}}{m+m+m} = \frac{x_{1i} + x_{2i} + x_{3i}}{3} = \frac{1\text{ m} + 2\text{ m} + 3\text{ m}}{3} = 2\text{ m}$$

$$x_f = \frac{x_{1f} + x_{2f} + x_{3f}}{3} = \frac{x_{1i} + x_{2i} + x_{3i} + 0.12\text{ m}}{3} = \frac{6\text{ m} + 0.12\text{ m}}{3}$$

$$\Delta x = x_f - x_i = \frac{6\text{ m} + 0.12\text{ m}}{3} - 2\text{ m} = \frac{0.12\text{ m}}{3} = 4.0\text{ cm}$$

The center of mass moves $\boxed{\text{4.0 cm in the positive } x\text{-direction}}$.

33. Strategy The x-coordinate of each three-dimensional shape is midway along its horizontal dimension.

Solution Find the x-component of the center of mass of the composite object.

$$x_{\text{CM}} = \frac{m_s x_s + m_c x_c + m_r x_r}{m_s + m_c + m_r} = \frac{(200\text{ g})(5.0\text{ cm}) + (450\text{ g})(10\text{ cm} + 17/2\text{ cm}) + (325\text{ g})(10\text{ cm} + 17\text{ cm} + 16/2\text{ cm})}{200\text{ g} + 450\text{ g} + 325\text{ g}}$$

$$= \boxed{21\text{ cm}}$$

35. Strategy The total momentum of the system is equal to the total mass of the system times the velocity of the center of mass.

Solution Find the total momentum.

$\vec{\mathbf{p}} = M\vec{\mathbf{v}}_{\text{CM}} = m_A\vec{\mathbf{v}}_A + m_B\vec{\mathbf{v}}_B$ since $\vec{\mathbf{p}} = \vec{\mathbf{p}}_A + \vec{\mathbf{p}}_B$. Thus, $\vec{\mathbf{v}}_{\text{CM}} = \dfrac{m_A\vec{\mathbf{v}}_A + m_B\vec{\mathbf{v}}_B}{m_A + m_B}$.

B
7 m/s
A
14 m/s
y
x

Find the components of $\vec{\mathbf{v}}_{\text{CM}}$.

$$v_{\text{CM}x} = \frac{m_A v_{Ax} + m_B v_{Bx}}{m_A + m_B} = \frac{(3\text{ kg})(14\text{ m/s}) + (4\text{ kg})(0)}{3\text{ kg} + 4\text{ kg}} = 6\text{ m/s and } v_{\text{CM}y} = \frac{(3\text{ kg})(0) + (4\text{ kg})(-7\text{ m/s})}{3\text{ kg} + 4\text{ kg}} = -4\text{ m/s}.$$

So, the components are $(v_{\text{CM}x}, v_{\text{CM}y}) = \boxed{(6\text{ m/s}, -4\text{ m/s})}$.

37. (a) Strategy Draw a diagram and use conservation of linear momentum.

Solution

4.0 m/s
5.0 m/s
y
45°
37°
x

$$\vec{\mathbf{p}} = M\vec{\mathbf{v}}_{\text{CM}} = \frac{M}{4}\vec{\mathbf{v}}_1 + \frac{M}{3}\vec{\mathbf{v}}_2 + \frac{5M}{12}\vec{\mathbf{v}}_3 = 0 \text{ Use components.}$$

$$\frac{M}{4}v_{1x} + \frac{M}{3}v_{2x} + \frac{5M}{12}v_{3x} = 3v_{1x} + 4v_{2x} + 5v_{3x} = 0\text{, so}$$

$$v_{3x} = -\frac{3v_{1x} + 4v_{2x}}{5} = -\frac{3(5.0\text{ m/s})\cos 37° + 4(4.0\text{ m/s})\cos 135°}{5} = -0.13\text{ m/s}.$$

$$\frac{M}{4}v_{1y} + \frac{M}{3}v_{2y} + \frac{5M}{12}v_{3y} = 3v_{1y} + 4v_{2y} + 5v_{3y} = 0\text{, so}$$

$$v_{3y} = -\frac{3v_{1y} + 4v_{2y}}{5} = -\frac{3(5.0\text{ m/s})\sin 37° + 4(4.0\text{ m/s})\sin 135°}{5} = -4.1\text{ m/s}.$$

The velocity components are $\boxed{(-0.13\text{ m/s}, -4.1\text{ m/s})}$.

(b) Strategy and Solution Due to the law of conservation of linear momentum,

$\boxed{\text{the center of mass of the system remains at the origin after the explosion}}$.

41. Strategy Use conservation of momentum.

Solution

(a) The collision is perfectly inelastic, so $v_{1f} = v_{2f} = v_f$. Find the speed of the two cars after the collision.

$$m_1v_{1i} + m_2v_{2i} = mv_{1i} + 4.0m(0) = m_1v_{1f} + m_2v_{2f} = mv_f + 4.0mv_f, \text{ so } v_f = \frac{v_{1i}}{5.0} = \frac{1.0\text{ m/s}}{5.0} = \boxed{0.20\text{ m/s}}.$$

(b) The cars are at rest after the collision, so $v_{1f} = v_{2f} = 0$.

$$mv_{1i} + 4.0mv_{2i} = 0, \text{ so } v_{2i} = -\frac{v_{1i}}{4.0} = -\frac{1.0\text{ m/s}}{4.0} = -0.25\text{ m/s}. \text{ The initial speed was } \boxed{0.25\text{ m/s}}.$$

43. Strategy Use conservation of momentum. The collision is perfectly inelastic, so $v_{1f} = v_{2f} = v_f$. Also, the block is initially at rest, so $v_{2i} = 0$.

Solution Find the speed of the block of wood and the bullet just after the collision.
$m_1v_{1f} + m_2v_{2f} = (m_1 + m_2)v_f = m_1v_{1i} + m_2v_{2i} = m_1v_{1i} + m_2(0)$, so

$$v_f = \frac{m_1}{m_1 + m_2}v_{1i} = \frac{0.050\text{ kg}}{0.050\text{ kg} + 0.95\text{ kg}}(100.0\text{ m/s}) = \boxed{5.0\text{ m/s}}.$$

45. Strategy The spring imparts the same (in magnitude) impulse to each block. (The same magnitude force is exerted on each block by the ends of the spring for the same amount of time.) So, each block has the same final magnitude of momentum. (The initial momentum is zero.)

Solution Find the mass of block B.

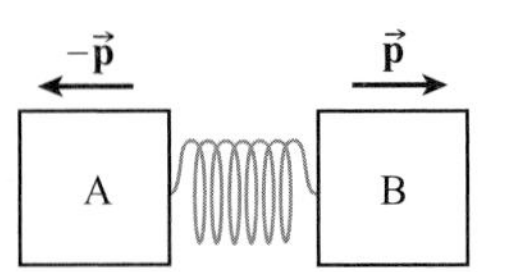

$m_Bv_B = m_Av_A$, so

$$m_B = \frac{v_A}{v_B}m_A = \frac{d_A/\Delta t}{d_B/\Delta t}m_A = \frac{d_A}{d_B}m_A = \frac{1.0\text{ m}}{3.0\text{ m}}(0.60\text{ kg}) = \boxed{0.20\text{ kg}}.$$

49. Strategy Use conservation of momentum. Let the positive direction be the initial direction of motion.

Solution Find the speed of the 5.0-kg body after the collision.
$m_1v_{1f} + m_2v_{2f} = m_1v_{1i} + m_2v_{2i}$, so

$$v_{2f} = \frac{m_1(v_{1i} - v_{1f}) + m_2v_{2i}}{m_2} = \frac{(1.0\text{ kg})[10.0\text{ m/s} - (-5.0\text{ m/s})] + (5.0\text{ kg})(0)}{5.0\text{ kg}} = \boxed{3.0\text{ m/s}}.$$

53. Strategy The collision is perfectly inelastic, so $v_{1f} = v_{2f} = v$. The block is initially at rest, so $v_{2i} = 0$ and $v_{1i} = v_i$. Use conservation of momentum and Eq. (4-9).

Solution Find the speed of the bullet and block system.

$$(m_{bul} + m_{blk})v = m_{bul}v_i + m_{blk}(0), \text{ so } v = \frac{m_{bul}}{m_{bul} + m_{blk}}v_i.$$

Determine the time it takes the system to hit the floor.

$$\Delta y = -h = v_{iy}\Delta t - \frac{1}{2}g(\Delta t)^2 = 0 - \frac{1}{2}g(\Delta t)^2, \text{ so } \Delta t = \sqrt{\frac{2h}{g}}.$$

Find the horizontal distance traveled.

$$\Delta x = v_{ix}\Delta t = v\Delta t = \frac{m_{bul}}{m_{bul} + m_{blk}}v_i\sqrt{\frac{2h}{g}} = \frac{0.010\text{ kg}}{0.010\text{ kg} + 4.0\text{ kg}}(400.0\text{ m/s})\sqrt{\frac{2(1.2\text{ m})}{9.80\text{ m/s}^2}} = \boxed{0.49\text{ m}}$$

57. Strategy Use conservation of linear momentum.

Solution Find $v_{\text{Bf}x}$.

$p_{\text{i}x} = mv_{\text{Ai}x} = p_{\text{f}x} = mv_{\text{Af}x} + mv_{\text{Bf}x}$, so $v_{\text{Bf}x} = v_{\text{Ai}x} - v_{\text{Af}x}$.

Find $v_{\text{Bf}y}$.

$p_{\text{i}y} = mv_{\text{Ai}y} = 0 = p_{\text{f}y} = mv_{\text{Af}y} + mv_{\text{Bf}y}$, so $v_{\text{Bf}y} = -v_{\text{Af}y}$.

Compute the final speed of puck B.

$$v_{\text{Bf}} = \sqrt{(v_{\text{Bf}x})^2 + (v_{\text{Bf}y})^2} = \sqrt{(v_{\text{Ai}x} - v_{\text{Af}x})^2 + (-v_{\text{Af}y})^2}$$
$$= \sqrt{[2.0\text{ m/s} - (1.0\text{ m/s})\cos 60°]^2 + [-(1.0\text{ m/s})\sin 60°]^2} = 1.7\text{ m/s}$$

Compute the direction of puck B.

$$\theta = \tan^{-1}\frac{v_{\text{Bf}y}}{v_{\text{Bf}x}} = \tan^{-1}\frac{-(1.0\text{ m/s})\sin 60°}{2.0\text{ m/s} - (1.0\text{ m/s})\cos 60°} = -30°$$

Thus, the speed and direction of puck B after the collision is $\boxed{1.7\text{ m/s at }30°\text{ below the }x\text{-axis}}$.

59. Strategy Use conservation of momentum. Refer to Practice Problem 7.11.

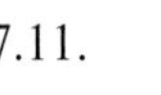

Solution

(a) Find the momentum change of the ball of mass m_1.

$$\Delta p_{1x} = -\Delta p_{2x} = m_2 v_{2\text{i}x} - m_2 v_{2\text{f}x} = m_2(0 - v_{2\text{f}x}) = -m_2 v_{2\text{f}x} = -5m_1\left[\frac{1}{4}v_\text{i}\cos(-36.9°)\right] = \boxed{-1.00 m_1 v_\text{i}}$$

$$\Delta p_{1y} = -\Delta p_{2y} = m_2 v_{2\text{i}y} - m_2 v_{2\text{f}y} = 5m_1(0 - v_{2\text{f}y}) = -5m_1 v_{2\text{f}y} = -5m_1\left[\frac{1}{4}v_\text{i}\sin(-36.9°)\right] = \boxed{0.751 m_1 v_\text{i}}$$

(b) Find the momentum change of the ball of mass m_2.

$$\Delta p_{2x} = -\Delta p_{1x} = m_1(v_{1\text{i}x} - v_{1\text{f}x}) = m_1(v_\text{i} - 0) = \boxed{m_1 v_\text{i}}$$

$$\Delta p_{2y} = -\Delta p_{1y} = m_1(v_{1\text{i}y} - v_{1\text{f}y}) = m_1(0 - v_1) = -m_1 v_1 = -m_1(0.751 v_\text{i}) = \boxed{-0.751 m_1 v_\text{i}}$$

$\boxed{\text{The momentum changes for each mass are equal and opposite.}}$

61. Strategy Use conservation of momentum.

Solution Find $v_{2\text{f}}$ in terms of $v_{1\text{f}}$.

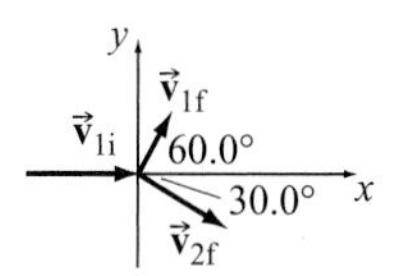

$mv_{1\text{f}y} + mv_{2\text{f}y} = mv_{1\text{f}}\sin\theta_1 + mv_{2\text{f}}\sin\theta_2 = mv_{1\text{i}y} + mv_{2\text{i}y} = 0 + 0$, so

$$v_{2\text{f}} = \frac{-\sin\theta_1}{\sin\theta_2}v_{1\text{f}} = \frac{-\sin 60.0°}{\sin(-30.0°)}v_{1\text{f}} = \boxed{1.73 v_{1\text{f}}}.$$

65. Strategy The collision is perfectly inelastic, so the final velocities of the cars are identical. Use conservation of momentum.

Solution Let the 1700-kg car be (1) and the 1300-kg car be (2).

$$p_{ix} = m_1 v_{1ix} + m_2 v_{2ix} = m_1 v_{1ix} + 0 = p_{fx} = (m_1 + m_2) v_{fx}, \text{ so } v_{fx} = \frac{m_1}{m_1 + m_2} v_{1ix}.$$

$$p_{iy} = m_1 v_{1iy} + m_2 v_{2iy} = p_{fy} = (m_1 + m_2) v_{fy}, \text{ so } v_{fy} = \frac{m_1 v_{1iy} + m_2 v_{2iy}}{m_1 + m_2}.$$

Compute the final speed and the direction.

$$v_f = \sqrt{v_{fx}^2 + v_{fy}^2} = \sqrt{\left(\frac{m_1 v_{1ix}}{m_1 + m_2}\right)^2 + \left(\frac{m_1 v_{1iy} + m_2 v_{2iy}}{m_1 + m_2}\right)^2}$$

$$= \frac{\sqrt{[(1700 \text{ kg})(14 \text{ m/s})\cos 45^\circ]^2 + [(1700 \text{ kg})(14 \text{ m/s})\sin 45^\circ + (1300 \text{ kg})(-18 \text{ m/s})]^2}}{1700 \text{ kg} + 1300 \text{ kg}} = 6.0 \text{ m/s}$$

$$\theta = \tan^{-1}\frac{v_{fy}}{v_{fx}} = \tan^{-1}\frac{(1700 \text{ kg})(14 \text{ m/s})\sin 45^\circ + (1300 \text{ kg})(-18 \text{ m/s})}{(1700 \text{ kg})(14 \text{ m/s})\cos 45^\circ} = -21^\circ$$

Thus, the final velocity of the cars is $\boxed{6.0 \text{ m/s at } 21^\circ \text{ S of E}}$.

69. Strategy Use conservation of momentum.

Solution Let swallow 1 and its coconut be (1) and swallow 2 and its coconut be (2) (before the collision). After the collision, let swallow 1's coconut be (3), swallow 2's coconut be (4), and the tangled-up swallows be (5).

$$p_{ix} = m_1 v_{1x} + m_2 v_{2x} = 0 + 0 = p_{fx} = m_3 v_{3x} + m_4 v_{4x} + m_5 v_{5x}, \text{ so } v_{5x} = -\frac{m_3 v_{3x} + m_4 v_{4x}}{m_5}.$$

$$p_{iy} = m_1 v_{1y} + m_2 v_{2y} = m_1 v_1 + m_2 v_2 = p_{fy} = m_3 v_{3y} + m_4 v_{4y} + m_5 v_{5y}, \text{ so}$$

$$v_{5y} = \frac{m_1 v_1 + m_2 v_2 - m_3 v_{3y} - m_4 v_{4y}}{m_5}.$$

Compute the final speed of the tangled swallows, v_5.

$$v_5 = \sqrt{v_{5x}^2 + v_{5y}^2} = \frac{1}{m_5}\sqrt{[-(m_3 v_{3x} + m_4 v_{4x})]^2 + (m_1 v_1 + m_2 v_2 - m_3 v_{3y} - m_4 v_{4y})^2}$$

$$= \frac{\sqrt{\begin{array}{l}[(0.80 \text{ kg})(13 \text{ m/s})\cos 260^\circ + (0.70 \text{ kg})(14 \text{ m/s})\cos 60^\circ]^2 \\ +[(1.07 \text{ kg})(20 \text{ m/s}) + (0.92 \text{ kg})(-15 \text{ m/s}) - (0.80 \text{ kg})(13 \text{ m/s})\sin 260^\circ - (0.70 \text{ kg})(14 \text{ m/s})\sin 60^\circ]^2\end{array}}}{0.270 \text{ kg} + 0.220 \text{ kg}}$$

$$= 20 \text{ m/s}$$

Compute the direction.

$$\theta = \tan^{-1}\frac{v_{5y}}{v_{5x}}$$

$$= \tan^{-1}\frac{(1.07 \text{ kg})(20 \text{ m/s}) + (0.92 \text{ kg})(-15 \text{ m/s}) - (0.80 \text{ kg})(13 \text{ m/s})\sin 260^\circ - (0.70 \text{ kg})(14 \text{ m/s})\sin 60^\circ}{-(0.80 \text{ kg})(13 \text{ m/s})\cos 260^\circ - (0.70 \text{ kg})(14 \text{ m/s})\cos 60^\circ}$$

$$= -72^\circ$$

Since $v_{5x} < 0$ and $v_{5y} > 0$, the velocity vector is located in the second quadrant, so the angle is $180^\circ - 72^\circ = 108^\circ$ from the positive x-axis or 18° west of north. Thus, the velocity of the birds immediately after the collision is $\boxed{20 \text{ m/s at } 18^\circ \text{ W of N}}$.

73. Strategy Use the impulse-momentum theorem.

Solution Find the average force exerted by the ground on the ball.

y x
54 m/s 53 m/s
22° 18°

$$F_{\text{av}}=\frac{\Delta p}{\Delta t}=\frac{m\sqrt{(\Delta v_x)^2+(\Delta v_y)^2}}{\Delta t}=\frac{0.060\text{ kg}}{0.065\text{ s}}\sqrt{[(53\text{ m/s})\cos 18°-(54\text{ m/s})\cos(-22°)]^2+[(53\text{ m/s})\sin 18°-(54\text{ m/s})\sin(-22°)]^2}=\boxed{34\text{ N}}$$

77. Strategy Use the impulse-momentum theorem.

Solution Compute the average forces imparted to the two gloved hands during the catches.

$$\text{Inexperienced: } F_{\text{av}}=\frac{\Delta p}{\Delta t}=\frac{m\Delta v}{\Delta t}=(0.14\text{ kg})\frac{(130\text{ km/h})\left(\frac{10^3\text{ m}}{\text{km}}\right)\left(\frac{1\text{ h}}{3600\text{ s}}\right)}{10^{-3}\text{ s}}=\boxed{5000\text{ N}}$$

$$\text{Experienced: } F_{\text{av}}=(0.14\text{ kg})\frac{(130\text{ km/h})\left(\frac{10^3\text{ m}}{\text{km}}\right)\left(\frac{1\text{ h}}{3600\text{ s}}\right)}{10\times10^{-3}\text{ s}}=\boxed{500\text{ N}}$$

81. Strategy Use conservation of momentum. The collision is perfectly inelastic, so $v_{1\text{f}}=v_{2\text{f}}=v_{\text{f}}$.

Solution Find the speed of the cars just after the collision.

$$m_1v_{1\text{f}}+m_2v_{2\text{f}}=(m_1+m_2)v_{\text{f}}=m_1v_{1\text{i}}+m_2v_{2\text{i}}=m_1v_{\text{i}}+0,\text{ so}$$

$$v_{\text{f}}=\frac{m_1g}{(m_1+m_2)g}v_{\text{i}}=\frac{13.6\text{ kN}}{13.6\text{ kN}+9.0\text{ kN}}(17.0\text{ m/s})=\boxed{10.2\text{ m/s}}.$$

85. Strategy We must determine the initial speeds of the two cars. The collision is perfectly inelastic, so the final velocities of the cars are identical. Use conservation of momentum and the work-kinetic energy theorem.

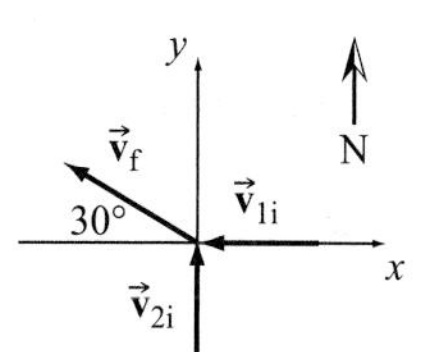

Solution Let the 1100-kg car be (1) and the 1300-kg car be (2). Use the work-kinetic energy theorem to determine the kinetic energy and, thus, the initial speed of the wrecked cars, which is the final speed of the collision.

$$W=F\Delta r=f_{\text{k}}\Delta r=-\mu_{\text{k}}mg\Delta r=\Delta K=0-\frac{1}{2}mv_{\text{i}}^2,\text{ so }v_{\text{i}}=\sqrt{2\mu_{\text{k}}g\Delta r}.$$

Thus, the final speed of the collision is $v_{\text{f}}=\sqrt{2\mu_{\text{k}}g\Delta r}$.

Find the initial speeds.

$$p_{\text{i}x}=m_1v_{1\text{i}x}+m_2v_{2\text{i}x}=m_1v_{1\text{i}}+0=p_{\text{f}x}=(m_1+m_2)v_{\text{f}x},\text{ so}$$

$$v_{1\text{i}}=\frac{m_1+m_2}{m_1}v_{\text{f}x}=\frac{m_1+m_2}{m_1}\sqrt{2\mu_{\text{k}}g\Delta r}\cos 150°=\frac{2400\text{ kg}}{1100\text{ kg}}\sqrt{2(0.80)(9.80\text{ m/s}^2)(17\text{ m})}\cos 150°\left(\frac{1\text{ km/h}}{0.2778\text{ m/s}}\right)=-110\text{ km/h}.$$

$$p_{\text{i}y}=m_1v_{1\text{i}y}+m_2v_{2\text{i}y}=0+m_2v_{2\text{i}}=p_{\text{f}y}=(m_1+m_2)v_{\text{f}y},\text{ so}$$

$$v_{2\text{i}}=\frac{m_1+m_2}{m_2}v_{\text{f}y}=\frac{m_1+m_2}{m_2}\sqrt{2\mu_{\text{k}}g\Delta r}\sin 150°=\frac{2400\text{ kg}}{1300\text{ kg}}\sqrt{2(0.80)(9.80\text{ m/s}^2)(17\text{ m})}\sin 150°\left(\frac{1\text{ km/h}}{0.2778\text{ m/s}}\right)=54\text{ km/h}.$$

Since $110>70$, $\boxed{\text{the lighter car was speeding}}$.

89. Strategy Use conservation of energy and momentum. Let $2m = m_B = 2m_A$.

Solution Find the maximum kinetic energy of A alone and, thus, its speed just before it strikes B.

$\Delta K = \frac{1}{2}mv_1^2 - 0 = -\Delta U = mgh - 0$, so $v_1 = \sqrt{2gh}$.

Use conservation of momentum to find the speed of the combined bobs just after impact. The collision is perfectly inelastic, so $v_{Af} = v_{Bf} = v_2$.

$m_A v_{Af} + m_B v_{Bf} = (m + 2m)v_2 = m_A v_{Ai} + m_B v_{Bi} = mv_1 + 0$, so $v_2 = \frac{1}{3}v_1$.

Find the maximum height.

$\Delta K = 0 - \frac{1}{2}mv_2^2 = -\frac{1}{2}m\left(\frac{1}{3}\sqrt{2gh}\right)^2 = -\Delta U = 0 - mgh_2$, so $h_2 = \boxed{\frac{1}{9}h}$.

93. Strategy Use conservation of momentum and Eq. (6-6) for the kinetic energies. Since the radium nucleus is at rest, $\vec{\mathbf{p}}_i = \vec{\mathbf{p}}_{Ra} = 0$.

Solution

(a) Find the ratio of the speed of the alpha particle to the speed of the radon nucleus.

$p_f = m_{Rn}v_{Rn} + m_\alpha v_\alpha = p_i = 0$, so $m_\alpha v_\alpha = -m_{Rn}v_{Rn}$. Therefore,

$\frac{v_\alpha}{v_{Rn}} = \frac{m_{Rn}}{m_\alpha} = \frac{222\text{ u}}{4\text{ u}} = \frac{222}{4} = \boxed{\frac{111}{2}}$, where the negative was dropped because speed is nonnegative.

(b) Since the initial momentum is zero, $\vec{\mathbf{p}}_{Rn} = -\vec{\mathbf{p}}_\alpha$; therefore, $\frac{|\vec{\mathbf{p}}_\alpha|}{|\vec{\mathbf{p}}_{Rn}|} = \frac{p_\alpha}{p_{Rn}} = \boxed{1}$.

(c) Find the ratio of the kinetic energies.

$$\frac{K_\alpha}{K_{Rn}} = \frac{\frac{1}{2}m_\alpha v_\alpha^2}{\frac{1}{2}m_{Rn}v_{Rn}^2} = \frac{m_\alpha}{m_{Rn}}\left(\frac{v_\alpha}{v_{Rn}}\right)^2 = \frac{4\text{ u}}{222\text{ u}}\left(\frac{111}{2}\right)^2 = \boxed{\frac{111}{2}}$$

Chapter 8

TORQUE AND ANGULAR MOMENTUM

Problems

1. Strategy and Solution I has units $\text{kg}\cdot\text{m}^2$. ω^2 has units $(\text{rad/s})^2$. So, $\frac{1}{2}I\omega^2$ has units $\text{kg}\cdot\text{m}^2\cdot\text{rad}^2/\text{s}^2 = \text{kg}\cdot\text{m}^2/\text{s}^2 = \text{J}$, which is a unit of energy.

5. Strategy $I = \frac{2}{5}MR^2$ for a solid sphere and mass density is $\rho = M/V$.

Solution

(a) $M = \rho V = \rho\frac{4}{3}\pi R^3$ for a solid sphere. Form a proportion.

$$\frac{M_{\text{child}}}{M_{\text{adult}}} = \left(\frac{R_{\text{child}}}{R_{\text{adult}}}\right)^3 = \left(\frac{1}{2}\right)^3 = \frac{1}{8},\ \text{so the mass is } \boxed{\text{reduced by a factor of 8}}.$$

(b) Form a proportion.

$$\frac{I_{\text{child}}}{I_{\text{adult}}} = \frac{1}{8}\left(\frac{R_{\text{child}}}{R_{\text{adult}}}\right)^2 = \frac{1}{8}\left(\frac{1}{2}\right)^2 = \frac{1}{32}$$

The rotational inertia is $\boxed{\text{reduced by a factor of 32}}$.

9. (a) Strategy and Solution Since a significant fraction of the wheel's kinetic energy is rotational, to model it as if it were sliding without friction would be unjustified. So, the answer is $\boxed{\text{no}}$.

(b) Strategy Use Eq. (8-1) and form a proportion.

Solution Find the fraction of the total kinetic energy that is rotational.

$$\frac{K_{\text{rot}}}{K_{\text{total}}} = \frac{4\left(\frac{1}{2}I\omega^2\right)}{\frac{1}{2}Mv^2 + 4\left(\frac{1}{2}I\omega^2\right)} = \frac{1}{\frac{Mv^2}{4I\omega^2}+1} = \frac{1}{1+\frac{Mv^2}{4I(v^2/R^2)}} = \frac{1}{1+\frac{MR^2}{4I}} = \frac{1}{1+\frac{(1300\ \text{kg})(0.35\ \text{m})^2}{4(0.705\ \text{kg}\cdot\text{m}^2)}} = \boxed{0.017}$$

13. Strategy Use Eq. (8-3) to compute the torque in each case.

Solution

(a) The force is applied perpendicularly to the door, so $\tau = rF = (1.26\ \text{m})(46.4\ \text{N}) = \boxed{58.5\ \text{N}\cdot\text{m}}$.

(b) The force is applied at $43.0°$ from the door's surface, so $|\tau| = rF_{\perp} = rF\sin\theta = (1.26\ \text{m})(46.4\ \text{N})\sin 43.0° = \boxed{39.9\ \text{N}\cdot\text{m}}$.

(c) Since the force is applied such that its line of action passes through the axis of the door hinges—the axis of rotation—there is no perpendicular component of the force and the torque is $\boxed{0}$.

17. Strategy Use Eq. (8-3).

Solution Find the magnitude of the torque.

$$|\tau| = F_\perp r = mgr = (40.0\text{ kg})(9.80\text{ N/kg})(2.0\text{ m}) = \boxed{780\text{ N}\cdot\text{m}}$$

21. Strategy The center of gravity is located at the center of mass. Let the origin be at the center of the door.

Solution Due to symmetry, $y_{CM} = 0$.

$$x_{CM} = \frac{m_1x_1 + m_2x_2}{M} = \frac{m_1(0) + m_2(x)}{M} = \frac{W_2x}{W} = \frac{(5.0\text{ N})(-0.75\text{ m})}{5.0\text{ N} + 300.0\text{ N}} = -0.012\text{ m}$$

The center of gravity is located $\boxed{\text{1.2 cm toward the doorknob as measured from the center of the door.}}$

25. (a) Strategy Use Eqs. (5-18) and (8-4) and Newton's second law.

Solution Find the torque.

$\omega_f = 120$ rpm

0.62 m

$$\omega_f = \omega_i + \alpha\Delta t = 0 + \alpha\Delta t = \alpha\Delta t \text{ and } \tau = Fr_\perp = ma_t r_\perp = m(r_\perp\alpha)r_\perp = mr_\perp{}^2\alpha, \text{ so}$$

$$\tau = mr_\perp{}^2\frac{\omega_f}{\Delta t} = \frac{(182\text{ kg})(0.62\text{ m})^2}{30.0\text{ s}}(120\text{ rpm})\left(\frac{2\pi\text{ rad}}{\text{rev}}\right)\left(\frac{1\text{ min}}{60\text{ s}}\right) = \boxed{29\text{ N}\cdot\text{m}}.$$

(b) Strategy α is constant, so $\omega_{av} = (\omega_f + \omega_i)/2$. Use Eq. (8-6).

Solution Find the work done.

$$W = \tau\Delta\theta = \tau\omega_{av}\Delta t = \tau\left(\frac{\omega_f + \omega_i}{2}\right)\Delta t = \tau\left(\frac{\omega_f + 0}{2}\right)\Delta t = \frac{(29.3\text{ N}\cdot\text{m})(120)(2\pi)(30.0\text{ s})}{2(60\text{ s})} = \boxed{5.5\text{ kJ}}$$

29. Strategy A system balances if its center of mass is above its base of support. Use Eq. (7-9) to find the center of mass of the metersticks.

Solution Let the left end of the lowest meterstick be the origin.

$$x_{CM} = \frac{mx_1 + mx_2 + mx_3 + mx_4}{4m} = \frac{x_1 + x_2 + x_3 + x_4}{4}$$

$$= \frac{0.5000 + (0.5000 + 0.3333) + (0.5000 + 0.3333 + 0.1667) + (0.5000 + 0.3333 + 0.1667 + 0.0833)}{4}\text{ m}$$

$$= 0.8542\text{ m}$$

Since $\boxed{\text{the center of mass} = 0.8542\text{ m} < 0.8600\text{ m, so the system balances}}$.

31. Strategy Choose the axis of rotation at the fulcrum. Use Eqs. (8-8).

Solution Find F.

$$\Sigma\tau = 0 = -F(3.0\text{ m}) + (1200\text{ N})(0.50\text{ m}), \text{ so } F = \frac{(1200\text{ N})(0.50\text{ m})}{3.0\text{ m}} = \boxed{200\text{ N}}.$$

33. Strategy Use Eqs. (8-8).

Solution Choose the axis of rotation at the point of contact between the driveway and the ladder.

$\Sigma F_x = 0 = f - N_w$, so $f = N_w$.

$$\Sigma\tau = 0 = N_w(4.7\text{ m}) - W_l(2.5\text{ m})\cos\theta - W_p\left(\frac{3.0\text{ m}}{4.7\text{ m}}\right)(5.0\text{ m})\cos\theta, \text{ so } N_w = \frac{\cos\theta}{4.7\text{ m}}\left[W_l(2.5\text{ m}) + W_p\left(\frac{15\text{ m}}{4.7}\right)\right].$$

Find θ.

$4.7\text{ m} = (5.0\text{ m})\sin\theta$, so $\theta = \sin^{-1}\dfrac{4.7}{5.0}$.

Calculate *f*.

$$f = N_w = \frac{\cos\sin^{-1}\frac{4.7}{5.0}}{4.7\text{ m}}\left[(120\text{ N})(2.5\text{ m}) + (680\text{ N})\left(\frac{15\text{ m}}{4.7}\right)\right] = 180\text{ N}$$

So, the force of friction is $\boxed{180\text{ N toward the wall}}$.

35. Strategy Use Eqs. (8-8).

Solution Choose the axis of rotation at the hinge.

$\Sigma\tau = 0 = Wl\cos\theta - Tl\sin\theta + mg\frac{l}{2}\cos\theta$, so $T = \boxed{\dfrac{mg/2+W}{\tan\theta}}$.

$\boxed{\text{For } \theta = 0, T \to \infty, \text{ and for } \theta = 90°, T \to 0.}$

37. Strategy Use Eqs. (8-8). Choose the axis of rotation at the point where the beam meets the store.

Solution The tension in the cable cannot exceed 417 N. Sum the torques.

$\Sigma\tau = 0 = T\sin\theta(1.50\text{ m}) - (50.0\text{ N})(0.75\text{ m}) - (200.0\text{ N})(1.00\text{ m})$

Solve for θ and substitute 417 N (the breaking strength) for *T*.

$$\theta = \sin^{-1}\frac{(50.0\text{ N})(0.75\text{ m}) + (200.0\text{ N})(1.00\text{ m})}{(417\text{ N})(1.50\text{ m})} = 22.3°$$

The minimum angle is $\boxed{22.3°}$.

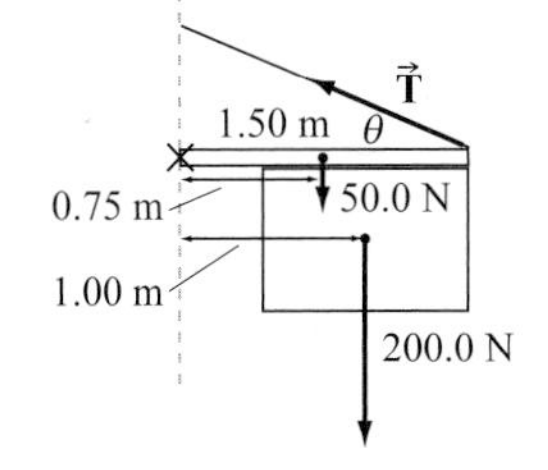

41. Strategy Use Eqs. (8-8). Choose the axis of rotation at the shoulder joint. One arm supports half of the person's weight, so $F_p = \frac{1}{2}(700\text{ N}) = 350\text{ N}$.

Solution Find the force each muscle exerts.

$\Sigma\tau = 0 = F_m(12\text{ cm})\sin 15° - F_g(27.5\text{ cm}) - F_p(60\text{ cm})$, so

$$F_m = \frac{F_g(27.5\text{ cm}) + F_p(60\text{ cm})}{(12\text{ cm})\sin 15°} = \frac{(30.0\text{ N})(27.5\text{ cm}) + (350\text{ N})(60\text{ cm})}{(12\text{ cm})\sin 15°} = \boxed{7.0\text{ kN}}.$$

43. Strategy Use Eqs. (8-8). Choose the axis of rotation at the elbow.

Solution Find the force exerted by the biceps muscle.

$\Sigma\tau = 0 = -W_m(35.0\text{ cm}) - W_a(16.5\text{ cm}) + F_b(5.00\text{ cm})\sin\theta$, so

$$F_b = \frac{W_m(35.0\text{ cm}) + W_a(16.5\text{ cm})}{(5.00\text{ cm})\sin\theta} = \frac{(9.9\text{ N})(35.0\text{ cm}) + (18.0\text{ N})(16.5\text{ cm})}{(5.00\text{ cm})\frac{30.0\text{ cm}}{\sqrt{(30.0\text{ cm})^2 + (5.00\text{ cm})^2}}} = \boxed{130\text{ N}}.$$

45. Strategy Assuming F_b is (nearly) straight down, F_s is simply equal to the magnitude of the sum of the forces due to gravity on your friend and the package.

Solution Find F_s.

$F_s = Mg + mg = (M+m)g = (55\text{ kg} + 10\text{ kg})(9.80\text{ N/kg}) = \boxed{640\text{ N}}$.

49. Strategy Use the rotational form of Newton's second law and Eq. (5-21).

Solution Find the torque that the motor must deliver.

$I = \frac{1}{2}MR^2$ for a uniform disk, so

$$\Sigma\tau = I\alpha = \frac{1}{2}MR^2\left(\frac{\omega_f^2 - \omega_i^2}{2\Delta\theta}\right) = \frac{MR^2\omega_f^2}{4\Delta\theta} = \frac{(0.22\text{ kg})\left(\frac{0.305\text{ m}}{2}\right)^2(3.49\text{ rad/s})^2}{4(2.0\text{ rev})(2\pi\text{ rad/rev})} = \boxed{0.0012\text{ N}\cdot\text{m}}.$$

53. Strategy Use the rotational form of Newton's second law and the definition of rotational inertia.

Solution Find the torque required to cause the angular acceleration.

$I = \sum_{i=A}^{D} m_i r_i^2 = (m_A + m_B + m_C + m_D)r^2$, since all four masses are (0.75 m)/2 from the axis.

$$\Sigma\tau = I\alpha = (4.0\text{ kg} + 3.0\text{ kg} + 5.0\text{ kg} + 2.0\text{ kg})[(0.75\text{ m})/2]^2(0.75\text{ rad/s}^2) = \boxed{1.5\text{ N}\cdot\text{m}}$$

55. (a) Strategy The rotational inertia of the merry-go-round is $I = \frac{1}{2}MR^2$ and that of the children is $I = 2MR^2$. Use the rotational form of Newton's second law.

Solution Find the torque on the merry-go-round.

$$\Sigma\tau = I\alpha = \left(\frac{1}{2}MR^2 + 2mR^2\right)\frac{\Delta\omega}{\Delta t}$$

$$= \left[\frac{1}{2}(350.0\text{ kg})(1.25\text{ m})^2 + 2(30.0\text{ kg})(1.25\text{ m})^2\right]\left(\frac{25\text{ rpm}}{20.0\text{ s}}\right)\left(\frac{2\pi\text{ rad}}{\text{rev}}\right)\left(\frac{1\text{ min}}{60\text{ s}}\right) = \boxed{48\text{ N}\cdot\text{m}}$$

(b) Strategy Let F be the magnitude of the tangential force with which each child must push the rim.

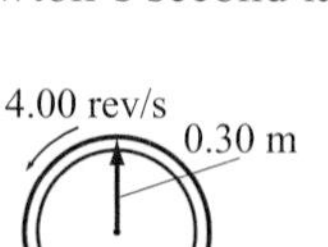

Solution Find F.

$$FR + FR = \Sigma\tau, \text{ so } F = \frac{\Sigma\tau}{2R} = \frac{48\text{ N}\cdot\text{m}}{2(1.25\text{ m})} = \boxed{19\text{ N}}.$$

57. Strategy The rotational inertia of the wheel is $I = MR^2$. Use the rotational form of Newton's second law.

Solution Find the magnitude of the average torque.

$$|\Sigma\tau_{av}| = I\alpha = MR^2\left|\frac{\Delta\omega}{\Delta t}\right| = (2\text{ kg})(0.30\text{ m})^2\left(\frac{4.00\text{ rev/s}}{50\text{ s}}\right)\left(\frac{2\pi\text{ rad}}{\text{rev}}\right) = \boxed{0.09\text{ N}\cdot\text{m}}$$

61. Strategy Use Eqs. (6-6) and (8-1).

Solution Find the total kinetic energies of each object.

Solid sphere:

$$K_{tr} + K_{rot} = \frac{1}{2}mv^2 + \frac{1}{2}I\omega^2 = \frac{1}{2}mv^2 + \frac{1}{2}\left(\frac{2}{5}mr^2\right)\left(\frac{v}{r}\right)^2 = \frac{1}{2}mv^2 + \frac{1}{5}mv^2 = \frac{7}{10}mv^2$$

Solid cylinder:

$$K_{tr} + K_{rot} = \frac{1}{2}mv^2 + \frac{1}{2}I\omega^2 = \frac{1}{2}mv^2 + \frac{1}{2}\left(\frac{1}{2}mr^2\right)\left(\frac{v}{r}\right)^2 = \frac{1}{2}mv^2 + \frac{1}{4}mv^2 = \frac{3}{4}mv^2$$

Hollow cylinder:

$$K_{tr}+K_{rot}=\frac{1}{2}mv^2+\frac{1}{2}I\omega^2=\frac{1}{2}mv^2+\frac{1}{2}mr^2\left(\frac{v}{r}\right)^2=\frac{1}{2}mv^2+\frac{1}{2}mv^2=mv^2$$

In order from smallest to largest, the total kinetic energies are

$$\boxed{\text{solid sphere: } K=\frac{7}{10}mv^2;\ \text{solid cylinder: } K=\frac{3}{4}mv^2;\ \text{hollow cylinder: } K=mv^2}.$$

65. (a) Strategy and Solution The drilled cylinder takes more time because it converts a larger fraction of its potential energy to rotational kinetic energy and a smaller fraction to translational kinetic energy than the solid cylinder; the drilled cylinder takes more time because its rotational inertia is larger.

(b) Strategy Use conservation of energy and the result for the acceleration from Example 8.13.

Solution Find the speeds of the solid and drilled cylinders.
Solid cylinder:

$$mgh=K_{tr}+K_{rot}=\frac{1}{2}mv^2+\frac{1}{2}I\omega^2=\frac{1}{2}mv^2+\frac{1}{2}\left(\frac{1}{2}mr^2\right)\left(\frac{v^2}{r^2}\right)=\frac{1}{2}mv^2+\frac{1}{4}mv^2=\frac{3}{4}mv^2,\ \text{so } v=2\sqrt{\frac{gh}{3}}.$$

Drilled cylinder:

$$mgh=\frac{1}{2}mv^2+\frac{1}{2}\left\{\frac{1}{2}m\left[r^2+(0.50r)^2\right]\right\}\left(\frac{v^2}{r^2}\right)=\frac{1}{2}mv^2+0.3125mv^2=0.8125mv^2,\ \text{so } v=\sqrt{\frac{gh}{0.8125}}.$$

The acceleration of the center of mass for a rolling ball is found in Example 8.13. The result is also valid for a rolling cylinder with the appropriate choice of I.

$$a_{CM}=\frac{g\sin\theta}{1+I/(mr^2)}$$

Solid cylinder: $a_{CM}=\dfrac{g\sin\theta}{1+\frac{1}{2}mr^2/(mr^2)}=\dfrac{2g\sin\theta}{3}$

Drilled cylinder: $a_{CM}=\dfrac{g\sin\theta}{1+0.625mr^2/(mr^2)}=\dfrac{g\sin\theta}{1.625}$

Since $\Delta v=v_f-v_i=v_f-0=a_{CM}\Delta t$,

$$\frac{\Delta t_d}{\Delta t_s}=\frac{\frac{v_d}{a_d}}{\frac{v_s}{a_s}}=\left(\frac{v_d}{v_s}\right)\left(\frac{a_s}{a_d}\right)=\left(\frac{\sqrt{\frac{gh}{0.8125}}}{2\sqrt{\frac{gh}{3}}}\right)\left(\frac{\frac{2g\sin\theta}{3}}{\frac{g\sin\theta}{1.625}}\right)=\frac{1}{2}\sqrt{\frac{3}{0.8125}}\left[\frac{2(1.625)}{3}\right]=1.04$$

The time for the drilled cylinder to roll down the incline is $\boxed{4\%\text{ longer}}$ than that for the solid cylinder.

67. Strategy Use conservation of energy and the relationship between speed and radial acceleration.

Solution At the top of the loop, the cylinder's speed must be at least the speed that results in a radial acceleration of g.

$$\frac{v^2}{r}=g,\ \text{so } v^2=gr.$$

The cylinder's kinetic energy is $\frac{1}{2}mv^2+\frac{1}{2}I\omega^2=\frac{1}{2}mv^2+\frac{1}{2}mr^2\left(\frac{v}{r}\right)^2=mv^2=mgr$, and it must equal the potential energy difference $mgh-mg(2r)$.

Thus, $mgr=mgh-2mgr=mg(h-2r)$, so $h=\boxed{3r}$.

69. (a) Strategy Let $r_1 = 0.00500$ m and $r_2 = 0.0200$ m. The tangential speed of the axle and the speed of the yo-yo are the same. Use conservation of energy.

Solution Find the speed of the yo-yo.

$$\Delta K = \frac{1}{2}mv^2 + \frac{1}{2}I\omega^2 = \frac{1}{2}mv^2 + \frac{1}{2}\left(\frac{1}{2}mr_2^2\right)\left(\frac{v^2}{r_1^2}\right) = \frac{1}{2}mv^2 + \frac{1}{4}m\left(\frac{r_2}{r_1}\right)^2 v^2 = -\Delta U = mgh, \text{ so}$$

$$v = \sqrt{\frac{4gh}{2+(r_2/r_1)^2}} = \sqrt{\frac{4(9.80 \text{ m/s}^2)(1.00 \text{ m})}{2+\left(\frac{0.0200 \text{ m}}{0.00500 \text{ m}}\right)^2}} = \boxed{1.5 \text{ m/s}}.$$

(b) Strategy Assume constant acceleration. Use Eq. (4-8).

Solution Find the time is takes the yo-yo to fall.

$$\Delta y = v_{av}\Delta t = \frac{1}{2}(v_{fy} + v_{iy})\Delta t = \frac{v}{2}\Delta t, \text{ so } \Delta t = \frac{2\Delta y}{v} = \frac{2(1.00 \text{ m})}{1.476 \text{ m/s}} = \boxed{1.36 \text{ s}}.$$

73. Strategy The initial rotational inertia is $I_i = \frac{1}{2}MR^2$, and the final rotational inertia is $I_f = \frac{1}{2}MR^2 + mr^2$, where M is the mass of the disk, m is the mass of the clod of clay, R is the radius of the disk, and r is the distance from the center of the disk (axis of rotation) to the center of the clod. Use conservation of angular momentum.

Solution Solve for the final angular speed.

$$L_i = I_i\omega_i = \frac{1}{2}MR^2\omega_i = L_f = I_f\omega_f = \left(\frac{1}{2}MR^2 + mr^2\right)\omega_f, \text{ so}$$

$$\omega_f = \frac{\frac{1}{2}MR^2}{\frac{1}{2}MR^2 + mr^2}\omega_i = \left(1+\frac{2mr^2}{MR^2}\right)^{-1}\omega_i = \left[1+\frac{2(0.12 \text{ g})(0.0800 \text{ m})^2}{(0.80 \text{ kg})(0.170 \text{ m})^2}\right]^{-1}(18.0 \text{ Hz})$$

$$= \boxed{16.9 \text{ Hz}}.$$

ω_f 17.0 cm 8.00 cm

75. Strategy Since the torque is constant, it is equal to the change in angular momentum divided by the time interval.

Solution Find the time to stop the spinning wheel

$$\tau = \frac{\Delta L}{\Delta t}, \text{ so } \Delta t = \frac{\Delta L}{\tau} = \frac{-6.40 \text{ kg}\cdot\text{m}^2/\text{s}}{-4.00 \text{ N}\cdot\text{m}} = \boxed{1.60 \text{ s}}.$$

77. Strategy Use conservation of angular momentum, Eq. (8-1), Eq. (8-14), and the relationship between period and angular velocity.

Solution

(a) Since the angular momentum is conserved, the ratio is $\boxed{1}$.

(b) Since the rotational inertia is proportional to the square of the radius, $\omega = \frac{L}{I} \propto \frac{L}{r^2}$.

Find the ratio of the angular velocities.

$$\frac{\omega_f}{\omega_i} = \frac{r_i^2}{r_f^2} = \left(\frac{1}{1.0\times10^{-4}}\right)^2 = \boxed{1.0\times10^8}$$

(c) The rotational kinetic energy is $K_{\text{rot}} = \frac{1}{2}I\omega^2 = \frac{1}{2}\left(\frac{L}{\omega}\right)\omega^2 = \frac{1}{2}L\omega.$

Find the ratio of the rotational kinetic energies.

$$\frac{K_\text{f}}{K_\text{i}} = \frac{\omega_\text{f}}{\omega_\text{i}} = \boxed{1.0\times10^8}$$

(d) The period is related to the angular velocity by $T = \frac{2\pi}{\omega}.$

Find the period of the star after collapse.

$$\frac{T_\text{f}}{T_\text{i}} = \frac{\omega_\text{i}}{\omega_\text{f}}, \text{ so } T_\text{f} = \frac{\omega_\text{i}}{\omega_\text{f}}T_\text{i} = (1.0\times10^{-8})(1.0\times10^7\text{ s}) = \boxed{0.10\text{ s}}.$$

79. Strategy Use conservation of angular momentum and Eq. (8-14).

Solution Find the skater's final angular velocity.

$$L_\text{i} = I_\text{i}\omega_\text{i} = L_\text{f} = I_\text{f}\omega_\text{f}, \text{ so } \omega_\text{f} = \frac{I_\text{i}}{I_\text{f}}\omega_\text{i} = \frac{2.50}{1.60}(10.0\text{ rad/s}) = \boxed{15.6\text{ rad/s}}.$$

81. Strategy Use conservation of angular momentum and Eq. (8-14).

Solution Find the diver's initial angular velocity.

$$L_\text{i} = I_\text{i}\omega_\text{i} = L_\text{f} = I_\text{f}\omega_\text{f}, \text{ so } \omega_\text{i} = \frac{I_\text{f}}{I_\text{i}}\omega_\text{f} = \frac{1}{3.00}\left(\frac{2.00\text{ rev}}{1.33\text{ s}}\right)\left(\frac{2\pi\text{ rad}}{\text{rev}}\right) = \boxed{3.15\text{ rad/s}}.$$

85. Strategy Use Eqs. (8-8). Choose the axis of rotation at the hinge attaching the crane to the cab (the pivot).

Solution Find T_1.

$$\Sigma\tau = 0 = T_2(12.2\text{ m})\sin 10.0° + T_1(12.2\text{ m})\sin 5.0° - (18\text{ kN})(6.1\text{ m})\sin 40.0° - (67\text{ kN})(12.2\text{ m})\sin 40.0° \text{ and}$$

$$\Sigma F_y = T_1 - 67\text{ kN} = 0, \text{ so } \boxed{T_1 = 67\text{ kN}}.$$

Find T_2.

$$T_2 = \frac{[(18\text{ kN})(6.1\text{ m}) + (67\text{ kN})(12.2\text{ m})]\sin 40.0° - (67\text{ kN})(12.2\text{ m})\sin 5.0°}{(12.2\text{ m})\sin 10.0°}, \text{ so } \boxed{T_2 = 250\text{ kN}}.$$

At the pivot:

$$\Sigma F_y = F_{\text{p}y} - 18\text{ kN} - 67\text{ kN} - T_1\cos 45.0° - T_2\cos 50.0° = 0, \text{ so}$$

$$F_{\text{p}y} = 18\text{ kN} + 67\text{ kN} + (247.7\text{ kN})\cos 50.0° + (67\text{ kN})\cos 45.0° = 291.6\text{ kN}.$$

$$\Sigma F_x = F_{\text{p}x} - T_1\sin 45.0° - T_2\sin 50.0° = 0, \text{ so}$$

$$F_{\text{p}x} = (247.7\text{ kN})\sin 50.0° + (67\text{ kN})\sin 45.0° = 237.1\text{ kN}.$$

Find the magnitude.

$$F_\text{p} = \sqrt{(237.1\text{ kN})^2 + (291.6\text{ kN})^2} = 380\text{ kN}$$

Find the direction.

$$\theta = \tan^{-1}\frac{291.6}{237.1} = 51°$$

So, $\boxed{\vec{\mathbf{F}}_\text{p} = 380\text{ kN at } 51° \text{ with the horizontal}}$.

89. Strategy The rotational inertia of each blade (uniform rod) is $I = \frac{1}{3}ML^2$, where L is the length of each blade. Find the angular acceleration of the fan using the definition; and use Eq. (8-9) to find the torque applied to the fan by the motor.

Solution The angular acceleration is $\alpha = \Delta\omega/\Delta t$. Find the torque.

$$\Sigma\tau = I\alpha = 4\left(\frac{1}{3}ML^2\right)\frac{\Delta\omega}{\Delta t} = \frac{4ML^2\Delta\omega}{3\Delta t} = \frac{4(0.35\text{ kg})(0.60\text{ m})^2(1.8\text{ rev/s})}{3(4.35\text{ s})}\left(\frac{2\pi\text{ rad}}{\text{rev}}\right) = \boxed{0.44\text{ N}\cdot\text{m}}$$

93. (a) Strategy The rotational inertia of a uniform solid sphere is $I = \frac{2}{5}MR^2$. Use Eq. (8-1).

Solution Find the kinetic energy of the Earth.

$$K_{\text{rot}} = \frac{1}{2}I\omega^2 = \frac{1}{2}\left(\frac{2}{5}MR^2\right)\omega^2 = \frac{1}{5}(5.974\times10^{24}\text{ kg})(6.371\times10^6\text{ m})^2\left(\frac{2\pi\text{ rad}}{24\text{ h}}\right)^2\left(\frac{1\text{ h}}{3600\text{ s}}\right)^2 = \boxed{2.6\times10^{29}\text{ J}}$$

(b) Strategy and Solution $T = \frac{2\pi}{\omega}$ and $K_{\text{rot}} \propto \omega^2$, so $\omega \propto \sqrt{K_{\text{rot}}}$ and $\frac{T_\text{f}}{T_\text{i}} = \frac{\omega_\text{i}}{\omega_\text{f}} = \sqrt{\frac{K_\text{i}}{K_\text{f}}}$. The change in the period is $T_\text{f} - T_\text{i} = \left(\sqrt{\frac{K_\text{i}}{K_\text{f}}} - 1\right)T_\text{i} = \left(\sqrt{\frac{1}{0.990}} - 1\right)(24\text{ h})\left(\frac{60\text{ min}}{1\text{ h}}\right) = 7\text{ min.}$

$\boxed{\text{The length of the day would increase by 7 minutes.}}$

(c) Strategy Divide 1.0% of the Earth's rotational kinetic energy by the world's energy usage.

Solution One percent of the Earth's rotational kinetic energy would supply the world's energy needs (at today's usage) for $\frac{0.010(2.6\times10^{29}\text{ J})}{1.0\times10^{21}\text{ J/yr}} = \boxed{2.6\text{ million years}}$.

97. Strategy The rotational inertia of the rod is $I = \frac{1}{3}mL^2$. Use conservation of energy.

Solution Find the speed of the lower end of the uniform rod when moving at its lowest point.

$$\Delta K = K_{\text{rot}} = \frac{1}{2}I\omega^2 = \frac{1}{2}\left(\frac{1}{3}mL^2\right)\left(\frac{v}{L}\right)^2 = \frac{1}{6}mv^2 = -\Delta U = mgh = mg\frac{L}{2},\text{ so } v = \boxed{\sqrt{3gL}}.$$

101. (a) Strategy The rotational inertia of a uniform thin rod is $I = \frac{1}{3}ML^2$.

Solution Compute the rotational inertia of the limb.

$$I = \frac{1}{3}ML^2 = \frac{1}{3}(0.0280\text{ kg})(0.0380\text{ m})^2 = \boxed{1.35\times10^{-5}\text{ kg}\cdot\text{m}^2}$$

(b) Strategy Use Eq. (8-9).

Solution Compute the muscular force required to achieve the blow.

$$\Sigma\tau = Fr = I\alpha = \frac{1}{3}ML^2\frac{\Delta\omega}{\Delta t},\text{ so } F = \frac{(0.0280\text{ kg})(0.0380\text{ m})^2(175\text{ rad/s})}{3(1.50\times10^{-3}\text{ s})(3.00\times10^{-3}\text{ m})} = \boxed{524\text{ N}}.$$

105. Strategy Since the mass is concentrated at the tip, $I = MR^2$. Use Eq. (8-14).

Solution Compute the angular momenta of the second and hour hands of the clock.

(a) $$L = I\omega = MR^2\omega = (0.10\text{ kg})(0.300\text{ m})^2\left(\frac{1\text{ rev}}{60\text{ s}}\right)\left(\frac{2\pi\text{ rad}}{\text{rev}}\right) = \boxed{9.4\times10^{-4}\text{ kg}\cdot\text{m}^2/\text{s}}$$

(b) $$L = (0.20\text{ kg})(0.200\text{ m})^2\left(\frac{1\text{ rev}}{12\text{ h}}\right)\left(\frac{1\text{ h}}{3600\text{ s}}\right)\left(\frac{2\pi\text{ rad}}{\text{rev}}\right) = \boxed{1.2\times10^{-6}\text{ kg}\cdot\text{m}^2/\text{s}}$$

109. (a) Strategy Use conservation of angular momentum, since no external torques act on the two-disk system.

Solution Find the final angular velocity.

$$L_\text{f} = I_\text{f}\omega_\text{f} = L_\text{i} = I_\text{i}\omega_\text{i},\text{ so }\omega_\text{f} = \frac{I_\text{i}\omega_\text{i}}{I_\text{f}} = \frac{I_\text{i}\omega_\text{i}}{I_\text{i} + \frac{1}{2}mr^2} = \frac{\omega_\text{i}}{1 + \frac{mr^2}{2\left(\frac{1}{2}MR^2\right)}} = \boxed{\frac{\omega_\text{i}}{1 + \frac{mr^2}{MR^2}}}.$$

(b) Strategy and Solution

$\boxed{\text{The total angular momentum does not change, since no external torques act on the system.}}$

(c) Strategy Compute the initial and final total kinetic energies and compare their values.

Solution

$$K_\text{i} = \frac{1}{2}I_\text{i}\omega_\text{i}^2 = \frac{1}{2}\left(\frac{1}{2}MR^2\right)\omega_\text{i}^2 = \frac{1}{4}MR^2\omega_\text{i}^2$$

$$K_\text{f} = \frac{1}{2}I_\text{f}\omega_\text{f}^2 = \frac{1}{2}\left(\frac{1}{2}MR^2 + \frac{1}{2}mr^2\right)\left(\frac{\omega_\text{i}}{1 + \frac{mr^2}{MR^2}}\right)^2 = \frac{1}{2}\left(\frac{1}{2}MR^2 + \frac{1}{2}mr^2\right)\left(\frac{\frac{1}{2}MR^2\omega_\text{i}}{\frac{1}{2}MR^2 + \frac{1}{2}mr^2}\right)^2$$

$$= \frac{\frac{1}{2}\left(\frac{1}{2}MR^2\right)^2\omega_\text{i}^2}{\frac{1}{2}MR^2 + \frac{1}{2}mr^2} = \frac{\frac{1}{4}MR^2\omega_\text{i}^2}{1 + \frac{mr^2}{MR^2}} = \frac{K_\text{i}}{1 + \frac{mr^2}{MR^2}}$$

So, $K_\text{f} \neq K_\text{i}$. Therefore, the answer is $\boxed{\text{yes; the kinetic energy changes.}}$

113. Strategy Use conservation of energy.

Solution

(a) Find the speed with which the roustabout reaches the ground.

$$\Delta K = \frac{1}{2}mv^2 = -\Delta U = mgL,\text{ so }v = \boxed{\sqrt{2gL}}.$$

(b) Find the speed with which the roustabout reaches the ground.

$$\Delta K = \frac{1}{2}I\omega^2 = \frac{1}{2}\left(\frac{1}{3}ML^2\right)\left(\frac{v}{L}\right)^2 = \frac{1}{6}Mv^2 = -\Delta U = Mg\frac{L}{2},\text{ so }v = \boxed{\sqrt{3gL}}.$$

(c) Since $\sqrt{2gL} < \sqrt{3gL}$, $\boxed{\text{the roustabout should jump}}$.

117. (a) Strategy Refer to Example 8.7. The system is in equilibrium until the ladder begins to slip.

Solution Use Newton's second law.

$\Sigma F_x = N_w - f = 0$, so $f = N_w$.

At the person's highest point, the frictional force has its maximum possible magnitude, $f = \mu_s N_f$.

Thus, $N_w = \mu_s N_f$.

$\Sigma F_y = N_f - Mg - mg = 0$, so $N_w = \mu_s g(M+m)$.

Choose the axis of rotation at the contact point between the ladder and the floor.

$\Sigma\tau = 0 = -N_w L\sin\theta + mg\left(\frac{1}{2}L\cos\theta\right) + Mgd\cos\theta$, so

$$d = \frac{N_w L\sin\theta - \frac{1}{2}mgL\cos\theta}{Mg\cos\theta} = \frac{\mu_s g(M+m)L\sin\theta - \frac{1}{2}mgL\cos\theta}{Mg\cos\theta} = \frac{\left[\mu_s(M+m)\sin\theta - \frac{m}{2}\cos\theta\right]L}{M\cos\theta}$$

$$= \boxed{\left(\mu_s\frac{M+m}{M}\tan\theta - \frac{m}{2M}\right)L}$$

(b) Strategy and Solution Since $\tan\theta$ increases as θ increases on the interval $0 \le \theta < 90°$, and since d increases if $\tan\theta$ increases [which is evident from the equation found in part (a)], placing the ladder at a larger angle θ allows a person to climb farther up the ladder without having it slip.

(c) Strategy Set $d = L$.

Solution Find the minimum angle that enables the person to climb all the way to the top of the ladder.

$$L = \left(\mu_s\frac{M+m}{M}\tan\theta - \frac{m}{2M}\right)L$$

$$1 + \frac{m}{2M} = \mu_s\frac{M+m}{M}\tan\theta$$

$$\frac{2M+m}{2} = \mu_s(M+m)\tan\theta$$

$$\frac{2M+m}{2\mu_s(M+m)} = \tan\theta$$

$$\theta = \tan^{-1}\frac{2M+m}{2\mu_s(M+m)} = \tan^{-1}\frac{2(60.0\text{ kg}) + 15.0\text{ kg}}{2(0.45)(60.0\text{ kg} + 15.0\text{ kg})} = \boxed{63°}$$

REVIEW AND SYNTHESIS: CHAPTERS 6–8

Review Exercises

1. (a) Strategy Multiply the extension per mass by the mass to find the maximum extension required.

Solution

$$\left(\frac{1.0 \text{ mm}}{25 \text{ g}}\right)(5.0 \text{ kg})\left(\frac{1000 \text{ g}}{1 \text{ kg}}\right)\left(\frac{1 \text{ m}}{1000 \text{ mm}}\right) = \boxed{0.20 \text{ m}}$$

(b) Strategy Set the weight of the mass equal to the magnitude of the force due to the spring scale. Use Hooke's law.

Solution

$$\text{Weight} = mg = kx, \text{ so } k = \frac{mg}{x} = \frac{(5.0 \text{ kg})(9.80 \text{ N/kg})}{0.20 \text{ m}} = \boxed{250 \text{ N/m}}.$$

5. Strategy The work done by the muscles is 22% of the energy expended. The gravitational potential energy gained by the person is equal to the work done by the muscles.

Solution $0.22E = W = \Delta U = mgh$, so

$$E = \frac{mgh}{0.22} = \frac{(80.0 \text{ kg})(9.80 \text{ N/kg})(15 \text{ m})}{0.22} = \boxed{53 \text{ kJ}}.$$

9. Strategy Use conservation of energy and Newton's second law.

Solution Find the normal force on the crate.

$\Sigma F_y = N - mg\cos\theta = 0$, so $N = mg\cos\theta$.

Find the force of sliding friction.

$\Sigma F_x = -f_k + mg\sin\theta = -\mu_k mg\cos\theta + mg\sin\theta = ma_x$, so

$a_x = -\mu_k g\cos\theta + g\sin\theta = -0.70(9.80 \text{ m/s}^2)\cos 53° + (9.80 \text{ m/s}^2)\sin 53° = 3.7 \text{ m/s}^2$.

Therefore, the acceleration of the block is $\boxed{3.7 \text{ m/s}^2 \text{ down the ramp}}$.

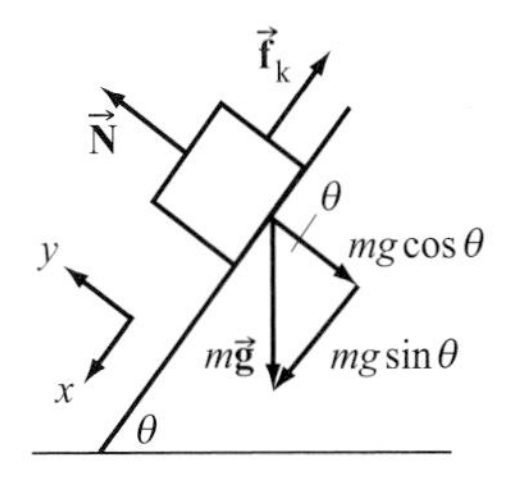

13. Strategy Use conservation of energy. Let $d = 2.05$ m. Then, the ramp rises $h = d\sin 5.00°$. The rotational inertia of a uniform sphere is $\frac{2}{5}mr^2$.

Solution Find the speed of the ball when it reaches the top of the ramp.

d
h
5.00°

$$0 = \Delta K + \Delta U = \frac{1}{2}mv_f^2 + \frac{1}{2}I\omega_f^2 - \frac{1}{2}mv_i^2 - \frac{1}{2}I\omega_i^2 + mgh$$

$$= \frac{1}{2}mv_f^2 + \frac{1}{2}\left(\frac{2}{5}mr^2\right)\left(\frac{v_f}{r}\right)^2 - \frac{1}{2}mv_i^2 - \frac{1}{2}\left(\frac{2}{5}mr^2\right)\left(\frac{v_i}{r}\right)^2 + mgh$$

$$= \frac{7}{10}mv_f^2 - \frac{7}{10}mv_i^2 + mgh, \text{ so}$$

$$v_f = \sqrt{v_i^2 - \frac{10}{7}gh} = \sqrt{(2.20 \text{ m/s})^2 - \frac{10}{7}(9.80 \text{ m/s}^2)(2.05 \text{ m})\sin 5.00°} = \boxed{1.53 \text{ m/s}}.$$

17. Strategy Use conservation of energy. The energy delivered to the fluid in the beaker plus the kinetic energies of the pulley, spool, axle, paddles, and the block are equal to the work done by gravity on the block, which is negative the change in the block's gravitational potential energy. The rotational inertia of the pulley (uniform solid disk) is $\frac{1}{2}m_\text{p}r^2$.

Solution Let the energy delivered to the fluid be E, the distance the block falls be h, and the rotational inertia of the spool, axle, and paddles be $I_\text{s} = 0.00140\ \text{kg}\cdot\text{m}^2$. Since the radii of the pulley and the spool are the same (r), their tangential speeds are the same, so let $v_\text{p} = v_\text{s} = v$.

$$m_\text{b}gh = \frac{1}{2}m_\text{b}v_\text{b}^2 + \frac{1}{2}I_\text{p}\omega_\text{p}^2 + \frac{1}{2}I_\text{s}\omega_\text{s}^2 + E = \frac{1}{2}m_\text{b}v_\text{b}^2 + \frac{1}{2}\left(\frac{1}{2}m_\text{p}r^2\right)\left(\frac{v}{r}\right)^2 + \frac{1}{2}I_\text{s}\left(\frac{v}{r}\right)^2 + E$$

The tangential speeds of the pulley and spool are equal to the speed of the block.

$$m_\text{b}gh = \frac{1}{2}m_\text{b}v_\text{b}^2 + \frac{1}{4}m_\text{p}v^2 + \frac{1}{2}I_\text{s}\frac{v^2}{r^2} + E = \frac{1}{2}m_\text{b}v^2 + \frac{1}{4}m_\text{p}v^2 + \frac{1}{2}I_\text{s}\frac{v^2}{r^2} + E, \text{ so}$$

$$E = m_\text{b}gh - \frac{v^2(2m_\text{b} + m_\text{p} + 2I_\text{s}/r^2)}{4}$$

$$= (0.870\ \text{kg})(9.80\ \text{m/s}^2)(2.50\ \text{m}) - \frac{(3.00\ \text{m/s})^2[2(0.870\ \text{kg}) + 0.0600\ \text{kg} + 2(0.00140\ \text{kg}\cdot\text{m}^2)/(0.0300\ \text{m})^2]}{4}$$

$$= \boxed{10.3\ \text{J}}.$$

21. Strategy Use energy conservation to find the speed of Jones just before he grabs Smith. Then, use momentum conservation to find the speed of both just after. Finally, again use energy conservation to find the final height.

Solution Find Jones's speed, v_J.

$$\frac{1}{2}m_\text{J}v_\text{J}^2 = m_\text{J}gh_\text{J}, \text{ so } v_\text{J} = \sqrt{2gh_\text{J}}.$$

Find the speed of both, v.

$$p_\text{i} = m_\text{J}v_\text{J} = p_\text{f} = (m_\text{J} + m_\text{S})v, \text{ so } v = \frac{m_\text{J}v_\text{J}}{m_\text{J} + m_\text{S}} = \frac{m_\text{J}\sqrt{2gh_\text{J}}}{m_\text{J} + m_\text{S}}.$$

Find the final height, h.

$$(m_\text{J} + m_\text{S})gh = \frac{1}{2}(m_\text{J} + m_\text{S})\left(\frac{m_\text{J}\sqrt{2gh_\text{J}}}{m_\text{J} + m_\text{S}}\right)^2, \text{ so } h = \frac{m_\text{J}^2h_\text{J}}{(m_\text{J} + m_\text{S})^2} = \frac{(78.0\ \text{kg})^2(3.70\ \text{m})}{(78.0\ \text{kg} + 55.0\ \text{kg})^2} = \boxed{1.27\ \text{m}}.$$

25. Strategy Use conservation of linear momentum.

Solution

$$p_{\text{i}x} = m_\text{b}v_\text{b} = p_{\text{f}x} = (m_\text{b} + m_\text{c})v_{\text{f}x}, \text{ so } v_{\text{f}x} = \frac{m_\text{b}v_\text{b}}{m_\text{b} + m_\text{c}}.$$

$$p_{\text{i}y} = m_\text{c}v_\text{c} = p_{\text{f}y} = (m_\text{b} + m_\text{c})v_{\text{f}y}, \text{ so } v_{\text{f}y} = \frac{m_\text{c}v_\text{c}}{m_\text{b} + m_\text{c}}.$$

N
y
x
$\vec{v}_\text{c}$
$\vec{v}_\text{b}$
$\vec{v}_\text{f}$

Compute the magnitude of the final velocity.

$$v = \sqrt{v_{\text{f}x}^2 + v_{\text{f}y}^2} = \sqrt{\left(\frac{m_\text{b}v_\text{b}}{m_\text{b} + m_\text{c}}\right)^2 + \left(\frac{m_\text{c}v_\text{c}}{m_\text{b} + m_\text{c}}\right)^2} = \frac{\sqrt{(m_\text{b}v_\text{b})^2 + (m_\text{c}v_\text{c})^2}}{m_\text{b} + m_\text{c}}$$

$$= \frac{\sqrt{[(2.00\ \text{kg})(2.70\ \text{m/s})]^2 + [(1.50\ \text{kg})(-3.20\ \text{m/s})]^2}}{2.00\ \text{kg} + 1.50\ \text{kg}} = 2.06\ \text{m/s}$$

Compute the angle.

$$\theta = \tan^{-1}\frac{v_{\text{fy}}}{v_{\text{fx}}} = \tan^{-1}\frac{\frac{m_\text{c}v_\text{c}}{m_\text{b}+m_\text{c}}}{\frac{m_\text{b}v_\text{b}}{m_\text{b}+m_\text{c}}} = \tan^{-1}\frac{m_\text{c}v_\text{c}}{m_\text{b}v_\text{b}} = \tan^{-1}\frac{(1.50\text{ kg})(-3.20\text{ m/s})}{(2.00\text{ kg})(2.70\text{ m/s})} = -41.6°$$

The velocity of the block and the clay after the collision is $\boxed{2.06\text{ m/s at }41.6°\text{ S of E}}$.

29. (a) Strategy Use conservation of energy.

Solution Let d be the distance moved along the incline by m_2. Both masses move the same distance and have the same speed, since they are connected by a rope.

$0 = \Delta K + \Delta U = \frac{1}{2}m_1v^2 + \frac{1}{2}m_2v^2 - 0 - 0 + m_1g(-d\sin\theta) + m_2gd\sin\phi$, so

$$v = \sqrt{\frac{2gd(m_1\sin\theta - m_2\sin\phi)}{m_1+m_2}} = \sqrt{\frac{2(9.80\text{ m/s}^2)(2.00\text{ m})[(6.00\text{ kg})\sin 36.9° - (4.00\text{ kg})\sin 45.0°]}{6.00\text{ kg} + 4.00\text{ kg}}}$$
$$= \boxed{1.7\text{ m/s}}.$$

(b) Strategy Use Newton's second law and Eq. (4-5).

Solution Let the positive direction be along the inclines from left to right.
For m_1: $\Sigma F = -T + m_1g\sin\theta = m_1a$, so $T = m_1g\sin\theta - m_1a$.
For m_2: $\Sigma F = T - m_2g\sin\phi = m_2a$, so $T = m_2g\sin\phi + m_2a$.
Solve for the acceleration.
$m_2g\sin\phi + m_2a = m_1g\sin\theta - m_1a$, so

$$a = \frac{g(m_1\sin\theta - m_2\sin\phi)}{m_1+m_2} = \frac{(9.80\text{ m/s}^2)[(6.00\text{ kg})\sin 36.9° - (4.00\text{ kg})\sin 45.0°]}{6.00\text{ kg}+4.00\text{ kg}} = \boxed{0.76\text{ m/s}^2}.$$

Find the speed.
$v_\text{f}^2 - v_\text{i}^2 = v^2 - 0 = 2ad$, so $v = \sqrt{2ad} = \sqrt{2(0.76\text{ m/s}^2)(2.00\text{ m})} = \boxed{1.7\text{ m/s}}$.

MCAT Review

1. Strategy Use conservation of momentum.

Solution
$p_\text{i} = mv_\text{i} = p_\text{f} = mv_\text{f} + p_\text{wall}$, so $p_\text{wall} = m(v_\text{i} - v_\text{f}) = (0.2\text{ kg})[2.0\text{ m/s} - (-1.0\text{ m/s})] = 0.6\text{ kg}\cdot\text{m/s}$.
The correct answer is $\boxed{\text{D}}$.

2. Strategy Use Hooke's law.

Solution Let up be the positive direction. The gravitational force on the mass is
$F = mg = (0.10\text{ kg})(-9.80\text{ m/s}^2) = -0.98\text{ N}$. Solving for the spring constant in Hooke's law, we have
$k = -\frac{F}{x} = -\frac{-0.98\text{ N}}{0.15\text{ m}} = 6.5\text{ N/m}$. Thus, the correct answer is $\boxed{\text{D}}$.

3. **Strategy** The net torque is zero.

Solution

$\Sigma\tau = 0 = F(0.60\text{ m}) - (1.0\times10^{-7}\text{ kg})(9.80\text{ m/s}^2)(0.40\text{ m})$, so

$$F = \frac{(1.0\times10^{-7}\text{ kg})(9.80\text{ m/s}^2)(0.40\text{ m})}{0.60\text{ m}} = 6.5\times10^{-7}\text{ N}.$$

The correct answer is B.

4. **Strategy** Determine the speed of the first ball just before in collides with the second. The collision is completely inelastic; that is, the balls stick together. Use conservation of momentum to find the speed of the balls after the collision.

Solution Find the speed of the first ball just before the collision.

$v_{fx} - v_{ix} = v_1 - 0 = a_x\Delta t$, so $v_1 = (10\text{ m/s}^2)(2.0\text{ s}) = 20\text{ m/s}$.

Find the speed v of the balls just after the collision.

$$p_i = m_1v_1 = p_f = (m_1+m_2)v, \text{ so } v = \frac{m_1v_1}{m_1+m_2} = \frac{(0.50\text{ kg})(20\text{ m/s})}{0.50\text{ kg}+1.0\text{ kg}} = 6.7\text{ m/s}.$$

The correct answer is B.

5. **Strategy** Use Newton's second law and Eq. (6-27).

Solution The gravitational force working against the motion of the car as it climbs the hill is $mg\sin 10°$, so the additional power required is

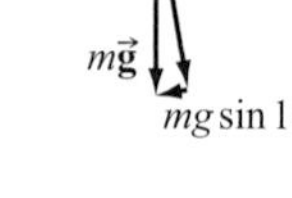

$$P_{\text{car}} = -P_{\text{grav}} = -Fv\cos 180° = (mg\sin 10°)v = (1000\text{ kg})(10\text{ m/s}^2)\sin 10°(15\text{ m/s})$$
$$= 1.5\times10^5\times\sin 10°\text{ W}.$$

The correct answer is D.

6. **Strategy** Find the vertical distance the patient would have climbed had the treadmill been stationary (and very long). Then, find the work done by the patient on the treadmill.

Solution The "distance" walked along the incline is $(2\text{ m/s})(600\text{ s}) = 1200\text{ m}$.

Thus, the vertical distance climbed is $(1200\text{ m})\sin 30° = 600\text{ m}$. The work done is

$W = Fd = mgd = (90\text{ kg})(10\text{ m/s}^2)(600\text{ m}) = 0.54\text{ MJ}$.

The correct answer is C.

7. **Strategy** Find the angle between the force exerted by the patient and the patient's velocity. Use Eq. (6-27).

Solution The force due to gravity is down, so the force exerted by the patient is up. The velocity is directed at the angle of the incline, or 30° above the horizontal, so the angle between the force and the velocity is 60°. Compute the mechanical power output of the patient.

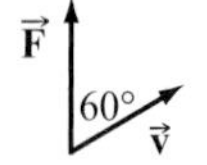

$P = Fv\cos\theta = mgv\cos\theta = (100\text{ kg})(10\text{ m/s}^2)(3\text{ m/s})\cos 60° = 1500\text{ W}$

The correct answer is B.

8. **Strategy and Solution** The force pushing each friction pad is normal to the wheel; that is, it is the normal force in $f_k = \mu_k N$. Solve for the normal force.

$$N = \frac{f_k}{\mu_k} = \frac{20\text{ N}}{0.4} = 50\text{ N}$$

This is the total force. The force pushing each friction pad is half this, or 25 N. The correct answer is $\boxed{\text{B}}$.

9. **Strategy** Find the average tangential speed at the friction pads. Then, use the relationship between tangential speed and radial acceleration.

Solution

The average tangential speed is $v = \frac{4800\text{ m}}{20\text{ min}} \times \frac{1\text{ min}}{60\text{ s}} = 4.0\text{ m/s}$. The radial acceleration is

$a_r = \frac{v^2}{r} = \frac{(4.0\text{ m/s})^2}{0.3\text{ m}} = 50\text{ m/s}^2$. The correct answer is $\boxed{\text{D}}$.

10. **Strategy** Use the work-kinetic energy theorem.

Solution The work done by friction on the wheel is $W = -f_k d$, where d is the linear distance the wheel passes between the pads before it stops. Relate d to the kinetic energy of the wheel.

$W_{\text{total}} = -f_k d = \Delta K = 0 - K_i$, so $d = \frac{K_i}{f_k}$.

Divide d by the circumference of a circle with radius 0.3 m to find the number of rotations.

$$\frac{d}{2\pi r} = \frac{K_i}{2\pi r f_k} = \frac{30\text{ J}}{2\pi(0.3\text{ m})(20\text{ N})} = 0.8\text{ rotations}$$

Since $0.8 < 1$, the correct answer is $\boxed{\text{A}}$.

11. **Strategy** Compute the average mechanical power output of the cyclist and compare it to the power consumed by the wheel at the friction pads.

Solution The metabolic power available for work is $535\text{ W} - 85\text{ W} = 450\text{ W}$. Since the efficiency is 20%, the average mechanical power output of the cyclist is $0.20 \times 450\text{ W} = 90\text{ W}$. The average tangential speed of the wheel is $v = \frac{4800\text{ m}}{20\text{ min}} \times \frac{1\text{ min}}{60\text{ s}} = 4.0\text{ m/s}$. Therefore, the power consumed by the friction pads is $P = f_k v = (20\text{ N})(4.0\text{ m/s}) = 80\text{ W}$. Thus, the difference between the average mechanical power output of the cyclist and the power consumed by the wheel at the friction pads is $90\text{ W} - 80\text{ W} = 10\text{ W}$.

The correct answer is $\boxed{\text{B}}$.

12. **Strategy and Solution** Increasing the force on the friction pads would increase the power consumed by the wheel at the friction pads (because $P = Fv$). So, if the cyclist is pedaling at the same rate and the power consumed by the friction pads increases, the difference between the two decreases and the fraction of mechanical power output of the cyclist consumed by the wheel at the friction pad increases. Thus, the correct answer is $\boxed{\text{D}}$.

13. Strategy Relate the cyclist's average metabolic rate to the energy released per volume of oxygen consumed, the time on the bike, and volume of oxygen consumed.

Solution The cyclist's average metabolic rate while riding is 535 W. The total energy used during 20 minutes is $(535\text{ W})(20\text{ min})\dfrac{60\text{ s}}{1\text{ min}} = 642{,}000\text{ J}$. The total energy released by the consumption of oxygen is $(20{,}000\text{ J/L})V$, where V is the volume of oxygen consumed. Equating these two expressions and solving for V gives the number of liters of oxygen the cyclist consumes.

$(20{,}000\text{ J/L})V = 642{,}000\text{ J}$, so $V = \dfrac{642{,}000\text{ J}}{20{,}000\text{ J/L}} = 32\text{ L} \approx 30\text{ L}$. The correct answer is $\boxed{\text{B}}$.

14. Strategy and Solution Since the force has been reduced by 50% and the distance has been doubled, the cyclist does the same amount of work $[W = 0.50F(2\Delta x) = F\Delta x]$. So, the energy transmitted in the second workout is equal to the energy transmitted in the first. The correct answer is $\boxed{\text{C}}$.

15. Strategy The circumference of a circle is $C = 2\pi r$. A wheel moves a distance equal to its circumference during each rotation. The wheel rotates twice during each rotation of the pedals.

Solution The circumference of a circle with a radius of 0.15 m is $2\pi(0.15\text{ m})$. The circumference of a circle with a radius of 0.3 m is $2\pi(0.3\text{ m})$. During each rotation of the pedals, a point on the wheel at a radius of 0.3 m moves a distance $2[2\pi(0.3\text{ m})]$. The ratio of the distance moved by a pedal to the distance moved by a point on the wheel located at a radius of 0.3 m in the same amount of time is $\dfrac{2\pi(0.15\text{ m})}{2[2\pi(0.3\text{ m})]} = 0.25$.

The correct answer is $\boxed{\text{A}}$.

16. Strategy Use the definition of power.

Solution

$P = \dfrac{\Delta E}{\Delta t}$, so $\Delta t = \dfrac{\Delta E}{P} = \left(\dfrac{300\text{ kcal}}{500\text{ W}}\right)\left(\dfrac{4186\text{ J}}{1\text{ kcal}}\right)\left(\dfrac{1\text{ min}}{60\text{ s}}\right) = 41.9\text{ min}$. The correct answer is $\boxed{\text{D}}$.

17. Strategy Consider the distance a point on the wheel travels for each situation.

Solution The circumference of a circle with a radius of 0.3 m is $2\pi(0.3\text{ m})$. The circumference of a circle with a radius of 0.4 m is $2\pi(0.4\text{ m})$. During each rotation, a point on a wheel travels a distance equal to the circumference. The force on the wheel is the same in each case, but the distance traveled by a point on the wheel is greater for a greater radius. In this case, the distance is $0.4\text{ m}/(0.3\text{ m}) = 1.33$ times farther or 33%. Since work is equal to the product of force times distance, the work done on the wheel per revolution is 33% more. Thus, the correct answer is $\boxed{\text{C}}$.

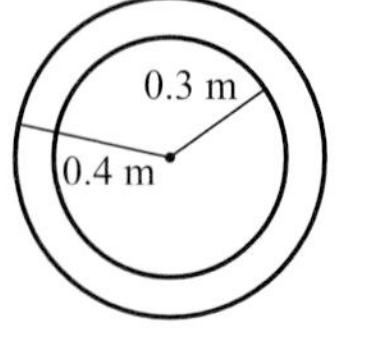

Chapter 9

FLUIDS

Problems

1. Strategy Use the definition of average pressure.

Solution Compute the average pressure.

$$P_{av} = \frac{F}{A} = \frac{500\text{ N}}{1.0\text{ cm}^2}\left(\frac{100\text{ cm}}{1\text{ m}}\right)^2\left(\frac{1\text{ atm}}{101.3\times 10^3\text{ Pa}}\right) = \boxed{49\text{ atm}}$$

5. Strategy The average pressure is the force applied to the floor divided by the contact area.

Solution

The baby applies a pressure of $P_b = \frac{F}{A} = \frac{m_b g}{3\left(\frac{1}{4}\pi d_s^2\right)} = \frac{4m_b g}{3\pi d_s^2}$.

The adult applies a pressure of $P_a = \frac{F}{A} = \frac{m_a g}{4\left(\frac{1}{4}\pi d_c^2\right)} = \frac{m_a g}{\pi d_c^2}$.

The ratio of these two pressures is $\frac{P_b}{P_a} = \frac{4m_b g}{3\pi d_s^2}\left(\frac{m_a g}{\pi d_c^2}\right)^{-1} = \frac{4m_b d_c^2}{3m_a d_s^2} = \frac{4(10\text{ kg})(0.060\text{ m})^2}{3(60\text{ kg})(0.020\text{ m})^2} = 2.0.$

$\boxed{\text{The baby applies 2.0 times as much pressure as the adult.}}$

9. (a) Strategy The pressure is the same throughout the hydraulic fluid, so $P = F_s/A_s = F_l/A_l$.

Solution Find the force that must be applied to the small piston.

$$F_a = F_s = \frac{A_s}{A_l}F_l = \frac{r_s^2}{r_l^2}F_l = \left(\frac{2.50\text{ cm}}{10.0\text{ cm}}\right)^2(10.0\text{ kN}) = \boxed{625\text{ N}}$$

(b) Strategy The work done at each piston must be equal.

Solution

$$W_l = F_l d_l = W_s = F_s d_s,\text{ so } d_l = \frac{F_s}{F_l}d_s = \frac{625\text{ N}}{10.0\text{ kN}}(10.0\text{ cm}) = \boxed{6.25\text{ mm}}.$$

(c) Strategy Compute the ratio of the weight to the applied force.

Solution Compute the mechanical advantage.

$$\frac{W}{F_a} = \frac{10.0\text{ kN}}{0.625\text{ kN}} = \boxed{16.0}$$

13. Strategy Use Eq. (9-4).

Solution Compute the pressure on the fish.

$$P = P_{\text{atm}} + \rho g d = 1.0 \text{ atm} + (1025 \text{ kg/m}^3)(9.80 \text{ m/s}^2)(10 \text{ m})\left(\frac{1 \text{ atm}}{1.013\times10^5 \text{ Pa}}\right) = \boxed{2.0 \text{ atm}}$$

15. (a) Strategy Use Eq. (9-4).

Solution Compute the pressure increase.

$$\Delta P = \rho_{\text{water}} g d = (1.00\times10^3 \text{ kg/m}^3)(9.80 \text{ m/s}^2)(35.0 \text{ m}) = \boxed{343 \text{ kPa}}$$

(b) Strategy Use Eq. (9-3).

Solution Compute the pressure decrease.

$$\Delta P = -\rho_{\text{air}} g h = -(1.20 \text{ kg/m}^3)(9.80 \text{ m/s}^2)(35 \text{ m}) = -410 \text{ Pa}$$

The pressure decreases by $\boxed{410 \text{ Pa}}$.

17. Strategy Use the definition of average pressure and Eq. (9-4).

Solution Find the force that the Dutch boy must exert to save the town.

$$\Delta P = \frac{F}{A} \text{ and } P - P_{\text{atm}} = \Delta P = \rho g d, \text{ so}$$

$$F = A\rho g d = (1.0\times10^{-4} \text{ m}^2)(1.0\times10^3 \text{ kg/m}^3)(9.80 \text{ m/s}^2)(3.0 \text{ m}) = \boxed{2.9 \text{ N}}.$$

21. Strategy Use the appropriate conversion factors to convert the gauge pressure into the various pressure units.

Solution

(a) $$(32 \text{ lb/in}^2)\left(\frac{1.013\times10^5 \text{ Pa}}{14.7 \text{ lb/in}^2}\right) = \boxed{2.2\times10^5 \text{ Pa}}$$

(b) $$(32 \text{ lb/in}^2)\left(\frac{760.0 \text{ torr}}{14.7 \text{ lb/in}^2}\right) = \boxed{1700 \text{ torr}}$$

(c) $$(32 \text{ lb/in}^2)\left(\frac{1 \text{ atm}}{14.7 \text{ lb/in}^2}\right) = \boxed{2.2 \text{ atm}}$$

25. Strategy The amount the fluid rises is one-half the difference of levels, or $\Delta h_{\text{oil}}/2$. Use Eq. (9-5).

Solution

(a) Find the amount that the fluid level rises.

$$\Delta P = \rho_{\text{Hg}} g \Delta h_{\text{Hg}} = \rho_{\text{oil}} g \Delta h_{\text{oil}}, \text{ so } \frac{\Delta h_{\text{oil}}}{2} = \frac{\rho_{\text{Hg}}}{2\rho_{\text{oil}}}\Delta h_{\text{Hg}} = \frac{13.6 \text{ g/cm}^3}{2(0.90 \text{ g/cm}^3)}(0.74 \text{ cm Hg}) = \boxed{5.6 \text{ cm}}.$$

(b) $$\frac{\Delta h_{\text{oil}}}{2}\left(\frac{\rho_{\text{oil}}}{\rho_{\text{Hg}}}\right) = (5.6 \text{ cm})\frac{0.90 \text{ g/cm}^3}{13.6 \text{ g/cm}^3} = \boxed{0.37 \text{ cm}}$$

27. (a) Strategy The relationship between the fraction of a floating object's volume that is submerged to the ratio of the object's density to the fluid in which it floats is $V_f/V_o = \rho_o/\rho_f$. Since the water contains ice, use the density of water at 0°C.

Solution Find the percent of the volume of ice that is submerged when it floats in water.

$$\frac{V_{submerged}}{V_{ice}} = \frac{\rho_{ice}}{\rho_{water}} = \frac{917\ \text{kg/m}^3}{999.87\ \text{kg/m}^3} = 0.917, \text{ or } \boxed{91.7\%}$$

(b) Strategy and Solution The specific gravity and the fraction of the object submerged in water are the same for objects that float, so the specific gravity of ice is $\boxed{0.917}$.

29. Strategy and Solution Since the goose has 25% of its volume submerged, its density is 25% of water's, or about $\boxed{250\ \text{kg/m}^3}$.

33. Strategy Compare the densities of a block of ebony and ethanol.

Solution The density of a block of ebony is between 1000 and 1300 kg/m^3. The density of ethanol is 790 kg/m^3. Since the density of ebony is more than that of ethanol, the block will sink; therefore, $\boxed{100\%}$ of the volume of the block of ebony is submerged.

35. Strategy Let the $+y$-direction be upward. Use Newton's second law and Eq. (9-7).

Solution

(a) $\Sigma F_y = F_B - mg = ma$, so

$$a = \frac{F_B}{m} - g = \frac{\rho_w gV}{m} - g = g\left(\frac{\rho_w V}{\rho V} - 1\right) = (9.80\ \text{m/s}^2)\left(\frac{1.00\ \text{g/cm}^3}{0.50\ \text{g/cm}^3} - 1\right) = 9.8\ \text{m/s}^2.$$

Thus, $\vec{\mathbf{a}} = \boxed{9.8\ \text{m/s}^2\ \text{upward}}$.

(b) $a = (9.80\ \text{m/s}^2)\left(\frac{1.00\ \text{g/cm}^3}{0.750\ \text{g/cm}^3} - 1\right) = 3.3\ \text{m/s}^2$, so $\vec{\mathbf{a}} = \boxed{3.3\ \text{m/s}^2\ \text{upward}}$.

(c) $a = (9.80\ \text{m/s}^2)\left(\frac{1.00\ \text{g/cm}^3}{0.125\ \text{g/cm}^3} - 1\right) = 68.6\ \text{m/s}^2$, so $\vec{\mathbf{a}} = \boxed{68.6\ \text{m/s}^2\ \text{upward}}$.

37. Strategy The weight of the alcohol displaced is equal to the buoyant force. Use Eqs. (9-2) and (9-9).

Solution Find the specific gravity of the alcohol.

$m_{alcohol}g = 1.03\ \text{N} - 0.730\ \text{N} = 0.30\ \text{N}$ and

$$\text{S.G.} = \frac{\rho_{alcohol}}{\rho_w} = \frac{m_{alcohol}}{\rho_w V_{alcohol}} = \frac{m_{alcohol}g}{\rho_w V_{alcohol}g} = \frac{0.30\ \text{N}}{(1.00\times10^3\ \text{kg/m}^3)(3.90\times10^{-5}\ \text{m}^3)(9.80\ \text{m/s}^2)} = \boxed{0.78}.$$

41. Strategy Use Eq. (9-13).

Solution Find the speed of the water as it passes through the nozzle.

$$A_2v_2 = A_1v_1,\ \text{so}\ v_2 = \frac{A_1}{A_2}v_1 = \frac{\pi r_1^2}{\pi r_2^2}v_1 = \left(\frac{1.0\ \text{cm}}{0.20\ \text{cm}}\right)^2(2.0\ \text{m/s}) = \boxed{50\ \text{m/s}}.$$

43. (a) Strategy Use Eq. (9-13).

Solution Find the speed of the water in the hose.

$$v_2 = \frac{A_1}{A_2}v_1 = \frac{\pi r_1^2}{\pi r_2^2}v_1 = \left(\frac{r_1}{r_2}\right)^2 v_1 = \left(\frac{1.00\ \text{mm}}{8.00\ \text{mm}}\right)^2(25.0\ \text{m/s}) = \boxed{39.1\ \text{cm/s}}$$

(b) Strategy Use Eq. (9-12).

Solution Compute the volume flow rate.

$$\frac{\Delta V}{\Delta t} = A_1v_1 = \pi(1.00\times10^{-3}\ \text{m})^2(25.0\ \text{m/s}) = \boxed{78.5\ \text{cm}^3/\text{s}}$$

(c) Strategy Use Eq. (9-11).

Solution Compute the mass flow rate.

$$\frac{\Delta m}{\Delta t} = \rho A_1v_1 = (1.00\ \text{g/cm}^3)(78.5\ \text{cm}^3/\text{s}) = \boxed{78.5\ \text{g/s}}$$

45. Strategy Use Eqs. (9-13) and (9-14). Since the pipe is horizontal, $y_1 = y_2$.

Solution Let the larger end be labeled 2. Find the speed of the water at the narrow end in terms of the speed at the larger end.

$$A_1v_1 = A_2v_2,\ \text{so}\ v_1 = \frac{A_2}{A_1}v_2.$$

Find the pressure at the narrow end of the segment of pipe.

$$P_1 + \frac{1}{2}\rho v_1^2 = P_2 + \frac{1}{2}\rho v_2^2,\ \text{so}\ P_1 = P_2 + \frac{1}{2}\rho\left[v_2^2 - \left(\frac{A_2}{A_1}v_2\right)^2\right]$$

$$= 1.20\times10^5\ \text{Pa} + \frac{1}{2}(1.00\times10^3\ \text{kg/m}^3)(0.040\ \text{m/s})^2\left[1 - \left(\frac{50.0\ \text{cm}^2}{0.500\ \text{cm}^2}\right)^2\right] = \boxed{1.12\times10^5\ \text{Pa}}.$$

47. Strategy Use Eq. (9-14).

Solution The potential energy difference is relatively small, so Bernoulli's equation becomes

$$P_1 + \frac{1}{2}\rho v_1^2 = P_2 + \frac{1}{2}\rho v_2^2,\ \text{or}\ P_1 - P_2 = \frac{1}{2}\rho v_2^2 - \frac{1}{2}\rho v_1^2.$$

Estimate the force.

$$F = \Delta PA = (P_1 - P_2)A = \left(\frac{1}{2}\rho v_2^2 - \frac{1}{2}\rho v_1^2\right)A = \frac{1}{2}A\rho(v_2^2 - v_1^2)$$

$$= \frac{1}{2}(28\ \text{m}^2)(1.3\ \text{kg/m}^3)[(190\ \text{m/s})^2 - (160\ \text{m/s})^2] = \boxed{1.9\times10^5\ \text{N}}$$

49. (a) Strategy The rate at which the well does work on the water is equal to the potential energy change of the water per unit time.

Solution Find the power output of the pump.

$$\frac{\Delta W}{\Delta t}=\frac{mg\Delta y}{\Delta t}=\rho g\Delta y\frac{\Delta V}{\Delta t}=(1.0\times10^{3}\text{ kg/m}^{3})(9.80\text{ m/s}^{2})(40.0\text{ m})(2.0\times10^{-4}\text{ m}^{3}\text{/s})=\boxed{78\text{ W}}$$

(b) Strategy Assume that the speed of the water at the top and bottom of the well is zero. Use Eq. (9-14).

Solution Find the pressure difference the pump must maintain.

$$\Delta P=\rho g\Delta y=(1.00\times10^{3}\text{ kg/m}^{3})(9.80\text{ m/s}^{2})(40.0\text{ m})=\boxed{392\text{ kPa}}$$

(c) Strategy and Solution Since ΔP is greater than atmospheric pressure, the pump must be $\boxed{\text{at the bottom}}$ so that it can push the water up.

53. Strategy Use Eq. (9-15).

Solution Show that viscosity has SI units of pascal-seconds.
Solve for η.

$$\frac{\Delta V}{\Delta t}=\frac{\pi}{8}\frac{\frac{\Delta P}{L}}{\eta}r^{4},\text{ so }\eta=\frac{\pi}{8}\frac{\Delta P}{L}\frac{\Delta t}{\Delta V}r^{4}.\text{ Thus, the units of }\eta\text{ are }\frac{\text{Pa}}{\text{m}}\cdot\frac{\text{s}}{\text{m}^{3}}\cdot\text{m}^{4}=\text{Pa}\cdot\text{s}.$$

57. Strategy Use Eq. (9-15). Form a ratio of the volume flow rates.

Solution Find the total flow rate in system C.

$$\frac{\frac{\Delta V}{\Delta t}_{\text{C}}}{\frac{\Delta V}{\Delta t}_{\text{B}}}=\frac{4\left[\frac{\pi\Delta P_{\text{C}}r^{4}}{8\eta\left(\frac{L}{2}\right)}\right]}{2\left(\frac{\pi\Delta P_{\text{B}}r^{4}}{8\eta L}\right)}=\frac{4\Delta P_{\text{C}}}{\Delta P_{\text{B}}},\text{ so }\frac{\Delta V}{\Delta t}_{\text{C}}=\frac{4\Delta P_{\text{C}}}{\Delta P_{\text{B}}}\left(\frac{\Delta V}{\Delta t}_{\text{B}}\right)=4\left(\frac{2.0\times10^{5}\text{ Pa}}{4.0\times10^{5}\text{ Pa}}\right)(0.020\text{ m}^{3}\text{/s})=\boxed{0.040\text{ m}^{3}\text{/s}}.$$

61. (a) Strategy Use Eq. (9-16).

Solution Find the drag force on the dinoflagellate in seawater.

$$F_{\text{D}}=6\pi\eta rv=6\pi(0.0010\text{ Pa}\cdot\text{s})(35.0\times10^{-6}\text{ m})\frac{1.0\times10^{-3}\text{ m}}{5.0\text{ s}}=\boxed{1.3\times10^{-10}\text{ N}}$$

(b) Strategy Assuming the flagellate travels the 1.0 mm with constant speed, the force with which it pushes on the water is equal in magnitude to the drag force. To find the power output, divide the work done by the flagellate by the time of travel.

Solution Find the power output of the flagellate.

$$P=\frac{W}{\Delta t}=\frac{F_{\text{D}}d}{\Delta t}=\frac{(1.3\times10^{-10}\text{ N})(1.0\times10^{-3}\text{ m})}{5.0\text{ s}}=\boxed{2.6\times10^{-14}\text{ W}}$$

63. Strategy Use Eq. (9-16) and Newton's second law.

Solution Find the viscosity of the second liquid.

$$\Sigma F_y = F_D + F_B - m_s g = 6\pi\eta rv + m_l g - m_s g = 6\pi\eta rv + (m_l - m_s)g = 6\pi\eta rv - \frac{4}{3}\pi r^3(\rho_s - \rho_l)g = 0,$$

so $\eta = \dfrac{\frac{4}{3}\pi r^3(\rho_s - \rho_l)g}{6\pi rv} = \dfrac{2r^2(\rho_s - \rho_l)g}{9v}.$

Find the viscosity of the second liquid by forming a proportion.

$$\frac{\eta_2}{\eta_1} = \frac{\frac{2r^2(\rho_s - \rho_l)g}{9v_2}}{\frac{2r^2(\rho_s - \rho_l)g}{9v_1}} = \frac{v_1}{v_2}, \text{ so } \eta_2 = \frac{v_1}{1.2v_1}\eta_1 = \frac{\eta_1}{1.2} = \frac{0.5 \text{ Pa}\cdot\text{s}}{1.2} = \boxed{0.4 \text{ Pa}\cdot\text{s}}.$$

65. Strategy Use Eq. (9-16) and Newton's second law.

Solution Find the terminal speed of the air bubble.

Aluminum sphere: $\Sigma F_y = F_{1D} + F_{1B} - m_1 g = 0$

Air bubble: $\Sigma F_y = F_{2B} - F_{2D} - m_2 g = 0$

Divide F_{2D} by F_{1D}.

$$\frac{F_{2D}}{F_{1D}} = \frac{6\pi\eta rv_2}{6\pi\eta rv_1} = \frac{F_{2B} - m_2 g}{m_1 g - F_{1B}} = \frac{m_w g - m_2 g}{m_1 g - m_w g}, \text{ so}$$

$$v_2 = \frac{m_w - m_2}{m_1 - m_w}v_1 = \frac{1 - \frac{m_2}{m_w}}{\frac{m_1}{m_w} - 1}v_1 = \frac{1 - \frac{\rho_a}{\rho_w}}{\frac{2.7\rho_w}{\rho_w} - 1}v_1 = \frac{1 - \frac{1.20}{1000}}{2.7 - 1}(5.0 \text{ cm/s}) = \boxed{2.9 \text{ cm/s}}.$$

67. Strategy For a viscous drag force, $v_t \propto m$. For a turbulent drag force, $v_t \propto \sqrt{m}$. For the data in the table, compute m/v_t and m/v_t^2.

Solution The data, m/v_t, and m/v_t^2 are organized in the table.

m (g)	8	12	16	20	24	28
v_t (cm/s)	1.0	1.5	2.0	2.5	3.0	3.5
$\frac{m}{v_t}$ (g·s/cm)	8	8.0	8.0	8.0	8.0	8.0
$\frac{m}{v_t^2}$ $(\text{g}\cdot\text{s}^2/\text{cm}^2)$	8	5.3	4.0	3.2	2.7	2.3

Since m/v_t is constant, the drag force is primarily viscous.

69. Strategy Use Eq. (9-17).

Solution Form a proportion to find the pressure inside the air bubble.

$$\frac{\Delta P_2}{\Delta P_1}=\frac{\frac{2\gamma}{r_2}}{\frac{2\gamma}{r_1}},\text{ so }\Delta P_2=\Delta P_1\frac{r_1}{r_2}=(10\text{ Pa})\frac{r_1}{2r_1}=\boxed{5\text{ Pa}}.$$

73. (a) Strategy The mass of the water is equal to its density times its volume.

Solution The weight of the water in the straw is

$$mg=(\rho V)g=\rho(\pi r^2 h)g=(1.00\times10^3\text{ kg/m}^3)\pi(0.00250\text{ m})^2(8.00\text{ m})(9.80\text{ m/s}^2)=\boxed{1.54\text{ N}}.$$

(b) Strategy Equate the pressures and solve for the force on the top of the barrel. Use the definition of pressure.

Solution Find the force with which the water in the barrel pushes up on the top of the barrel.

$$F_\text{b}=\frac{A_\text{b}}{A_\text{s}}F_\text{s}=\frac{r_\text{b}^2}{r_\text{s}^2}F_\text{s}=\frac{(25.0\text{ cm})^2}{(0.250\text{ cm})^2}(1.54\text{ N})=\boxed{1.54\times10^4\text{ N}}.$$

(c) Strategy Consider the nature of pressure in a column of fluid.

Solution For a given depth, the pressure is the same everywhere, so the very tall, narrow column of water is as effective as having a whole barrel of water filled to the same height and pushing upward on the barrel top.

77. (a) Strategy Assume that the change in the height of the water level in the vat is negligible and that the pressures at the top of the vat and at the outlet are the same. Use Bernoulli's equation.

Solution Let the top of the vat be labeled 1. With the above assumptions, Bernoulli's equation becomes

$$\rho g y_1=\rho g y_2+\frac{1}{2}\rho v_2^2,\text{ so }v_2=\sqrt{2g(y_1-y_2)}=\sqrt{2(9.80\text{ m/s}^2)(1.80\text{ m})}=\boxed{5.94\text{ m/s}}.$$

(b) Strategy and Solution The density "falls out" of Bernoulli's equation in our calculation of the speed, so as long as we can assume Bernoulli's equation applies, it doesn't matter what fluid is in the vat.

(c) Strategy and Solution Since the speed is directly proportional to the square root of the gravitational field strength, $\sqrt{1.6/9.80}=0.40$, the speed would be reduced by a factor of 0.40.

81. Strategy Use Eq. (9-14) with $v_1=v_2=0$.

Solution Find the height between the basement and the seventh floor.

$$P_1+\rho g y_1=P_2+\rho g y_2,\text{ so }y_1-y_2=\frac{P_2-P_1}{\rho g}=\frac{4.10\times10^5\text{ Pa}-1.85\times10^5\text{ Pa}}{(1.00\times10^3\text{ kg/m}^3)(9.80\text{ m/s}^2)}=\boxed{23.0\text{ m}}.$$

85. **(a)** **Strategy** Use Eq. (9-2).

Solution Find the weight of the beach ball.

$$Weight = m_b g + \frac{4}{3}\pi r^3 \rho_a g = (9.80\ \text{m/s}^2)\left[0.10\ \text{kg} + \frac{4}{3}\pi(0.200\ \text{m})^3(1.3\ \text{kg/m}^3)\right] = \boxed{1.4\ \text{N}}$$

(b) **Strategy** The buoyant force is equal to the mass of displaced air and is directed upward. Use Eq. (9-7).

Solution Find the buoyant force on the beach ball.

$$F_B = \rho g V = (1.3\ \text{kg/m}^3)(9.80\ \text{m/s}^2)\frac{4}{3}\pi(0.200\ \text{m})^3 = 0.43\ \text{N, so } \vec{\mathbf{F}}_B = \boxed{0.43\ \text{N upward}}.$$

(c) **Strategy** Use Newton's second law.

Solution Find the acceleration of the beach ball at the top of its trajectory.

$\Sigma F_y = F_B - mg = ma$, so

$$a = \frac{F_B - mg}{m} = \frac{F_B}{m} - g = \frac{\rho_a g V}{m_b + \rho_a V} - g = \frac{g}{\frac{m_b}{\rho_a V} + 1} - g = g\left[\left(1 + \frac{m_b}{\rho_a V}\right)^{-1} - 1\right]$$

$$= (9.80\ \text{m/s}^2)\left\{\left[1 + \frac{0.10\ \text{kg}}{(1.3\ \text{kg/m}^3)\frac{4}{3}\pi(0.200\ \text{m})^3}\right]^{-1} - 1\right\} = -6.8\ \text{m/s}^2.$$

Thus, $\vec{\mathbf{a}} = \boxed{6.8\ \text{m/s}^2\ \text{downward}}$.

$\vec{\mathbf{F}}_B$ y $\vec{\mathbf{v}} = 0$ $m\vec{\mathbf{g}}$

89. **Strategy** Use Eq. (9-7) and Newton's second law.

Solution Find an expression for d.

$$\Sigma F_y = F_B - mg = \rho g V - mg = \rho g A d - mg = \rho g(\pi r^2)d - mg = 0,\ \text{so } d = \frac{m}{\pi \rho r^2}.$$

$$\boxed{d \text{ is not a linear function of } \rho\text{: } d = \frac{m}{\pi \rho r^2}.}$$

93. **Strategy** Use Eqs. (9-13) and (9-14).

Solution

(a) Find the speed of the water as it exits the showerhead (v_1).

$$P_1 + \rho g y_1 + \frac{1}{2}\rho v_1^2 = P_2 + \rho g y_2 + \frac{1}{2}\rho v_2^2$$

$$\frac{1}{2}\rho(v_1^2 - v_2^2) = P_2 - P_1 + \rho g(y_2 - y_1)$$

$$v_1^2 - \frac{A_1^2}{A_2^2}v_1^2 = \frac{2P_{\text{gauge}}}{\rho} - 2gh$$

$$v_1 = \sqrt{\frac{\frac{2P_{\text{gauge}}}{\rho} - 2gh}{1 - \left[N\pi r_1^2/(\pi r_2^2)\right]^2}} = \sqrt{\frac{\frac{2(410\times10^3\ \text{Pa})}{1.00\times10^3\ \text{kg/m}^3} - 2(9.80\ \text{m/s}^2)(6.7\ \text{m})}{1 - 36^2\left(\frac{0.33\ \text{mm}}{6.3\ \text{mm}}\right)^4}} = \boxed{26\ \text{m/s}}$$

(b) Find the speed of the water as it moves through the output pipe of the pump.

$$v_2 = \frac{A_1}{A_2} v_1 = \frac{N\pi r_1^2}{\pi r_2^2} v_1 = 36\left(\frac{0.33 \text{ mm}}{6.3 \text{ mm}}\right)^2 (26 \text{ m/s}) = \boxed{2.6 \text{ m/s}}$$

97. Strategy Use the relationships between pressure, density, force, area, and height.

Solution Find the density of the liquid.

$$\Delta P = \frac{W_1}{A} = \frac{\rho_1 g V_1}{A} = \rho_w g \Delta y_w, \text{ so}$$

$$\rho_1 = \frac{\rho_w \Delta y_w A}{V_1} = \frac{\rho_w \Delta y_w \pi r^2}{\pi r^2 h} = \frac{\rho_w \Delta y_w}{h} = \frac{(1.0 \text{ g/cm}^3)[0.45 \text{ m} - (0.50 \text{ m} - 0.30 \text{ m})]}{0.30 \text{ m}} = \boxed{0.83 \text{ g/cm}^3}.$$

101. Strategy Use Eq. (9-15).

Solution Find the absolute pressure at the bug's end of the feeding tube.

$$\frac{\pi \Delta P r^4}{8\eta L} = \frac{\Delta V}{\Delta t}$$

$$P_{\text{arm}} - P_{\text{bug}} = \left(\frac{8\eta L}{\pi r^4}\right)\frac{\Delta V}{\Delta t}$$

$$P_{\text{bug}} = P_{\text{arm}} - \left(\frac{8\eta L}{\pi r^4}\right)\frac{\Delta V}{\Delta t}$$

$$P_{\text{bug}} = 105 \text{ kPa} - \frac{8(0.0013 \text{ Pa}\cdot\text{s})(0.20\times10^{-3} \text{ m})}{\pi(5.0\times10^{-6} \text{ m})^4}\left(\frac{0.30 \text{ cm}^3}{25 \text{ min}}\right)\left(\frac{1 \text{ m}}{100 \text{ cm}}\right)^3\left(\frac{1 \text{ min}}{60 \text{ s}}\right) = \boxed{-110 \text{ kPa}}$$

Chapter 10

ELASTICITY AND OSCILLATIONS

Problems

1. Strategy The stress is proportional to the strain. Use Eq. (10-4).

Solution Find the vertical compression of the beam.

$$Y\frac{\Delta L}{L}=\frac{F}{A},\text{ so }\Delta L=\frac{FL}{YA}=\frac{(5.8\times10^4\text{ N})(2.5\text{ m})}{(200\times10^9\text{ Pa})(7.5\times10^{-3}\text{ m}^2)}=\boxed{0.097\text{ mm}}.$$

3. Strategy The stress is proportional to the strain. Use Eq. (10-4).

Solution Find how much the wire stretches.

$$\Delta L=\frac{FL}{YA}=\frac{(5.0\times10^3\text{ N})(2.0\text{ m})}{(9.2\times10^{10}\text{ Pa})(5.0\text{ mm}^2)\left(\frac{1\text{ m}}{1000\text{ mm}}\right)^2}=\boxed{2.2\text{ cm}}$$

5. (a) Strategy The average power required is equal to the kinetic energy change divided by the elapsed time it takes for the flea to reach its peak velocity.

Solution Find the average power required.

$$P_{\text{av}}=\frac{\Delta K}{\Delta t}=\frac{\frac{1}{2}m(v_\text{f}^2-v_\text{i}^2)}{\Delta t}=\frac{mv_\text{f}^2}{2\Delta t}=\frac{(0.45\times10^{-6}\text{ kg})(0.74\text{ m/s})^2}{2(1.0\times10^{-3}\text{ s})}=\boxed{1.2\times10^{-4}\text{ W}}$$

(b) Strategy and Solution Compute the power output of the flea.

$$(60\text{ W/kg})(0.45\times10^{-6}\text{ kg})(0.20)=\boxed{5.4\times10^{-6}\text{ W}}<1.2\times10^{-4}\text{ W}$$

$\boxed{\text{No}}$, the flea's muscle cannot provide the power needed.

(c) Strategy There are two pads, so the total energy stored is $E=2U=2[\frac{1}{2}k(\Delta L)^2]=k(\Delta L)^2$. Use Eq. (10-5).

Solution Find the energy stored in the resilin pads.

$$E=k(\Delta L)^2=Y\frac{A}{L}(\Delta L)^2=Y\frac{L^2}{L}L^2=YL^3=(1.7\times10^6\text{ N/m}^2)(6.0\times10^{-5}\text{ m})^3=\boxed{3.7\times10^{-7}\text{ J}}$$

(d) Strategy Use the definition of average power.

Solution Compute the power provided by the resilin pads.

$$P_{\text{av}}=\frac{\Delta E}{\Delta t}=\frac{3.7\times10^{-7}\text{ J}}{1.0\times10^{-3}\text{ s}}=\boxed{3.7\times10^{-4}\text{ W}}>1.2\times10^{-4}\text{ W}$$

$\boxed{\text{Yes}}$, enough power is provided for the jump.

9. Strategy Refer to Fig. 10.4c. The stress is proportional to the strain.

Solution Calculate Young's moduli for tension and compression of bone.
Tension:
For tensile stress and strain, the graph is far from being linear, but for relatively small values of stress and strain, it is approximately linear. So, for small values of tensile stress and strain, Young's Modulus is

$$Y = \frac{\text{stress}}{\text{strain}} = \frac{5.0\times10^7\ \text{N/m}^2}{0.0033} = \boxed{1.5\times10^{10}\ \text{N/m}^2}.$$

Compression:

Similarly, for small values of compressive stress and strain, $Y = \frac{-4.5\times10^7\ \text{N/m}^2}{-0.0050} = \boxed{9.0\times10^9\ \text{N/m}^2}.$

11. Strategy Set the stress equal to the tensile strength of the hair to find the diameter of the hair.

Solution Find the diameter of the hair.

$$\text{tensile strength} = \frac{F}{A} = \frac{F}{\frac{1}{4}\pi d^2}, \text{ so } d = \sqrt{\frac{4F}{\pi(\text{tensile strength})}} = \sqrt{\frac{4(1.2\ \text{N})}{\pi(2.0\times10^8\ \text{Pa})}} = \boxed{8.7\times10^{-5}\ \text{m}}.$$

13. Strategy The stress on the copper wire must be less than its elastic limit.

Solution Find the maximum load that can be suspended from the copper wire.

$$\frac{F}{A} < \text{elastic limit, so } F < \pi r^2(\text{elastic limit}) = \pi(0.0010\ \text{m})^2(2.0\times10^8\ \text{Pa}) = \boxed{630\ \text{N}}.$$

17. Strategy Set the stresses equal to the compressive strengths to determine the effective cross-sectional areas.

Solution Find the effective cross-sectional areas.

$$\text{Human: } \frac{F}{A} = 1.6\times10^8\ \text{Pa, so } A = \frac{5\times10^4\ \text{N}}{1.6\times10^8\ \text{Pa}} = \boxed{3\ \text{cm}^2}. \qquad \text{Horse: } A = \frac{10\times10^4\ \text{N}}{1.4\times10^8\ \text{Pa}} = \boxed{7.1\ \text{cm}^2}$$

21. Strategy Use Hooke's law for volume deformations.

Solution Find the change in volume of the sphere.

$$\Delta P = -B\frac{\Delta V}{V}, \text{ so } \Delta V = -\frac{V\Delta P}{B} = -\frac{(1.00\ \text{cm}^3)(9.12\times10^6\ \text{Pa})}{160\times10^9\ \text{Pa}} = -57\times10^{-6}\ \text{cm}^3.$$

$\boxed{\text{The volume of the steel sphere would decrease by } 57\times10^{-6}\ \text{cm}^3.}$

25. Strategy Use Hooke's law for shear deformations.

Solution Find the magnitude of the tangential force.

$$\frac{F}{A} = \frac{F}{L^2} = S\frac{\Delta x}{L}, \text{ so } F = S\Delta xL = (940\ \text{Pa})(0.64\times10^{-2}\ \text{m})(0.050\ \text{m}) = \boxed{0.30\ \text{N}}.$$

29. Strategy At the maximum extension of the spring, $x = A$ and the magnitude of the acceleration is maximum. Use Eq. (10-22).

Solution Find the magnitude of the acceleration at the point of maximum extension of the spring.

$$a_m = \omega^2 A = \frac{4\pi^2 A}{T^2} = \frac{4\pi^2(0.050\ \text{m})}{(0.50\ \text{s})^2} = \boxed{7.9\ \text{m/s}^2}$$

33. Strategy Replace each quantity with its SI units.

Solution a has units m/s^2. $-\omega^2 x$ has units $\text{s}^{-2}\cdot\text{m} = \text{m/s}^2$. So, $a = -\omega^2 x$ is consistent for units.

ω has units s^{-1}. $\sqrt{k/m}$ has units $\sqrt{\dfrac{\text{N/m}}{\text{kg}}} = \sqrt{\dfrac{\text{kg}\cdot\text{m/s}^2}{\text{m}\cdot\text{kg}}} = \sqrt{\dfrac{1}{\text{s}^2}} = \text{s}^{-1}$. So, $\sqrt{k/m}$ has the same units as ω.

35. Strategy The angular frequency of oscillation is inversely proportional to the square root of the mass. Form a proportion.

Solution Find the new value of ω.

$$\omega \propto \sqrt{\frac{1}{m}}, \text{ so } \frac{\omega_\text{f}}{\omega_\text{i}} = \frac{\sqrt{1/m_\text{f}}}{\sqrt{1/m_\text{i}}} = \sqrt{\frac{m_\text{i}}{m_\text{f}}} = \sqrt{\frac{1}{4.0}} = \frac{1}{2.0}. \text{ Therefore, } \omega_\text{f} = \frac{\omega_\text{i}}{2.0} = \frac{10.0\ \text{rad/s}}{2.0} = \boxed{5.0\ \text{rad/s}}.$$

37. Strategy Use Eqs. (10-21) and (10-22).

Solution

(a) Find v_m and a_m in terms of f. Then compare high- and low-frequency sounds.

$v_\text{m} = \omega A = 2\pi f A \propto f$ and $a_\text{m} = \omega^2 A = 4\pi^2 f^2 A \propto f^2$, so v_m and a_m are greatest for $\boxed{\text{high frequency}}$.

(b) $v_\text{m} = 2\pi(20.0\ \text{Hz})(1.0\times10^{-8}\ \text{m}) = \boxed{1.3\times10^{-6}\ \text{m/s}}$

$a_\text{m} = 4\pi^2(20.0\ \text{Hz})^2(1.0\times10^{-8}\ \text{m}) = \boxed{1.6\times10^{-4}\ \text{m/s}^2}$

(c) $v_\text{m} = 2\pi(20.0\times10^3\ \text{Hz})(1.0\times10^{-8}\ \text{m}) = \boxed{0.0013\ \text{m/s}}$

$a_\text{m} = 4\pi^2(20.0\times10^3\ \text{Hz})^2(1.0\times10^{-8}\ \text{m}) = \boxed{160\ \text{m/s}^2}$

39. Strategy Use Eqs. (10-21) and (10-22) and Newton's second law.

Solution Find the radio's maximum displacement and maximum speed, and the maximum net force exerted on it.

(a) $a_\text{m} = \omega^2 A$, so $A = \dfrac{a_\text{m}}{\omega^2} = \dfrac{98\ \text{m/s}^2}{4\pi^2(120\ \text{Hz})^2} = \boxed{1.7\times10^{-4}\ \text{m}}$.

(b) $v_\text{m} = \omega A = \omega\dfrac{a_\text{m}}{\omega^2} = \dfrac{a_\text{m}}{\omega} = \dfrac{98\ \text{m/s}^2}{2\pi(120\ \text{Hz})} = \boxed{0.13\ \text{m/s}}$

(c) According to Newton's second law, $F_\text{m} = ma_\text{m} = (5.24\ \text{kg})(98\ \text{m/s}^2) = \boxed{510\ \text{N}}$.

41. (a) Strategy Use Newton's second law and Eq. (10-22).

Solution Find the maximum force acting on the diaphragm.

$F_\text{m} = ma_\text{m} = m\omega^2 A = 4\pi^2(0.0500\ \text{kg})(2.0\times10^3\ \text{Hz})^2(1.8\times10^{-4}\ \text{m}) = \boxed{1.4\ \text{kN}}$.

(b) Strategy The maximum elastic potential energy of the diaphragm is equal to the total mechanical energy.

Solution Find the mechanical energy of the diaphragm.

$$E = U = \frac{1}{2}m\omega^2 A^2 = 2\pi^2(0.0500\text{ kg})(2.0\times10^3\text{ Hz})^2(1.8\times10^{-4}\text{ m})^2 = \boxed{0.13\text{ J}}.$$

43. (a) Strategy The speed is maximum when the spring and mass system is at its equilibrium point. Use Newton's second law.

Solution Find the extension of the spring.

$$\Sigma F_y = kx - mg = 0,\ \text{so } x = \frac{mg}{k} = \frac{(0.60\text{ kg})(9.80\text{ N/kg})}{15\text{ N/m}} = \boxed{0.39\text{ m}}.$$

kx

mg

(b) Strategy Use Eqs. (10-20a) and (10-21).

Solution Find the maximum speed of the body.

$$v_\text{m} = \omega A = \sqrt{\frac{k}{m}}x = \sqrt{\frac{15\text{ N/m}}{0.60\text{ kg}}}(0.39\text{ m}) = \boxed{2.0\text{ m/s}}$$

45. Strategy Use Hooke's law, Newton's second law, and Eq. (10-20c).

Solution Find the "spring constant" of the boat. At equilibrium,

$$\Sigma F_y = kx - m_\text{person}g = 0,\ \text{so } k = \frac{m_\text{person}g}{x}.$$

kx

$m_\text{person}g$

Compute the period of oscillation.

$$T = 2\pi\sqrt{\frac{m_\text{total}}{k}} = 2\pi\sqrt{\frac{m_\text{total}}{m_\text{person}g/x}} = 2\pi\sqrt{\frac{m_\text{total}x}{m_\text{person}g}} = 2\pi\sqrt{\frac{(47\text{ kg}+92\text{ kg})(0.080\text{ m})}{(92\text{ kg})(9.80\text{ m/s}^2)}} = \boxed{0.70\text{ s}}$$

49. Strategy The maximum kinetic energy occurs at the equilibrium point where $v = v_\text{m} = \omega A.$ For SHM, $\omega = \sqrt{k/m}.$

Solution Find the maximum kinetic energy of the body.

$$K_\text{max} = \frac{1}{2}mv_\text{m}^2 = \frac{1}{2}m\omega^2A^2 = \frac{1}{2}m\left(\frac{k}{m}\right)A^2 = \frac{1}{2}kA^2 = \frac{1}{2}(2.5\text{ N/m})(0.040\text{ m})^2 = \boxed{2.0\text{ mJ}}$$

51. Strategy Use the definition of average speed and Eq. (10-21). In (d), graph v_x on the vertical axis and t on the horizontal axis.

Solution

(a) The average speed is the total distance traveled divided by the time of travel.

$$v_\text{av} = \frac{\Delta x}{\Delta t} = \frac{4A}{T} = \frac{4A}{2\pi/\omega} = \boxed{\frac{2}{\pi}\omega A}$$

(b) The maximum speed for SHM is $v_\text{m} = \boxed{\omega A}.$

(c) $$\frac{v_\text{av}}{v_\text{m}} = \frac{\frac{2}{\pi}\omega A}{\omega A} = \boxed{\frac{2}{\pi}}$$

(d) Graph $v_x(t)$ and a line from the origin to v_m.

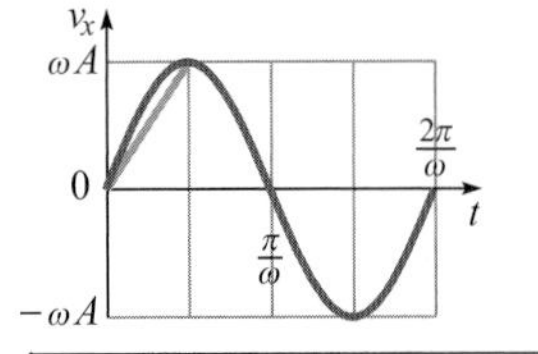

> If the acceleration were constant so that the speed varied linearly, the average speed would be 1/2 of the maximum velocity. Since the actual speed is always larger than what it would be for constant acceleration, the average speed must be larger.

53. (a) Strategy Use Eq. (10-20a) to find the spring constant. Then, find the elastic potential energy using $\frac{1}{2}kx^2$.

Solution Find the spring constant.

$$\omega=\sqrt{\frac{k}{m}},\text{ so } k=\omega^2 m=(2.00\text{ Hz})^2(2\pi\text{ rad/cycle})^2(0.2300\text{ kg})=36.3\text{ N/m}.$$

The equation for the elastic potential energy is

$$U(t)=\frac{1}{2}(36.3\text{ N/m})(0.0800\text{ m})^2\sin^2[(2.00\text{ Hz})(2\pi\text{ rad/cycle})t]=(116\text{ mJ})\sin^2\left[(4.00\pi\text{ s}^{-1})t\right].$$

Since the sine function is squared, the period of $U(t)$ is half that of a sine function or

$$T=\frac{\pi}{\omega}=\frac{\pi}{4.00\pi\text{ s}^{-1}}=250\text{ ms. Graph } U(t).$$

U (mJ)
116
0
0 250 500 t (ms)

(b) Strategy Find the kinetic energy using $\frac{1}{2}mv_x{}^2$.

Solution The equation for the kinetic energy is

$$K(t)=\frac{1}{2}(0.2300\text{ kg})(2.00\text{ Hz})^2\left(\frac{2\pi\text{ rad}}{\text{cycle}}\right)^2(0.0800\text{ m})^2\cos^2\left[(2.00\text{ Hz})\left(\frac{2\pi\text{ rad}}{\text{cycle}}\right)t\right]$$
$$=(116\text{ mJ})\cos^2\left[(4.00\pi\text{ s}^{-1})t\right].$$

Since the cosine function is squared, the period of $K(t)$ is half that of a cosine function or

$$T=\frac{\pi}{\omega}=\frac{\pi}{4.00\pi\text{ s}^{-1}}=250\text{ ms, which is the same as } U(t)\text{. Graph } K(t).$$

K (mJ)
116
0
0 250 500 t (ms)

(c) Strategy Add $U(t)$ and $K(t)$ and graph the result.

Solution

$$E(t)=U(t)+K(t)=(116\text{ mJ})\sin^2\left[(4.00\pi\text{ s}^{-1})t\right]+(116\text{ mJ})\cos^2\left[(4.00\pi\text{ s}^{-1})t\right]$$
$$=(116\text{ mJ})\left\{\sin^2\left[(4.00\pi\text{ s}^{-1})t\right]+\cos^2\left[(4.00\pi\text{ s}^{-1})t\right]\right\}=(116\text{ mJ})(1)=116\text{ mJ}$$

Graph $E(t)=U(t)+K(t)$.

E (mJ)
116
0
0 250 500 t (ms)

(d) Strategy and Solution Friction does nonconservative work on the object, thus, $\boxed{U,\ K,\text{ and }E\text{ would gradually be reduced to zero}}$.

57. Strategy and Solution According to Eq. (10-26b), $T=2\pi\sqrt{\dfrac{L}{g}}$, which does not depend upon the mass.

Therefore, $T=\boxed{1.5\text{ s}}$.

61. Strategy Use Eq. (10-26b) to find the length of the pendulum. Then, form a ratio of the lengths.

Solution Solve for L.

$$T=2\pi\sqrt{\frac{L}{g}},\text{ so }L=\frac{gT^2}{4\pi^2}.$$

Form a proportion.

$$\frac{L_2}{L_1}=\frac{T_2^2}{T_1^2}=\left(\frac{1.00\text{ s}}{0.950\text{ s}}\right)^2=\boxed{1.11}$$

65. Strategy The amplitude is $(20.0\text{ mm})/2=10.0\text{ mm}$ and the period is 2.00 s. Use Eqs. (10-21) and (10-26b) and conservation of energy.

Solution Find the maximum speed of the pendulum bob.
1st method:

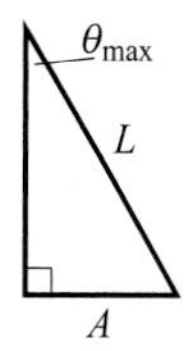

$$v_\text{m}=\omega A=\frac{2\pi}{T}A=\frac{2\pi(10.0\times10^{-3}\text{ m})}{2.00\text{ s}}=\boxed{3.14\text{ cm/s}}$$

2nd method:
Find the length L of the pendulum.

$$T=2\pi\sqrt{\frac{L}{g}},\text{ so }L=\frac{T^2g}{4\pi^2}=\frac{(2.00\text{ s})^2(9.80\text{ m/s}^2)}{4\pi^2}=0.993\text{ m}.$$

Find θ_max.

Since the displacement is small, $\theta_\text{max}\approx\sin\theta_\text{max}=\dfrac{A}{L}=\dfrac{10.0\times10^{-3}\text{ m}}{0.993\text{ m}}=1.007\times10^{-2}\text{ rad}.$

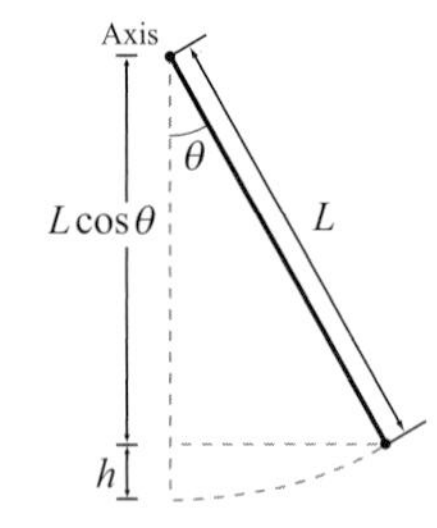

The height h of the pendulum bob above its lowest point is the difference between the length of the pendulum L and the vertical distance from the axis to the bob when it is at its maximum height, $L\cos\theta$. So, $h = L(1-\cos\theta)$.

Find v_m.

$$\Delta K = \frac{1}{2}mv_\text{m}^2 = -\Delta U = mgh, \text{ so}$$

$$v_\text{m} = \sqrt{2gh} = \sqrt{2gL(1-\cos\theta)} = \sqrt{2(9.80\text{ m/s}^2)(0.993\text{ m})(1-\cos 1.007\times 10^{-2})} = \boxed{3.14\text{ cm/s}}.$$

69. Strategy $E = \frac{1}{2}m\omega^2 A^2 \propto A^2$ for a pendulum. Form a proportion.

Solution Find by what factor the energy has decayed.

$$\frac{E_2}{E_1} = \frac{A_2^2}{A_1^2} = \frac{(A_1/20.0)^2}{A_1^2} = \frac{1}{20.0^2} = \frac{1}{400}$$

$\boxed{\text{The energy has decayed by a factor of 400}}$.

73. Strategy Use Equation (10-5) and conservation of energy.

Solution Find the "spring constant" of the bar.

$$k = Y\frac{A}{L} = \frac{Y\pi r^2}{L} = \frac{(2.0\times 10^{11}\text{ Pa})\pi(0.75\times 10^{-2}\text{ m})^2}{0.50\text{ m}} = \boxed{7.1\times 10^7\text{ N/m}}$$

The gravitational potential energy is converted into elastic potential energy.

$$\frac{1}{2}k\Delta L^2 = \frac{1}{2}\frac{Y\pi r^2}{L}\Delta L^2 = mgh, \text{ so } \Delta L = \sqrt{\frac{2Lmgh}{Y\pi r^2}} = \sqrt{\frac{2(0.50\text{ m})(0.70\text{ kg})(9.80\text{ m/s}^2)(1.0\text{ m})}{(2.0\times 10^{11}\text{ Pa})\pi(0.75\times 10^{-2}\text{ m})^2}} = \boxed{0.44\text{ mm}}.$$

77. Strategy Refer to Eqs. (10-20). Use conservation of energy and the fact that $E \propto A^2$.

Solution Analyze the mass and spring system.

(a) The frequency and period don't vary with amplitude, they only vary with m and k. Since these two values remain constant, so do the frequency and period.

(b) Since the total energy of a spring is directly proportional to the square of the amplitude, the total energy for an amplitude of $2D$ is four times that for an amplitude of D.

(c) The initial speed will essentially result in a greater initial displacement; therefore, it will have a greater amplitude. Since the frequency and period don't vary with amplitude, the frequency and period are still the same.

(d) The energy is greater when given an initial push, since it has an amplitude greater than $2D$. The increase in energy is $\frac{1}{2}mv_\text{i}^2$, due to the initial kinetic energy.

81. Strategy The inertia of the system is $I = \frac{1}{3}m_1L^2 + m_2L^2$. Use Eq. (10-28b) and the definition of center of mass.

Solution

(a) The distance to the center of mass from the rotation axis is

$$d = \frac{m_1\frac{L}{2} + m_2L}{m_1 + m_2} = \frac{\frac{m_1}{2} + m_2}{m}L.$$

Find the period of this physical pendulum.

$$T = 2\pi\sqrt{\frac{I}{mgd}} = 2\pi\sqrt{\frac{\frac{1}{3}m_1L^2 + m_2L^2}{mgd}} = 2\pi\sqrt{\frac{L^2\left(\frac{m_1}{3} + m_2\right)}{mg\left[\left(\frac{m_1}{2} + m_2\right)\Big/m\right]L}} = \boxed{2\pi\sqrt{\frac{L\left(\frac{m_1}{3} + m_2\right)}{g\left(\frac{m_1}{2} + m_2\right)}} = 2\pi\sqrt{\frac{2L(m_1 + 3m_2)}{3g(m_1 + 2m_2)}}}$$

(b) For each case, replace the smaller mass with zero. Then

$$\boxed{\text{for } m_1 >> m_2,\ T = 2\pi\sqrt{\frac{2L}{3g}},\ \text{and for } m_1 << m_2,\ T = 2\pi\sqrt{\frac{L}{g}}}.$$

The former is the period for the uniform rod alone and the latter is the period for the block alone.

85. Strategy $I = mL^2$ for a simple pendulum of length L and mass m. Use Eq. (10-28a).

Solution

$$\omega = \sqrt{\frac{mgd}{I}} = \sqrt{\frac{mgd}{mL^2}} = \sqrt{\frac{gd}{L^2}} = \sqrt{\frac{g(L)}{L^2}} = \sqrt{\frac{g}{L}},$$ which is the angular frequency of a simple pendulum.

89. (a) Strategy Use conservation of energy.

Solution Let the maximum displacement be d. Find d.

$$E = K + U = \frac{1}{2}mv^2 + \frac{1}{2}kx^2 = \frac{1}{2}kd^2, \text{ so}$$

$$d = \sqrt{\frac{mv^2}{k} + x^2} = \sqrt{\frac{(1.24\text{ kg})(0.543\text{ m/s})^2}{9.82\text{ N/m}} + (0.345\text{ m})^2} = \boxed{0.395\text{ m}}.$$

(b) Strategy The maximum speed of the block occurs when the block is at it equilibrium position, $x = 0$. Use conservation of energy.

Solution Find the maximum speed of the block.

$$E = K + U = \frac{1}{2}mv^2 + \frac{1}{2}kx^2 = \frac{1}{2}kd^2, \text{ so}$$

$$v_{\text{max}} = \sqrt{\frac{k(d^2 - x^2)}{m}} = \sqrt{\frac{(9.82\text{ N/m})[(0.395\text{ m})^2 - 0^2]}{1.24\text{ kg}}} = \boxed{1.11\text{ m/s}}.$$

(c) Strategy Use the result from part (b).

Solution

$$v = \sqrt{\frac{k(d^2 - x^2)}{m}} = \sqrt{\frac{(9.82\text{ N/m})[(0.3953\text{ m})^2 - (0.200\text{ m})^2]}{1.24\text{ kg}}} = \boxed{0.960\text{ m/s}}$$

91. Strategy Use Eq. (10-2) to find the tensile stress. Then compare the tensile stress to the elastic limit of steel piano wire.

Solution Find the tensile stress in the piano wire in Problem 90.

$$\text{tensile stress} = \frac{F}{A} = \frac{T}{\frac{1}{4}\pi d^2} = \frac{4T}{\pi d^2} = \frac{4(402\text{ N})}{\pi(0.80\times10^{-3}\text{ m})^2} = 8.0\times10^8\text{ Pa} < 8.26\times10^8\text{ Pa}$$

The tensile stress is $\boxed{8.0\times10^8\text{ Pa; it is just under the elastic limit.}}$

93. Strategy Treat the swinging gibbon as a physical pendulum. Use Eq. (10-28a).

Solution Estimate the frequency of oscillation of the gibbon.

$$f \approx \frac{1}{2\pi}\sqrt{\frac{mgd}{I}} = \frac{1}{2\pi}\sqrt{\frac{(9.80\text{ m/s}^2)(0.40\text{ m})}{0.25\text{ m}^2}} = \boxed{0.63\text{ Hz}}$$

97. (a) Strategy Gravitational potential energy is converted into elastic potential energy in the bungee cord. Assume the cord obeys Hooke's law. Assume SHM and use Eq. (10-20c).

Solution Find k for the cord.

$$\frac{1}{2}ky^2 = mgh,\text{ so } k = \frac{2mgh}{y^2}.$$

Find the period of oscillations of the bungee cord.

$$T = 2\pi\sqrt{\frac{m}{k}} = 2\pi\sqrt{\frac{m}{2mgh/y^2}} = \pi y\sqrt{\frac{2}{gh}} = \pi(50.0\text{ m} - 33.0\text{ m})\sqrt{\frac{2}{(9.78\text{ m/s}^2)(50.0\text{ m})}} = \boxed{3.42\text{ s}}$$

(b) Strategy Use conservation of energy and the quadratic formula.

Solution Find the extension of the bungee cord y_2.

$$\frac{1}{2}ky_2^2 = \frac{1}{2}\left(\frac{2m_1gh}{y_1^2}\right)y_2^2 = m_2gh = m_2g(y_2 + 33.0\text{ m}),\text{ so } 0 = y_2^2 - \frac{m_2y_1^2}{m_1h}y_2 - \frac{m_2y_1^2(33.0\text{ m})}{m_1h}.$$

Solve for y_2.

$$y_2 = \frac{\frac{m_2y_1^2}{m_1h} \pm \sqrt{\left(-\frac{m_2y_1^2}{m_1h}\right)^2 - 4(1)\left[-\frac{m_2y_1^2(33.0\text{ m})}{m_1h}\right]}}{2(1)}$$

$$= \frac{(80.0\text{ kg})(17.0\text{ m})^2}{2(60.0\text{ kg})(50.0\text{ m})} \pm \frac{1}{2}\sqrt{\left[\frac{(80.0\text{ kg})(17.0\text{ m})^2}{(60.0\text{ kg})(50.0\text{ m})}\right]^2 + \frac{(80.0\text{ kg})(17.0\text{ m})^2(132\text{ m})}{(60.0\text{ kg})(50.0\text{ m})}}$$

$$= 20.3\text{ m or } -12.6\text{ m}$$

$y_2 > 0$, so $y_2 = 20.3$ m, and 33.0 m + 20.3 m = 53.3 m > 50.0 m.

$\boxed{\text{No}}$, he should not use the same cord because his greater mass will stretch if too much and he will hit the water.

Chapter 11

WAVES

Problems

1. (a) Strategy It takes half the time for the sound to cross the valley as it does to make the round trip.

Solution The cliff is $\Delta x = v\Delta t = (343\ \text{m/s})(0.75\ \text{s}) = \boxed{260\ \text{m}}$ away.

(b) Strategy Treat the radio and echo as isotropic sources. Use Eq. (11-1) and form a proportion.

Solution Find the intensity of the music arriving at the cliff.

$$\frac{I_{\text{cliff}}}{I_{\text{radio}}} = \frac{P/(4\pi r_{\text{c}}^2)}{P/(4\pi r_{\text{r}}^2)} = \frac{r_{\text{r}}^2}{r_{\text{c}}^2},\ \text{so}\ I_{\text{cliff}} = \left(\frac{r_{\text{r}}}{r_{\text{c}}}\right)^2 I_{\text{radio}} = \left(\frac{1}{257.25}\right)^2 (1.0\times10^{-5}\ \text{W/m}^2) = \boxed{1.5\times10^{-10}\ \text{W/m}^2}.$$

3. Strategy Form a proportion with the intensities, treating the jet airplane as an isotropic source. Use Eq. (11-1).

Solution Find the intensity of the sound waves at the ears of the person.

$$\frac{I_2}{I_1} = \frac{P/(4\pi r_2^2)}{P/(4\pi r_1^2)} = \left(\frac{r_1}{r_2}\right)^2,\ \text{so}\ I_2 = \left(\frac{r_1}{r_2}\right)^2 I_1 = \left(\frac{5.0\ \text{m}}{120\ \text{m}}\right)^2 (1.0\times10^2\ \text{W/m}^2) = \boxed{170\ \text{mW/m}^2}.$$

5. Strategy The power equals the intensity times the area.

Solution Find the rate at which the Sun emits electromagnetic waves.

$$P = IA = I(4\pi R_{\text{E}}^2) = 4\pi(1.4\times10^3\ \text{W/m}^2)(1.50\times10^{11}\ \text{m})^2 = \boxed{4.0\times10^{26}\ \text{W}}$$

9. Strategy Use Eq. (11-4).

Solution Find the speed of the transverse waves on the string.

$$v = \sqrt{\frac{F}{\mu}} = \sqrt{\frac{90.0\ \text{N}}{3.20\times10^{-3}\ \text{kg/m}}} = \boxed{168\ \text{m/s}}$$

13. Strategy Use Eq. (11-5).

Solution Find the wavelength.

$$\lambda = vT = (75.0\ \text{m/s})(5.00\times10^{-3}\ \text{s}) = \boxed{0.375\ \text{m}}$$

15. Strategy Use Eq. (11-6).

Solution Find the frequencies.

(a) $f = \dfrac{v}{\lambda} = \dfrac{340\ \text{m/s}}{1.0\ \text{m}} = \boxed{340\ \text{Hz}}$

(b) $f = \dfrac{v}{\lambda} = \dfrac{3.0\times10^8\ \text{m/s}}{1.0\ \text{m}} = \boxed{3.0\times10^8\ \text{Hz}}$

17. Strategy Use Eq. (11-6).

Solution Find the frequency with which the buoy bobs up and down.

$$f=\frac{v}{\lambda}=\frac{2.5\text{ m/s}}{7.5\text{ m}}=\boxed{0.33\text{ Hz}}$$

21. Strategy The equation for a transverse sinusoidal wave moving in the negative x-direction can be written in the form $y(x, t) = A\sin(kx+\omega t)$. Use Eq. (11-7) to find the angular frequency and the wavenumber.

Solution $A = 0.120$ m, $\lambda = 0.300$ m, $v = 6.40$ m/s, and $y(x, t) = A\sin(\omega t + kx)$.

$$\omega=\frac{2\pi v}{\lambda}=\frac{2\pi(6.40\text{ m/s})}{0.300\text{ m}}=134\text{ s}^{-1}\text{ and } k=\frac{2\pi}{\lambda}=\frac{2\pi}{0.300\text{ m}}=20.9\text{ m}^{-1}.$$

Thus, the equation is $\boxed{y(x,t)=(0.120\text{ m})\sin[(134\text{ s}^{-1})t+(20.9\text{ m}^{-1})x]}$.

25. (a) Strategy and Solution Since the argument of the cosine function is $\omega t + kx$ (both terms are positive), the wave is moving $\boxed{\text{to the left}}$.

(b) Strategy The maximum y-value is the amplitude. The wave repeats every 4.0 cm, so the wavelength is 0.040 m. Use Eq. (11-7) to find the angular frequency and wavenumber.

Solution

$$y_{\text{max}}=2.0\text{ mm, so }\boxed{A=2.0\text{ mm}};\ \omega=\frac{2\pi v}{\lambda}=\frac{2\pi(10.0\text{ m/s})}{0.040\text{ m}}=\boxed{1600\text{ rad/s}};$$

$$k=\frac{2\pi}{\lambda}=\frac{2\pi}{0.040\text{ m}}=\boxed{160\text{ rad/m}}$$

(c) Strategy Choose the point $(x, y) = (0, 0)$. $0 = A\cos[\omega t + k(0)]$, so $0 = \cos\omega t$. $\cos\omega t = 0$ when $\omega t = n\pi/2$, where n is an odd integer. The smallest nonnegative n is 1 and it will give the smallest nonnegative time. Use Eq. (11-6).

Solution

$$t=\frac{\pi}{2\omega}=\frac{\pi}{4\pi f}=\frac{1}{4f}=\frac{\lambda}{4v}=\frac{0.040\text{ m}}{4(10.0\text{ m/s})}=1.0\text{ ms}$$

The period is $T=\frac{1}{f}=\frac{\lambda}{v}=\frac{0.040\text{ m}}{10.0\text{ m/s}}=4.0\text{ ms}$, so the three times are $\boxed{1.0\text{ ms, }5.0\text{ ms, and }9.0\text{ ms}}$.

29. Strategy Use Eqs. (10-21) and (10-22). v_y leads y by 1/4 cycle, so v_y is a cosine function. Plot the graphs.

Solution $y(x,t)=(1.2\text{ mm})\sin[(2.0\pi\text{ s}^{-1})t-(0.50\pi\text{ m}^{-1})x]$; calculate the maximum speed.

$v_\text{m}=\omega A=(2.0\pi\text{ s}^{-1})(1.2\text{ mm})=7.5\text{ mm/s}$, so $v_y(0,t)=(7.5\text{ mm/s})\cos(2.0\pi\text{ s}^{-1})t$ at $x=0$.

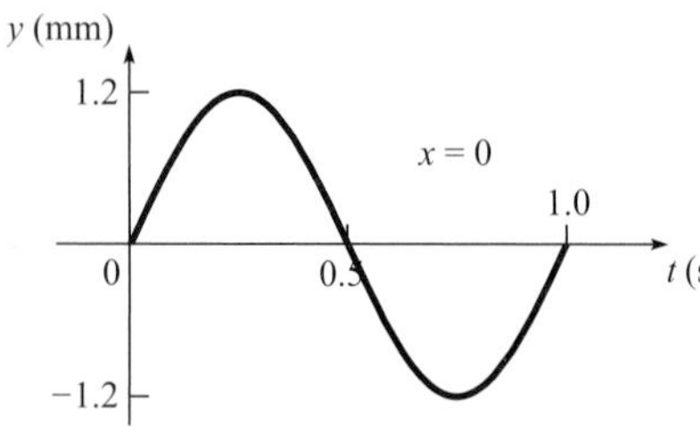

y (mm)
1.2
x = 0
1.0
0
0.5
t (s)
−1.2

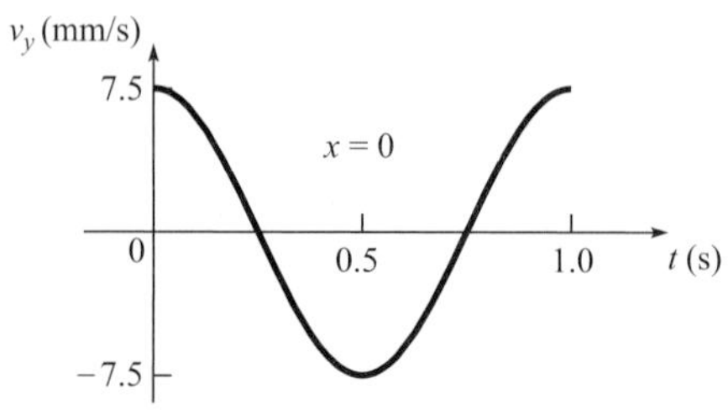

v_y (mm/s)
7.5
x = 0
0
0.5
1.0
t (s)
−7.5

31. Strategy Compute the positions of the peaks of each pulse for the given times. Then use the principle of superposition to graph the shape of the cord for each time.

Solution

t (s)	Short Pulse Position	Tall Pulse Position
0.15	$10\text{ cm}+(40\text{ cm/s})(0.15\text{ s})=16\text{ cm}$	$30\text{ cm}-(40\text{ cm/s})(0.15\text{ s})=24\text{ cm}$
0.25	$10\text{ cm}+(40\text{ cm/s})(0.25\text{ s})=20\text{ cm}$	$30\text{ cm}-(40\text{ cm/s})(0.25\text{ s})=20\text{ cm}$
0.30	$10\text{ cm}+(40\text{ cm/s})(0.30\text{ s})=22\text{ cm}$	$30\text{ cm}-(40\text{ cm/s})(0.30\text{ s})=18\text{ cm}$

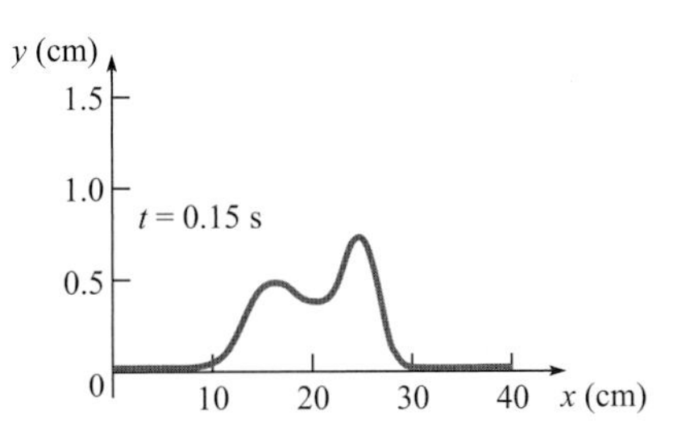

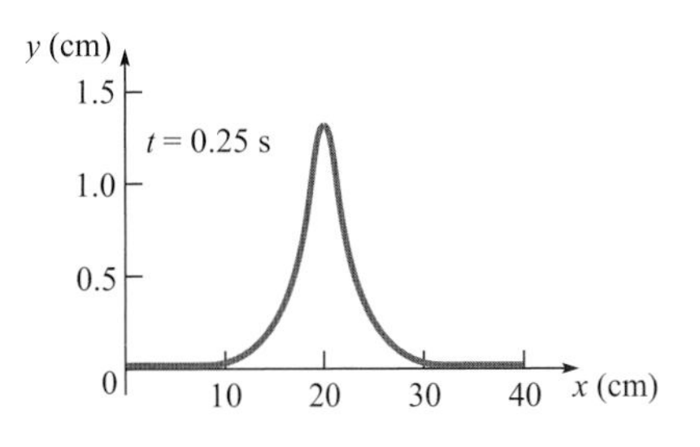

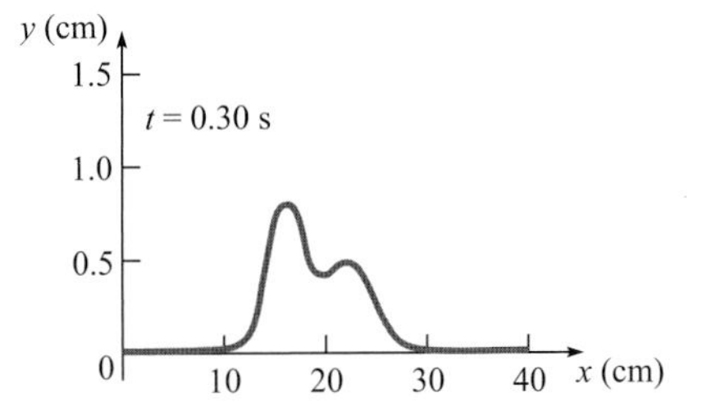

33. Strategy Use the principle of superposition and the trigonometric identity
$\sin\alpha+\sin\beta=2\sin\left(\frac{\alpha+\beta}{2}\right)\cos\left(\frac{\alpha-\beta}{2}\right).$

Solution Find the traveling sine wave.

$$y=y_1+y_2=A\sin(\omega t+kx)+A\sin(\omega t+kx-\phi)=2A\sin\left(\omega t+kx-\frac{\phi}{2}\right)\cos\frac{\phi}{2}=A'\sin\left(\omega t+kx-\frac{\phi}{2}\right)$$

where $A'=2A\cos\frac{\phi}{2}=6.69\text{ cm}$. Find ϕ.

$$\cos\frac{\phi}{2}=\frac{A'}{2A},\text{ so }\phi=2\cos^{-1}\frac{A'}{2A}=2\cos^{-1}\frac{6.69\text{ cm}}{2(5.00\text{ cm})}=\boxed{96.0^\circ}.$$

37. Strategy Use the law of refraction, Eq. (11-10).

Solution Find the speed of the S wave past the boundary.

$$\frac{v_1}{v_2}=\frac{\sin\theta_1}{\sin\theta_2},\text{ so }v_2=\frac{\sin\theta_2}{\sin\theta_1}v_1=\frac{\sin 34^\circ}{\sin 22^\circ}(3.2\text{ m/s})=\boxed{4.8\text{ m/s}}.$$

39. Strategy Refer to the figure. Use $\Delta x = v_x \Delta t$ and the principle of superposition.

Solution The pulse moves 1.80 m – 1.50 m = 0.30 m in 0.20 s. So, the speed of the wave is $v = \dfrac{0.30\text{ m}}{0.20\text{ s}} = 1.5\text{ m/s}$. When the pulse reaches the right endpoint, it is reflected and inverted. When exactly half of the pulse has been reflected and inverted, the superposition of the incident and reflected waves results in the cancellation of the waves $(y_1 + y_2 = 0)$. Thus, the string looks flat at $t = \dfrac{x}{v} = \dfrac{4.0\text{ m} - 1.5\text{ m}}{1.5\text{ m/s}} = \boxed{1.7\text{ s}}$.

41. Strategy The waves are coherent. Use the principle of superposition.

Solution

(a) The resulting wave will have its largest amplitude if the waves interfere constructively. The phase difference is $\boxed{0°}$, and the amplitude is $A_1 + A_2 = 5.0\text{ cm} + 3.0\text{ cm} = \boxed{8.0\text{ cm}}$.

(b) The resulting wave will have its smallest amplitude if the waves interfere destructively. The phase difference is $\boxed{180°}$, and the amplitude is $|A_1 - A_2| = |5.0\text{ cm} - 3.0\text{ cm}| = \boxed{2.0\text{ cm}}$.

(c) $8.0\text{ cm} : 2.0\text{ cm} = \boxed{4:1}$

43. Strategy Intensity is proportional to the amplitude squared. For constructive interference, the amplitude of the superposition is the sum of the original amplitudes.

Solution Find A_1/A_2.

$$\frac{A_1}{A_2} = \sqrt{\frac{I_1}{I_2}} = \sqrt{\frac{25}{15}} = \sqrt{\frac{5.0}{3.0}}$$

Find the amplitude of the superposition.

$$A = A_1 + A_2 = A_2\sqrt{\frac{5.0}{3.0}} + A_2 = A_2\left(1 + \sqrt{\frac{5.0}{3.0}}\right)$$

Find the intensity of the superposition.

$$\sqrt{\frac{I}{I_2}} = \frac{A}{A_2} = 1 + \sqrt{\frac{5.0}{3.0}},\text{ so } I = \left(1 + \sqrt{\frac{5.0}{3.0}}\right)^2 I_2 = \left(1 + \sqrt{\frac{5.0}{3.0}}\right)^2 (15\text{ mW/m}^2) = \boxed{79\text{ mW/m}^2}.$$

45. Strategy Intensity is proportional to the amplitude squared. For constructive interference, the amplitude of the superposition is the sum of the original amplitudes. For destructive interference, the amplitude of the superposition is the absolute value of the difference of the original amplitudes. For incoherent waves, the intensities add.

Solution

(a) Find A_1/A_2.

$$\frac{A_1}{A_2} = \sqrt{\frac{I_1}{I_2}} = \sqrt{\frac{0.040}{0.090}} = \frac{2.0}{3.0}$$

Find the amplitude of the superposition.

$$A = A_1 + A_2 = A_2\left(\frac{2.0}{3.0}\right) + A_2 = A_2\left(1+\frac{2.0}{3.0}\right)$$

Find the intensity of the superposition.

$$\sqrt{\frac{I}{I_2}} = \frac{A}{A_2} = 1+\frac{2.0}{3.0}, \text{ so } I = \left(1+\frac{2.0}{3.0}\right)^2 I_2 = \left(1+\frac{2.0}{3.0}\right)^2 (0.090\ \text{W/m}^2) = \boxed{0.25\ \text{W/m}^2}.$$

(b) Find the amplitude of the superposition.

$$A = |A_1 - A_2| = A_2\left(1-\frac{2.0}{3.0}\right)$$

Find the intensity of the superposition.

$$I = \left(1-\frac{2.0}{3.0}\right)^2 I_2 = \left(1-\frac{2.0}{3.0}\right)^2 (0.090\ \text{W/m}^2) = \boxed{0.010\ \text{W/m}^2}$$

(c) Find the intensity of the superposition.

$$I = I_1 + I_2 = 0.040\ \text{W/m}^2 + 0.090\ \text{W/m}^2 = \boxed{0.130\ \text{W/m}^2}$$

49. Strategy Use Eqs. (11-2) and (11-13).

Solution

$f_n = \dfrac{nv}{2L}$ and $v = \sqrt{\dfrac{TL}{m}}$. Find the total mass of the wire.

$$f_1 = \frac{v}{2L} = \frac{1}{2L}\sqrt{\frac{TL}{m}} = \sqrt{\frac{T}{4Lm}}, \text{ so } m = \frac{T}{4Lf_1^2} = \frac{300.0\ \text{N}}{4(2.0\ \text{m})(27.5\ \text{Hz})^2} = \boxed{0.050\ \text{kg}}.$$

53. Strategy The frequencies are given by $f_n = nv/(2L)$. The speed of the transverse waves is related to the tension by $v = \sqrt{T/\mu}$.

Solution

(a) Find the frequency of the fundamental oscillation.

$$f_1 = \frac{v}{2L} = \frac{1}{2L}\sqrt{\frac{T}{\mu}} = \frac{1}{2(1.5\ \text{m})}\sqrt{\frac{12\ \text{N}}{1.2\times10^{-3}\ \text{kg/m}}} = \boxed{33\ \text{Hz}}$$

(b) Find the tension.

$$f_3 = \frac{3v}{2L} = \frac{3}{2L}\sqrt{\frac{T}{\mu}}, \text{ so } T = \frac{4\mu L^2 f_3^2}{9} = \frac{4(1.2\times10^{-3}\ \text{kg/m})(1.5\ \text{m})^2(0.50\times10^3\ \text{Hz})^2}{9} = \boxed{300\ \text{N}}.$$

57. (a) Strategy and Solution All frequencies higher than the fundamental are integral multiples of the fundamental. Since there are no other frequencies between the two given, the fundamental is the difference between those two. Thus, the fundamental frequency is $1040\ \text{Hz} - 780\ \text{Hz} = \boxed{260\ \text{Hz}}$.

(b) Strategy Use Eqs. (11-2) and (11-13).

Solution Find the total mass of the string.

$$f_1 = \frac{v}{2L} = \frac{1}{2L}\sqrt{\frac{FL}{m}} = \sqrt{\frac{F}{4mL}}, \text{ so } m = \frac{F}{4f_1^2 L} = \frac{1200\ \text{N}}{4(260\ \text{Hz})^2(1.6\ \text{m})} = \boxed{2.8\ \text{g}}.$$

61. Strategy Use Eq. (11-6).

Solution Find the wavelength of the radio waves.

$$\lambda = \frac{v}{f} = \frac{3.0\times10^8\ \text{m/s}}{90\times10^6\ \text{Hz}} = \boxed{3.3\ \text{m}}$$

65. Strategy Use dimensional analysis.

Solution λ has units m. g has units m/s^2. $\lambda\cdot g$ has units m^2/s^2. $\sqrt{\lambda g}$ has units m/s. So, $\boxed{v \propto \sqrt{\lambda g}}$.

69. Strategy Speed is inversely proportional to the time of travel. Form a proportion and use $\Delta x = v\Delta t$.

Solution Relate the speeds to the times of travel.

$$\frac{v_\text{P}}{v_\text{S}} = \frac{10.0\ \text{km/s}}{8.0\ \text{km/s}} = \frac{5.00}{4.0} = \frac{t_\text{S}}{t_\text{P}}$$

Find the time for the S wave to travel from the source to the detector.

$$\Delta t = t_\text{S} - t_\text{P} = t_\text{S} - t_\text{S}\left(\frac{4.0}{5.00}\right) = t_\text{S}(1-0.80),\ \text{so}\ t_\text{S} = \frac{2.0\ \text{s}}{0.20} = 10\ \text{s}.$$

Calculate the distance between the source and the detector.

$$d = v_\text{S}t_\text{S} = (8.0\ \text{km/s})(10\ \text{s}) = \boxed{80\ \text{km}}$$

73. Strategy Use the principle of superposition.

Solution $\Delta x = 1.80\ \text{m} - 1.50\ \text{m} = 0.30\ \text{m}$ in $\Delta t = 0.20\ \text{s}$, so $v = 0.30\ \text{m}/(0.20\ \text{s}) = 1.5\ \text{m/s}$.

Find the position of the peak at $t = 1.6$ s.

$$x_\text{peak} = x_\text{i} + vt = 1.5\ \text{m} + (1.5\ \text{m/s})(1.6\ \text{s}) = 3.9\ \text{m}$$

The peak of the pulse is nearly to the end of the string. The reflected pulse is below the string, so most of the height of the original pulse is cancelled.

y (cm)
10
5
0
x = 0
x = 3.9 m
x = 4.0 m

77. Strategy The tension is 0.93 of the tensile strength. Use Eqs. (10-2), (11-2), and (11-13).

Solution Set the stress equal to the strength.

$$\frac{F}{A} = \text{strength} = \frac{\frac{T}{0.93}}{A},\ \text{so}\ T = 0.93A(\text{strength}).\ \ f_n = \frac{nv}{2L}\ \text{and}\ v = \sqrt{\frac{TL}{m}}.\ \text{Find}\ f_1.$$

$$f_1 = \frac{v}{2L} = \frac{1}{2L}\sqrt{\frac{TL}{m}} = \frac{1}{2}\sqrt{\frac{TL}{mL^2}} = \frac{1}{2}\sqrt{\frac{0.93A(\text{strength})L}{mL^2}} = \frac{1}{2}\sqrt{\frac{0.93(\text{strength})V}{mL^2}} = \frac{1}{2}\sqrt{\frac{0.93(\text{strength})}{\rho L^2}}$$

$$= \frac{1}{2}\sqrt{\frac{0.93(6.3\times10^8\ \text{Pa})}{(8500\ \text{kg/m}^3)(0.094\ \text{m})^2}} = \boxed{1.4\ \text{kHz}}$$

Chapter 12

SOUND

Problems

1. **Strategy** Use Eqs. (11-6) and (12-3).

   Solution Find the wavelength of the ultrasonic waves.

   $$\lambda = \frac{v}{f} \text{ and } v = v_0\sqrt{\frac{T}{T_0}}, \text{ so } \lambda = \frac{v}{f} = \frac{v_0}{f}\sqrt{\frac{T}{T_0}} = \frac{331 \text{ m/s}}{1.0\times10^5 \text{ Hz}}\sqrt{\frac{273.15 \text{ K}+15 \text{ K}}{273.15}} = \boxed{3.4 \text{ mm}}.$$

5. **Strategy** Replace each quantity with its SI units and simplify. In (a), use Eq. (12-1). In (b), analyze each combination of ρ and B.

   Solution

   (a) Show that Eq. (12-1) gives the speed of sound in m/s.

   $$v = \sqrt{\frac{B}{\rho}}, \text{ so } \sqrt{\frac{\text{N/m}^2}{\text{kg/m}^3}} = \sqrt{\frac{(\text{kg}\cdot\text{m/s}^2)/\text{m}^2}{\text{kg/m}^3}} = \sqrt{\frac{1/(\text{m}\cdot\text{s}^2)}{1/\text{m}^3}} = \sqrt{\frac{\text{m}^2}{\text{s}^2}} = \text{m/s}.$$

   (b) Show that no other combination of B and ρ of than $\sqrt{B/\rho}$ can give dimensions of speed.

   $$\frac{\rho}{B} \text{ has units } \frac{\text{kg}}{\text{m}^3}\cdot\frac{\text{m}^2}{\text{N}} = \frac{\text{kg}}{\text{m}\cdot\text{kg}\cdot\text{m/s}^2} = \frac{\text{s}^2}{\text{m}^2}; \; \rho B \text{ has units } \frac{\text{kg}}{\text{m}^3}\cdot\frac{\text{N}}{\text{m}^2} = \frac{\text{kg}^2\cdot\text{m/s}^2}{\text{m}^5} = \frac{\text{kg}^2}{\text{m}^4\cdot\text{s}^2}; \text{ and}$$

   $$\frac{1}{\rho B} \text{ has units } \frac{\text{m}^4\cdot\text{s}^2}{\text{kg}^2}.$$

   No power of the above three combinations (other than $-1/2$, which gives $\sqrt{B/\rho}$) will give the dimensions of speed; therefore, Eq. (12-1) must be correct except for the possibility of a dimensionless constant.

7. **Strategy** Use Eq. (12-1).

   Solution Find the speed of sound in mercury.

   $$v = \sqrt{\frac{B}{\rho}} = \sqrt{\frac{2.8\times10^{10} \text{ Pa}}{1.36\times10^4 \text{ kg/m}^3}} = \boxed{1.4 \text{ km/s}}$$

9. **Strategy** Use $\Delta x = v_x \Delta t$ and the speeds of light and sound. For $T = 20.0°\text{C}$, $v = 343$ m/s.

   Solution Verify the rule of thumb.

   $$\Delta t_{\text{light}} = \frac{\Delta x}{c} = \frac{1.6\times10^3 \text{ m}}{3\times10^8 \text{ m/s}} = 5 \text{ μs is negligible; } \Delta t_{\text{sound}} = \frac{\Delta x}{v} = \frac{1.6\times10^3 \text{ m}}{343 \text{ m/s}} = \boxed{4.7 \text{ s}}.$$

   4.7 s = 5 s to one significant figure. The rule of thumb is approximately correct.

13. Strategy Solve for the intensity in Eq. (12-8). The sound is incoherent, so add the three intensities; then solve for the combined intensity level of the three machines.

Solution Solve for the intensity.

$$\beta = (10\text{ db})\log_{10}\frac{I}{I_0},\text{ so } 10^{\frac{\beta}{10\text{ dB}}} = \frac{I}{I_0} \text{ and } I = I_0 10^{\frac{\beta}{10\text{ dB}}}.$$

Find the combined intensity level.

$$I_{\text{total}} = I_1 + I_2 + I_3 = I_0 10^{\frac{\beta_1}{10\text{ dB}}} + I_0 10^{\frac{\beta_2}{10\text{ dB}}} + I_0 10^{\frac{\beta_3}{10\text{ dB}}},\text{ so } \frac{I_{\text{total}}}{I_0} = 10^{\frac{\beta_1}{10\text{ dB}}} + 10^{\frac{\beta_2}{10\text{ dB}}} + 10^{\frac{\beta_3}{10\text{ dB}}}.\text{ Thus,}$$

$$\beta_{\text{total}} = (10\text{ dB})\log_{10}\frac{I_{\text{total}}}{I_0} = (10\text{ dB})\log_{10}\left(10^{\frac{\beta_1}{10\text{ dB}}} + 10^{\frac{\beta_2}{10\text{ dB}}} + 10^{\frac{\beta_3}{10\text{ dB}}}\right) = (10\text{ dB})\log_{10}\left(10^{\frac{85\text{ dB}}{10\text{ dB}}} + 10^{\frac{90\text{ dB}}{10\text{ dB}}} + 10^{\frac{93\text{ dB}}{10\text{ dB}}}\right)$$

$$= \boxed{95\text{ dB}}.$$

Since 95 dB is comparable to 93 dB, the intensity level of all three machines running is not much different than with only one machine running.

17. Strategy Use Eq. (12-8) to find an expression for the intensity. Then use each relationship given for the intensities to obtain the relationships for the intensity levels.

Solution Solve for the intensity.

$$\beta = (10\text{ db})\log\frac{I}{I_0}$$

$$10^{\frac{\beta}{10\text{ dB}}} = \frac{I}{I_0}$$

$$I = I_0 10^{\frac{\beta}{10\text{ dB}}}$$

(a) Show that if $I_2 = 10.0 I_1$, $\beta_2 = \beta_1 + 10.0$ dB.

$$I_2 = 10.0 I_1$$

$$I_0 10^{\frac{\beta_2}{10\text{ dB}}} = 10.0 I_0 10^{\frac{\beta_1}{10\text{ dB}}}$$

$$10^{\frac{\beta_2}{10\text{ dB}}} = (10.0)10^{\frac{\beta_1}{10\text{ dB}}}$$

$$\log 10^{\frac{\beta_2}{10\text{ dB}}} = \log\left[(10.0)10^{\frac{\beta_1}{10\text{ dB}}}\right] = \log 10.0 + \log 10^{\frac{\beta_1}{10\text{ dB}}}$$

$$\frac{\beta_2}{10\text{ dB}} = 1.00 + \frac{\beta_1}{10\text{ dB}}$$

$$\beta_2 = \beta_1 + 10.0\text{ dB}$$

(b) Show that if $I_2 = 2.0 I_1$, $\beta_2 = \beta_1 + 3.0$ dB.

$$I_2 = 2.0 I_1$$

$$I_0 10^{\frac{\beta_2}{10\text{ dB}}} = 2.0 I_0 10^{\frac{\beta_1}{10\text{ dB}}}$$

$$10^{\frac{\beta_2}{10\text{ dB}}} = (2.0)10^{\frac{\beta_1}{10\text{ dB}}}$$

$$\log 10^{\frac{\beta_2}{10\text{ dB}}} = \log\left[(2.0)10^{\frac{\beta_1}{10\text{ dB}}}\right] = \log 2.0 + \log 10^{\frac{\beta_1}{10\text{ dB}}}$$

$$\frac{\beta_2}{10\text{ dB}} = 0.30 + \frac{\beta_1}{10\text{ dB}}$$

$$\beta_2 = \beta_1 + 3.0\text{ dB}$$

21. (a) Strategy $f_n = nv/(2L)$ for a pipe open at both ends and $v = 343\text{ m/s}$ for $T = 20.0°\text{C}$.

Solution Find the length of the organ pipe.

$$f_1 = \frac{v}{2L}, \text{ so } L = \frac{v}{2f_1} = \frac{343\text{ m/s}}{2(261.5\text{ Hz})} = \boxed{65.6\text{ cm}}$$

(b) Strategy The frequency of the organ pipe is proportional to the speed of the waves and the speed is proportional to the square root of temperature, so $f \propto v \propto \sqrt{T}$.

Solution Find the fundamental frequency after the temperature drop.

$$f_{0.0°} = f_{20°}\sqrt{\frac{T_{0.0°}}{T_{20°}}} = (261.5\text{ Hz})\sqrt{\frac{273.15\text{ K} + 0.0\text{ K}}{273.15\text{ K} + 20.0\text{ K}}} = \boxed{252.4\text{ Hz}}$$

23. Strategy $f_n = nv/(2L)$ for a pipe open at both ends.

Solution Find the length of the organ pipe.

$$f_1 = \frac{v}{2L}, \text{ so } L = \frac{v}{2f_1} = \frac{331\text{ m/s}}{2(382\text{ Hz})} = \boxed{43.3\text{ cm}}.$$

25. Strategy $f_n = nv/(2L)$ for a pipe open at both ends and $v = v_0\sqrt{T/T_0}$.

Solution Find an expression for the temperature in terms of *n*.

$$f_n = \frac{nv}{2L} = \frac{nv_0}{2L}\sqrt{\frac{T}{T_0}}, \text{ so } T = T_0\left(\frac{2Lf_n}{nv_0}\right)^2 = \frac{273.15\text{ K}}{n^2}\left[\frac{2(2.0\text{ m})(702\text{ Hz})}{331\text{ m/s}}\right]^2 = \frac{19{,}658\text{ K}}{n^2}.$$

The assumed temperature range is 293 K (20°C) to 308 K (35°C). We need to find *n* such that *T* falls within this range. By trial and error, *n* is found to be 8. So, $T = 19{,}658\text{ K}/8^2 = 307\text{ K} = \boxed{34°\text{C}}$.

29. Strategy If the displacement antinode is at the end of the tube, $L = 30.0$ cm. For a pipe closed at one end, $\lambda_n = 4L/n$, where $n = 1, 3, 5, \ldots$. The resonance produced by the smallest value of *L* corresponds to the fundamental. In (c), use Eq. (11-6).

Solution

(a) Compute the wavelength of the sound.

$$\lambda_1 = 4L = 4(30.0\text{ cm}) = \boxed{1.20\text{ m}}$$

(b) The next larger value of *L* corresponds to $n = 3$.

$$L = \frac{3\lambda_1}{4} = \frac{3(1.20\text{ m})}{4} = \boxed{90.0\text{ cm}}$$

(c) Compute the speed of sound in the tube.

$$v = \lambda f = (1.20\text{ m})(282\text{ Hz}) = \boxed{338\text{ m/s}}$$

33. (a) Strategy Use Eqs. (11-2) and (11-13).

Solution Solve for the tension in the string.

$$v=\sqrt{\frac{FL}{m}} \text{ and } f_1=\frac{v}{2L}, \text{ so } F=4mLf_1^2=4(0.300\times10^{-3}\text{ kg})(0.655\text{ m})(330.0\text{ Hz})^2=\boxed{85.6\text{ N}}.$$

(b) Strategy and Solution The waves travel on the string with a speed of

$$v=2Lf_1=2(0.655\text{ m})(330.0\text{ Hz})=\boxed{432\text{ m/s}}.$$

(c) Strategy and Solution Since the other musician is lowering the frequency of the whistle, the frequency being played is 330.0 Hz + 5 Hz = $\boxed{335\text{ Hz}}$ when beats are first heard.

(d) Strategy Use Eq. (12-10b) for a pipe closed at one end.

Solution Find the length of the slide whistle.

$$f=\frac{v}{4L}, \text{ so } L=\frac{v}{4f}=\frac{343\text{ m/s}}{4(335\text{ Hz})}=\boxed{0.256\text{ m}}.$$

37. Strategy Since the observer is moving and the source is stationary, use Eq. (12-13).

Solution As Mandy walks toward one siren (1), $v_o<0$. As she recedes from the other siren (2), $v_o>0$. Find the beat frequency heard by Mandy.

$$f_1-f_2=\left(1+\frac{|v_o|}{v}\right)f_s-\left(1-\frac{|v_o|}{v}\right)f_s=\frac{2|v_o|f_s}{v}=\frac{2(1.56\text{ m/s})(698\text{ Hz})}{343\text{ m/s}}=\boxed{6.35\text{ Hz}}$$

39. Strategy Since the observer is moving and the source is stationary, use Eq. (12-13).

Solution Compute the frequencies of the sound observed by the moving observer.

(a) The observer is moving toward a stationary source ($v_o<0$).

$$f_o=\left(1-\frac{v_o}{v}\right)f_s=[1-(-0.50)](1.0\text{ kHz})=\boxed{1.5\text{ kHz}}$$

(b) The observer is now moving away from the source ($v_o>0$).

$$f_o=(1-0.50)(1.0\text{ kHz})=\boxed{500\text{ Hz}}$$

41. Strategy Since both source and observer are moving, use Eq. (12-14).

Solution Compute the frequencies of the sound observed by the moving observer.

(a) A source and an observer are traveling toward each other ($v_s>0$, $v_o<0$).

$$f_o=\frac{1-\frac{v_o}{v}}{1-\frac{v_s}{v}}f_s=\frac{1-(-0.50)}{1-0.50}(1.0\text{ kHz})=\boxed{3.0\text{ kHz}}$$

(b) A source and an observer are traveling away from each other ($v_s<0$, $v_o>0$).

$$f_o=\frac{1-0.50}{1+0.50}(1.0\text{ kHz})=\boxed{330\text{ Hz}}$$

(c) A source and an observer are traveling in the same direction ($v_s < 0$, $v_o < 0$).

$$f_o = \frac{1+0.50}{1+0.50}(1.0\text{ kHz}) = \boxed{1.0\text{ kHz}}$$

45. Strategy Use $d = v\Delta t$ for the distances traveled by the plane and the shock wave during a time interval Δt.

Solution Show that the sine of the angle θ that the shock wave makes with the direction of motion of the plane is equal to the ratio of the speed of sound to the speed of the plane.

$d_{\text{plane}} = v_{\text{plane}}\Delta t$ and $d_{\text{sound}} = v_{\text{sound}}\Delta t$.

A right triangle is formed by d_{plane} and d_{sound}. d_{plane} is the hypotenuse and d_{sound} is opposite θ. Therefore,

$$\sin\theta = \frac{\text{opposite}}{\text{hypotenuse}} = \frac{d_{\text{sound}}}{d_{\text{plane}}} = \frac{v_{\text{sound}}\Delta t}{v_{\text{plane}}\Delta t} = \frac{v_{\text{sound}}}{v_{\text{plane}}}.$$

47. Strategy The distance traveled (round trip) by the sound wave in time Δt is $v\Delta t$. The depth d of the lake is half this distance.

Solution Find the depth of the lake.

$$d = \frac{1}{2}v\Delta t = \frac{1}{2}(0.540\text{ s})(1493\text{ m/s}) = \boxed{403\text{ m}}$$

49. Strategy First treat the moth as a receiver moving away from the source (the bat); then treat the moth as the source moving away from an observer (the bat). Use Eqs. (12-3) and (12-14).

Solution Moth as receiver (v_m and $v_b > 0$; in the direction of propagation):

$$f_1 = \frac{1-\frac{v_m}{v}}{1-\frac{v_b}{v}}f_s$$

Moth as source (v_m and $v_b < 0$; opposite the direction of propagation):

$$f_2 = \frac{1+\frac{v_b}{v}}{1+\frac{v_m}{v}}f_1 = \frac{\left(1+\frac{v_b}{v}\right)\left(1-\frac{v_m}{v}\right)}{\left(1+\frac{v_m}{v}\right)\left(1-\frac{v_b}{v}\right)}f_s = \frac{(v+v_b)(v-v_m)}{(v+v_m)(v-v_b)}f_s$$

Find v at 10.0°C.

$$v = v_0\sqrt{\frac{T}{T_0}} = (331\text{ m/s})\sqrt{\frac{273.15\text{ K} + 10.0\text{ K}}{273.15\text{ K}}} = 337\text{ m/s}$$

Calculate the frequency, f_2.

$$f_2 = \frac{(337\text{ m/s}+4.40\text{ m/s})(337\text{ m/s}-1.20\text{ m/s})}{(337\text{ m/s}+1.20\text{ m/s})(337\text{ m/s}-4.40\text{ m/s})}(82.0\text{ kHz}) = \boxed{83.6\text{ kHz}}$$

53. (a) Strategy The distance traveled (round trip) by the sound of the firing pistol in time Δt is $v\Delta t$. The distance between the ship and one side of the fjord is half this distance. Use Eq. (12-3).

Solution Find the distance between the ship and one side of the fjord.

$$d = \frac{1}{2}v\Delta t = \frac{1}{2}\Delta t v_0\sqrt{\frac{T}{T_0}} = \frac{1}{2}(4.0\text{ s})(331\text{ m/s})\sqrt{\frac{273.15\text{ K} + 5.0\text{ K}}{273.15\text{ K}}} = \boxed{670\text{ m}}$$

(b) Strategy Let the distance to the closer side of the fjord be $d_1 = \frac{1}{2}vt_1$; then the distance to the other side of the fjord is $d_2 = \frac{1}{2}vt_2$.

Solution Find the time interval between the two echoes.

$$\Delta t = t_2 - t_1 = \frac{2d_2}{v} - t_1 = \frac{2d_2}{2d_1/t_1} - t_1 = \frac{d_2}{d_1}t_1 - t_1 = \left(\frac{1.80\text{ km} - 0.668\text{ km}}{0.668\text{ km}} - 1\right)(4.0\text{ s}) = \boxed{2.8\text{ s}}$$

57. Strategy $f_n = nv/(4L)$ for a pipe closed at one end, where $n = 1, 3, 5, \ldots$. The speed of sound at $T = 20.0°\text{C}$ is $v = 343\text{ m/s}$.

Solution Calculate the four lowest standing-wave frequencies for the organ pipe.

$$f_1 = \frac{343\text{ m/s}}{4(4.80\text{ m})} = \boxed{17.9\text{ Hz}},\ f_3 = \frac{3(343\text{ m/s})}{4(4.80\text{ m})} = \boxed{53.6\text{ Hz}},\ f_5 = \frac{5(343\text{ m/s})}{4(4.80\text{ m})} = \boxed{89.3\text{ Hz}},\text{ and}$$

$$f_7 = \frac{7(343\text{ m/s})}{4(4.80\text{ m})} = \boxed{125\text{ Hz}}.$$

61. (a) Strategy and Solution The intensity is inversely proportional to distance, so $I_1 \propto I_0/d^2$, where I_1 and I_0 are the intensity at the object causing the reflection and the original intensity, respectively, and d is the distance between the bat and the object. Similarly, for the intensity of the echo, $I_2 \propto I_1/d^2$ where I_2 is the intensity of the echo at the bat. Therefore, $I_2 \propto \frac{I_1}{d^2} \propto \frac{I_0/d^2}{d^2} \propto \frac{1}{d^4}$.

(b) Strategy Use the result of part (a).

Solution Find the percent increase in the intensity of the echo in terms of the distances.

$$\frac{\Delta I}{I_0}\times 100\% = \frac{I - I_0}{I_0}\times 100\% = \frac{\frac{1}{d^4} - \frac{1}{d_0^4}}{\frac{1}{d_0^4}}\times 100\% = \left[\left(\frac{d_0}{d}\right)^4 - 1\right]\times 100\%$$

Compute the percent increase for each object.

First object: $\left[\left(\frac{0.60}{0.50}\right)^4 - 1\right]\times 100\% = \boxed{110\%}$; second object: $\left[\left(\frac{1.10}{1.00}\right)^4 - 1\right]\times 100\% = \boxed{46\%}$

65. Strategy The greatest common factor of the frequencies is the fundamental. Compute ratios of the frequencies and use these ratios to determine the greatest common factor of the frequencies.

Solution Compute the ratios.

$$\frac{588}{392} = \frac{3}{2}\text{ and }\frac{980}{392} = \frac{5}{2},\text{ so }\frac{392}{2} = 196\times 1,\ 196\times 3 = 588,\text{ and }196\times 5 = 980.$$

The fundamental frequency is $\boxed{196\text{ Hz}}$.

69. (a) Strategy The maximum speed of an element of air in the sound wave is given by $v_m = \omega A = \omega s_0$. Use Eqs. (12-6) and (12-7).

Solution Find the maximum speed.

$$I = \frac{p_0^2}{2\rho v} = \frac{(\omega v\rho s_0)^2}{2\rho v} = \frac{(\omega s_0)^2\rho v}{2} = \frac{v_m{}^2\rho v}{2},\text{ so }v_m = \sqrt{\frac{2I}{\rho v}} = \sqrt{\frac{2(1.0\times 10^{-12}\text{ W/m}^2)}{(1.3\text{ kg/m}^3)(340\text{ m/s})}} = \boxed{6.7\times 10^{-8}\text{ m/s}}.$$

(b) Strategy The average kinetic energy of the eardrum is equal to half the kinetic energy of an air element with the maximum speed.

Solution Find the maximum kinetic energy.

$$K_{\text{av}} = \frac{K_{\text{m}}}{2} = \frac{1}{4}mv_{\text{m}}^2 = \frac{1}{4}(0.0001\text{ kg})(6.7\times10^{-8}\text{ m/s})^2 = \boxed{1\times10^{-19}\text{ J}}$$

(c) Strategy and Solution The average kinetic energy of the eardrum due to collisions with air molecules in the presence of a sound wave at the threshold of human hearing is about ten times that in the absence of a sound wave. $\boxed{\text{The ear is about as sensitive as it can be}}$.

73. Strategy The frequency in the fluid is the same as for the wave in air. Use Eq. (12-7).

Solution Find the ratio of the pressure amplitude of the wave in air to that in the fluid.

$$0.80I_{\text{air}} = \frac{0.80p_{\text{air}}^2}{2\rho_{\text{air}}v_{\text{air}}} = I_{\text{fluid}} = \frac{p_{\text{fluid}}^2}{2\rho_{\text{fluid}}v_{\text{fluid}}}, \text{ so } \frac{p_{\text{air}}}{p_{\text{fluid}}} = \sqrt{\frac{\rho_{\text{air}}v_{\text{air}}}{0.80\rho_{\text{fluid}}v_{\text{fluid}}}} = \sqrt{\frac{(1.20)(343)}{0.80(1001.80)(1493)}} = 0.019.$$

The ratio is $\boxed{0.019}$.

REVIEW AND SYNTHESIS: CHAPTERS 9–12

Review Exercises

1. **Strategy** The magnitude of the buoyant force on an object in water is equal to the weight of the water displaced by the object.

   Solution

   (a) Lead is much denser than aluminum, so for the same mass, its volume is much less. Therefore, $\boxed{\text{aluminum}}$ has the larger buoyant force acting on it; $\boxed{\text{since it is less dense it occupies more volume}}$.

   (b) Steel is denser than wood. Even though the wood is floating, it displaces more water than does the steel. Therefore, $\boxed{\text{wood}}$ has the larger buoyant force acting on it; $\boxed{\text{since it displaces more water than the steel}}$.

   (c) Lead: $\rho_w g V_{Pb} = \rho_w g \dfrac{m_{Pb}}{\rho_{Pb}} = (1.00\times10^3\ \text{kg/m}^3)(9.80\ \text{m/s}^2)\dfrac{1.0\ \text{kg}}{11{,}300\ \text{kg/m}^3} = \boxed{0.87\ \text{N}}$

   Aluminum: $\rho_w g V_{Al} = \rho_w g \dfrac{m_{Al}}{\rho_{Al}} = (1.00\times10^3\ \text{kg/m}^3)(9.80\ \text{m/s}^2)\dfrac{1.0\ \text{kg}}{2702\ \text{kg/m}^3} = \boxed{3.6\ \text{N}}$

   Steel: $\rho_w g V_{Steel} = \rho_w g \dfrac{m_{Steel}}{\rho_{Steel}} = (1.00\times10^3\ \text{kg/m}^3)(9.80\ \text{m/s}^2)\dfrac{1.0\ \text{kg}}{7860\ \text{kg/m}^3} = \boxed{1.2\ \text{N}}$

   Wood: $mg = (1.0\ \text{kg})(9.80\ \text{m/s}^2) = \boxed{9.8\ \text{N}}$ (Since the wood is floating, the buoyant force is equal to its weight.)

5. **Strategy** Use Eqs. (10-20a), (10-21), and (10-22), and Newton's second law.

   Solution The normal force on the 1.0-kg block m_1 is $N = m_1 g$. So, the force of friction on m_1 is $f = \mu N = \mu m_1 g$.

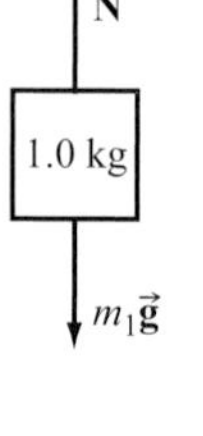

   Find the maximum acceleration that the top block can experience before it starts to slip.

   $\Sigma F = f = m_1 a$, so $a = \dfrac{f}{m_1} = \dfrac{\mu m_1 g}{m_1} = \mu g.$

   For SHM, the maximum acceleration is $a_m = \omega^2 A = \dfrac{k}{m} A$, which in this case is equal to $\dfrac{kA}{m_1 + m_2}$. Equate the accelerations and solve for A.

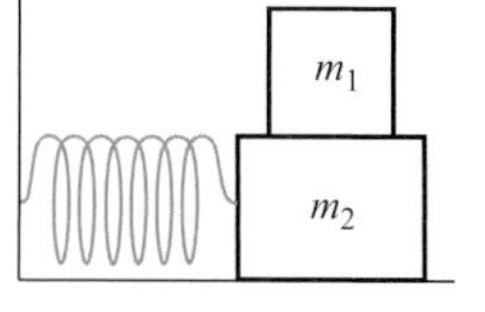

   $\dfrac{kA}{m_1 + m_2} = \mu g$, so $A = \dfrac{\mu(m_1 + m_2)g}{k}$.

   The maximum speed is $v_m = \omega A$. Compute the maximum speed that this set of blocks can have without the top block slipping.

   $$v_m = \omega A = \sqrt{\frac{k}{m_1 + m_2}}\left[\frac{\mu(m_1 + m_2)g}{k}\right] = \mu g\sqrt{\frac{m_1 + m_2}{k}} = 0.45(9.80\ \text{m/s}^2)\sqrt{\frac{1.0\ \text{kg} + 5.0\ \text{kg}}{150\ \text{N/m}}} = \boxed{0.88\ \text{m/s}}$$

9. (a) Strategy Use Eqs. (11-2) and (11-13).

Solution Find the tension in the guitar string.

$$f_1 = \frac{v}{2L} = \frac{1}{2L}\sqrt{\frac{FL}{m}} = \sqrt{\frac{F}{4Lm}}, \text{ so } F = 4Lmf_1^2 = 4(0.655 \text{ m})(0.00331 \text{ kg})(82 \text{ Hz})^2 = \boxed{58 \text{ N}}.$$

(b) Strategy Use Eqs. (11-4) and (11-13).

Solution Find the length of the lowest frequency string when it is fingered at the fifth fret.

$$f_1 = \frac{v}{2L} = \frac{1}{2L}\sqrt{\frac{F}{\mu}}, \text{ so } L = \frac{1}{2f_1}\sqrt{\frac{F}{\mu}} = \frac{1}{2(110 \text{ Hz})}\sqrt{\frac{58 \text{ N}}{0.00331 \text{ kg}/(0.655 \text{ m})}} = \boxed{49 \text{ cm}}.$$

13. Strategy The frequency of the sound is increased by a factor equal to the number of holes in the disk. Use Eq. (11-6).

Solution The frequency of the sound is

$f = 25(60.0 \text{ Hz}) = \boxed{1500 \text{ Hz}}$.

Compute the wavelength that corresponds to this frequency.

$$\lambda = \frac{v}{f} = \frac{343 \text{ m/s}}{1.50\times10^3 \text{ Hz}} = \boxed{22.9 \text{ cm}}$$

17. Strategy The source is moving in the direction of propagation of the sound, so $v_s > 0$. The observer is moving in the direction opposite the propagation of the sound, so $v_o < 0$. Use Eq. (12-14).

Solution Find the frequency heard by the passenger in the oncoming boat.

$$f_o = \frac{v - v_o}{v - v_s} f_s = \frac{343 \text{ m/s} - (-15.6 \text{ m/s})}{343 \text{ m/s} - 20.1 \text{ m/s}}(312 \text{ Hz}) = \boxed{346 \text{ Hz}}$$

MCAT Review

1. Strategy The buoyant force on the brick is equal in magnitude to the weight of the volume of water it displaces.

Solution The brick is completely submerged, so its volume is equal to that of the displaced water. The weight of the displaced water is $30 \text{ N} - 20 \text{ N} = 10 \text{ N}$. Find the volume of the brick.

$$\rho_w g V_w = \rho_w g V_{brick} = 10 \text{ N}, \text{ so } V_{brick} = \frac{10 \text{ N}}{\rho_w g} = \frac{10 \text{ N}}{(1000 \text{ kg/m}^3)(10 \text{ m/s}^2)} = 1\times10^{-3} \text{ m}^3$$

The correct answer is $\boxed{\text{A}}$.

2. Strategy The expansion of the cable obeys $F = k\Delta L$.

Solution Compute the expansion of the cable.

$$\Delta L = \frac{F}{k} = \frac{5000 \text{ N}}{5.0\times10^6 \text{ N/m}} = 10^{-3} \text{ m}$$

The correct answer is $\boxed{\text{A}}$.

3. **Strategy** Solve for the intensity in the definition of sound level.

   Solution Find the intensity of the fire siren.

   $\text{SL} = 10\log_{10}\frac{I}{I_0}$, so $I = I_0 10^{\text{SL}/10} = (1.0\times10^{-12}\ \text{W/m}^2)10^{100/10} = 1.0\times10^{-2}\ \text{W/m}^2$.

   The correct answer is $\boxed{\text{D}}$.

4. **Strategy and Solution** The two centimeters of liquid with a specific gravity of 0.5 are equivalent to one centimeter of water; that is, the 6-cm column of liquid is equivalent to a 5-cm column of water. Therefore, the new gauge pressure at the base of the column is five-fourths the original. The correct answer is $\boxed{\text{C}}$.

5. **Strategy** The wavelength is inversely proportional to the index n. Let $\lambda_n = 8$ m and $\lambda_{n+2} = 4.8$ m.

   Solution Find n.

   $$L = \frac{n\lambda_n}{4} = \frac{(n+2)\lambda_{n+2}}{4}, \text{ so } \frac{\lambda_n}{\lambda_{n+2}} = \frac{n+2}{n} = 1 + \frac{2}{n}, \text{ or } n = 2\left(\frac{\lambda_n}{\lambda_{n+2}} - 1\right)^{-1}.$$

   Compute L.

   $$L = \frac{n\lambda_n}{4} = 2\left(\frac{\lambda_n}{\lambda_{n+2}} - 1\right)^{-1}\frac{\lambda_n}{4} = \frac{1}{2}\left(\frac{8\text{ m}}{4.8\text{ m}} - 1\right)^{-1}(8\text{ m}) = 6\text{ m}$$

   The correct answer is $\boxed{\text{C}}$.

6. **Strategy** The wave may interfere within the range of possibility of totally constructive or totally destructive interference.

   Solution For totally constructive interference, the amplitude of the combined waves is $5+3=8$ units. For totally destructive interference, the amplitude of the combined waves is $5-3=2$ units. The correct answer is $\boxed{\text{B}}$.

7. **Strategy and Solution** As the bob repeatedly swings to and fro, it speeds up and slows down, as well as changes direction. Therefore, its linear acceleration must change in both magnitude and direction.
   The correct answer is $\boxed{\text{D}}$.

8. **Strategy** Since $K \propto v^2$ and $v \propto r^{-4}$, $K \propto r^{-8}$.

   Solution Compute the ratio of kinetic energies.

   $$\frac{K_2}{K_1} = \left(\frac{2}{1}\right)^{-8} = \frac{1}{256} = \frac{1}{4^4} \text{ or } K_2 : K_1 = 1 : 4^4.$$

   The correct answer is $\boxed{\text{B}}$.

9. **Strategy** The buoyant force on a ball is equal to the weight of the volume of water displaced by that ball.

   Solution Since B_1 is not fully submerged and B_2 and B_3 are, the buoyant force on B_1 is less than the buoyant forces on the other two. Since B_2 and B_3 are fully submerged, the buoyant forces on each are the same.

   The correct answer is $\boxed{\text{B}}$.

10. Strategy and Solution Ball 1 is floating, so its density is less than that of the water. Since Ball 2 is submerged within the water, its density is the same as that of the water. Ball 3 is sitting on the bottom of the tank, so its density is greater than that of the water. The correct answer is $\boxed{A}$.

11. Strategy The supporting force of the bottom of the tank is equal to the weight of Ball 3 less the buoyant force of the water.

Solution Compute the supporting force on Ball 3.

$$\rho_{\text{Ball 3}} g V_{\text{Ball 3}} - \rho_{\text{water}} g V_{\text{Ball 3}} = g V_{\text{Ball 3}}(\rho_{\text{Ball 3}} - \rho_{\text{water}})$$
$$= (9.80 \text{ m/s}^2)(1.0\times10^{-6} \text{ m}^3)(7.8\times10^3 \text{ kg/m}^3 - 1.0\times10^3 \text{ kg/m}^3) = 6.7\times10^{-2} \text{ N}$$

The correct answer is $\boxed{B}$.

12. Strategy The relationship between the fraction of a floating object's volume that is submerged to the ratio of the object's density to the fluid in which it floats is $V_\text{f}/V_\text{o} = \rho_\text{o}/\rho_\text{f}$. So, the fraction of the object that is not submerged is $V_\text{ns} = 1 - V_\text{f}/V_\text{o} = 1 - \rho_\text{o}/\rho_\text{f}$.

Solution Find the fraction of the volume of Ball 1 that is above the surface of the water.

$$1 - \frac{\rho_{\text{Ball 1}}}{\rho_{\text{water}}} = 1 - \frac{8.0\times10^2 \text{ kg/m}^3}{1.0\times10^3 \text{ kg/m}^3} = \frac{1}{5}$$

The correct answer is $\boxed{D}$.

13. Strategy The pressure difference at a depth d in water is given by $\rho_{\text{water}} g d$.

Solution Compute the approximate difference in pressure between the two balls.

$$\rho_{\text{water}} g d = (1.0\times10^3 \text{ kg/m}^3)(9.80 \text{ m/s}^2)(0.20 \text{ m}) = 2.0\times10^3 \text{ N/m}^2$$

The correct answer is $\boxed{C}$.

14. Strategy For the force exerted on Ball 3 by the bottom of the tank to be zero, Ball 3 must have the same density as the water.

Solution Find the volume of the hollow portion of Ball 3, V_H.

$$\rho_{\text{Ball 3}} = \frac{m_{\text{Ball 3}}}{V_{\text{Ball 3}}} = \frac{\rho_{\text{Fe}} V_{\text{Fe}}}{V_{\text{Ball 3}}} = \rho_{\text{water}} \text{ and } V_\text{H} = V_{\text{Ball 3}} - V_{\text{Fe}}, \text{ so}$$

$$V_\text{H} = V_{\text{Ball 3}} - \frac{\rho_{\text{water}}}{\rho_{\text{Fe}}} V_{\text{Ball 3}} = V_{\text{Ball 3}}\left(1 - \frac{\rho_{\text{water}}}{\rho_{\text{Fe}}}\right) = (1.0\times10^{-6} \text{ m}^3)\left(1 - \frac{1.0\times10^3 \text{ kg/m}^3}{7.8\times10^3 \text{ kg/m}^3}\right) = 0.87\times10^{-6} \text{ m}^3.$$

The correct answer is $\boxed{C}$.

Chapter 13

TEMPERATURE AND THE IDEAL GAS

Problems

1. Strategy Use Eqs. (13-2b) and (13-3).

Solution Convert the temperature.

(a) $T_C = \dfrac{T_F - 32°F}{1.8°F/°C} = \dfrac{84°F - 32°F}{1.8°F/°C} = \boxed{29°C}$

(b) $T = 29\text{ K} + 273.15\text{ K} = \boxed{302\text{ K}}$

5. Strategy and Solution There are $78 + 114 = 192$ degrees C and 144 degrees J between the freezing and boiling points of ethyl alcohol. Thus, the conversion factor is $144/192 = 0.750$. So, $T_J = (0.750\,°J/°C)T_C + A$, where A is the offset to be determined. Find A by setting both temperatures equal to their respective boiling temperatures. $144°J = (0.750\,°J/°C)(78°C) + A$, so $A = 144°J - (0.750\,°J/°C)(78°C) = 85.5°J$.

Thus, the conversion from °J to °C is given by $\boxed{T_J = (0.750\,°J/°C)T_C + 85.5°J}$.

9. Strategy Since each concrete slab expands along its entire length, only half of the expansion is considered for a particular gap. Two sections meet at a gap, so the gap should be as wide as the expansion of one concrete slab. Use Eq. (13-4).

Solution Find the sizes of the expansion gaps.

(a) $\Delta L = L_0\alpha\Delta T = (15\text{ m})(12\times10^{-6}\text{ K}^{-1})(40.0°C - 20.0°C) = \boxed{3.6\text{ mm}}$

(b) $\Delta L = L_0\alpha\Delta T = (15\text{ m})(12\times10^{-6}\text{ K}^{-1})(-20.0°C - 20.0°C) = -7.2\text{ mm}$

gap width $= 7.2\text{ mm} + 3.6\text{ mm} = \boxed{10.8\text{ mm}}$

11. Strategy The hole expands just as if it were a solid brass disk. Use Eq. (13-6).

Solution Find the increase in area of the hole.

$\Delta A = 2\alpha A_0\Delta T = 2(1.9\times10^{-5}\,°C^{-1})(1.00\text{ mm}^2)(30.0°C - 20.0°C) = \boxed{3.8\times10^{-4}\text{ mm}^2}$

13. Strategy A decrease in volume for a fixed mass increases the density, and vice versa. Use Eq. (13-7).

Solution

(a) $\dfrac{\Delta V}{V_0} = \beta\Delta T$ and $\dfrac{\Delta\rho}{\rho} = -\dfrac{\Delta V}{V_0}$. Thus, $\dfrac{\Delta\rho}{\rho} = -\beta\Delta T$, so $\Delta\rho = -\beta\rho\Delta T$.

(b) Compute the fractional change in density.

$\dfrac{\Delta\rho}{\rho} = -\beta\Delta T = -(57\times10^{-6}\text{ K}^{-1})(-10.0°C - 32°C) = \boxed{2.4\times10^{-3}}$

17. Strategy Use Eq. (13-4). The internal radius of the ring expands as if it were a solid piece of brass.

Solution Find the temperature at which the internal radius of the ring is 1.0010 cm.

$\frac{\Delta L}{L_0} = \alpha \Delta T$, so $\frac{\Delta L}{\alpha L_0} = T - T_0$. Compute the temperature.

$$T = \frac{\Delta L}{\alpha L_0} + T_0 = \frac{1.0010\text{ cm} - 1.0000\text{ cm}}{(19\times10^{-6}\text{ K}^{-1})(1.0000\text{ cm})} + 22.0°\text{C} = \boxed{75°\text{C}}$$

21. Strategy Use Eqs. (13-4) and (13-7) and the given initial volume.

Solution Find the fractional change.

$$\frac{\Delta V}{V_0} = \frac{V - V_0}{V_0} = \frac{(s_0 + \Delta s)^3 - s_0^3}{s_0^3} = \frac{s_0^3 + 2s_0^2\Delta s + s_0(\Delta s)^2 + s_0^2\Delta s + 2s_0(\Delta s)^2 + (\Delta s)^3 - s_0^3}{s_0^3} = \frac{3s_0^2\Delta s + 3s_0(\Delta s)^2 + (\Delta s)^3}{s_0^3}$$

Now, since $\Delta s << s_0$, $3s_0^2\Delta s >> s_0(\Delta s)^2$ and $(\Delta s)^3$. Thus, we have $\frac{\Delta V}{V_0} \approx \frac{3s_0^2\Delta s}{s_0^3} = 3\frac{\Delta s}{s_0} = 3\alpha\Delta T$ since $\frac{\Delta s}{s_0} = \alpha\Delta T$.

25. Strategy Use Eqs. (13-4) and (13-5).

Solution Find the new scale of the rule.

$$\frac{\Delta L_\text{r}}{L_\text{r0}} = \frac{L_\text{r} - L_\text{r0}}{L_\text{r0}} = \frac{L_\text{r}}{L_\text{r0}} - 1 = \alpha_\text{r}\Delta T, \text{ so } \frac{L_\text{r}}{L_\text{r0}} = 1 + \alpha_\text{r}\Delta T.$$

Divide the new length of the brick by the new scale.

$$\frac{L_\text{b}}{1+\alpha_\text{r}\Delta T} = \frac{L_\text{b0}(1+\alpha_\text{b}\Delta T)}{1+\alpha_\text{r}\Delta T} = \frac{(25.00\text{ cm})[1+(0.75\times10^{-6}\text{ K}^{-1})(60.00\text{ K})]}{1+(12\times10^{-6}\text{ K}^{-1})(60.00\text{ K})} = \boxed{24.98\text{ cm}}$$

29. Strategy Add the molecular masses of each element in carbon dioxide.

Solution Find the mass of carbon dioxide in kg.

$$\text{mass of CO}_2\text{ in kg} = m_\text{C} + 2m_\text{O} = [12.011\text{ u} + 2(15.9994\text{ u})](1.6605\times10^{-27}\text{ kg/u}) = \boxed{7.31\times10^{-26}\text{ kg}}$$

33. Strategy The number of SiO_2 molecules (and the number of Si atoms) N is roughly equal to the volume of a sand grain V_g divided by the volume of a SiO_2 molecule V_m.

Solution Find the order of magnitude of the number of silicon atoms in a grain of sand.

$$N = \frac{V_\text{g}}{V_\text{m}} = \frac{\frac{4}{3}\pi r_\text{g}^3}{\frac{4}{3}\pi r_\text{m}^3} = \left(\frac{d_\text{g}}{d_\text{m}}\right)^3 = \left(\frac{0.5\times10^{-3}\text{ m}}{0.5\times10^{-9}\text{ m}}\right)^3 = \boxed{10^{18}\text{ atoms}}$$

35. Strategy Divide the total mass by the molar mass of sucrose to find the number of moles. Then use Eq. (13-11) to find the number of hydrogen atoms.

Solution $m_{\text{C}_{12}\text{H}_{22}\text{O}_{11}} = 12(12.011\text{ g/mol}) + 22(1.00794\text{ g/mol}) + 11(15.9994\text{ g/mol}) = 342.30\text{ g/mol}$

There are 342.30 grams of sucrose per mole, so there are $\frac{684.6\text{ g}}{342.30\text{ g/mol}} = 2.000$ mol of sucrose.

There are 2.000(22) = 44.00 moles of hydrogen. Find the number of hydrogen atoms.

$$N = nN_\text{A} = (44.00\text{ mol})(6.022\times10^{23}\text{ mol}^{-1}) = \boxed{2.650\times10^{25}\text{ atoms}}$$

37. Strategy Divide the total mass of methane by its molar mass.

Solution Find the number of moles.

$$n_{CH_4} = \frac{\text{mass of } CH_4}{\text{molar mass of } CH_4} = \frac{144.36 \text{ g}}{12.011 \text{ g/mol} + 4(1.00794 \text{ g/mol})} = \boxed{8.9985 \text{ mol}}$$

41. Strategy Use the macroscopic form of the ideal gas law, Eq. (13-16).

Solution Find the new temperature of the air.

$$PV = nRT, \text{ so } \frac{T_f}{T_i} = \frac{\frac{P_f V_f}{nR}}{\frac{P_i V_i}{nR}} = \frac{P_f V_f}{P_i V_i}, \text{ or}$$

$$T_f = \frac{P_f V_f T_i}{P_i V_i} = \frac{20.0 P_i (0.111 V_i) T_i}{P_i V_i} = 20.0(0.111)(30 \text{ K} + 273.15 \text{ K}) = 673 \text{ K} = \boxed{400°\text{C}}.$$

45. Strategy The volume and moles of the gas are constant. Use Gay-Lussac's law.

Solution Find the pressure at the higher temperature.

$$P \propto T, \text{ so } P_f = \frac{T_f}{T_i} P_i = \frac{70.0 \text{ K} + 273.15 \text{ K}}{20.0 \text{ K} + 273.15 \text{ K}}(115 \text{ kPa}) = \boxed{135 \text{ kPa}}.$$

49. Strategy The number of moles of the gas is constant and $V = \frac{1}{6}\pi d^3$. Use the ideal gas law and Eq. (9-3).

Solution Find the diameter of the bubble when it reaches the surface.

$$\frac{P_f V_f}{T_f} = \frac{P_i V_i}{T_i}, \text{ so } V_f = \frac{1}{6}\pi d_f^3 = \frac{P_i T_f}{P_f T_i} V_i = \frac{(P_f + \rho g h) T_f}{P_f T_i}\left(\frac{1}{6}\pi d_i^3\right). \text{ Solve for } d_f.$$

$$d_f = d_i \sqrt[3]{\frac{(P_f + \rho g h) T_f}{P_f T_i}}$$

$$= (1.00 \text{ mm}) \sqrt[3]{\frac{[(1.0 \text{ atm})(1.013 \times 10^5 \text{ Pa/atm}) + (1.0 \times 10^3 \text{ kg/m}^3)(9.80 \text{ m/s}^2)(80.0 \text{ m})](273.15 \text{ K} + 18 \text{ K})}{(1.0 \text{ atm})(1.013 \times 10^5 \text{ Pa/atm})(273.15 \text{ K} + 4 \text{ K})}}$$

$$= \boxed{2.1 \text{ mm}}$$

53. Strategy The temperature and the number of moles of the gas are constant. Use Boyle's law and Eq. (9-3).

Solution Find the factor by which the diver's lungs expand.

$$P \propto \frac{1}{V}, \text{ so } \frac{V_f}{V_i} = \frac{P_i}{P_f} = \frac{P_f + \rho g d}{P_f} = 1 + \frac{\rho g d}{P_f} = 1 + \frac{(1.03 \times 10^3 \text{ kg/m}^3)(9.80 \text{ m/s}^2)(5.0 \text{ m})}{1.013 \times 10^5 \text{ Pa}} = \boxed{1.50}.$$

55. Strategy The number of moles of air is constant. Assume that the temperature is constant.
$V_\text{f} = \dfrac{\Delta V}{\Delta t}\Delta t$ where $\dfrac{\Delta V}{\Delta t} = 0.500\ \text{L/s} = 5.00\times10^2\ \text{cm}^3/\text{s}$ and $P_\text{f} = P_\text{i} + \rho gd$. Use Boyle's law.

Solution Find how long the tank of air will last for each depth.

(a) $\dfrac{V_\text{f}}{V_\text{i}} = \dfrac{\frac{\Delta V}{\Delta t}\Delta t}{V_\text{i}} = \dfrac{P_\text{i}}{P_\text{f}} = \dfrac{P_\text{i}}{P_\text{i} + \rho gd}$, so

$$\Delta t = \frac{P_\text{i}V_\text{i}}{\frac{\Delta V}{\Delta t}(P_\text{i} + \rho gd)}$$

$$= \frac{(1.0\times10^7\ \text{Pa})(0.010\ \text{m}^3)}{(5.00\times10^2\ \text{cm}^3/\text{s})(10^{-6}\ \text{m}^3/\text{cm}^3)(60\ \text{s/min})\left[1.013\times10^5\ \text{Pa} + (1.0\times10^3\ \text{kg/m}^3)(9.80\ \text{m/s}^2)(2.0\ \text{m})\right]}$$

$$= \boxed{28\ \text{min}}$$

(b) $$\Delta t = \frac{(1.0\times10^7\ \text{Pa})(0.010\ \text{m}^3)}{(5.00\times10^2\ \text{cm}^3/\text{s})(10^{-6}\ \text{m}^3/\text{cm}^3)(60\ \text{s/min})\left[1.013\times10^5\ \text{Pa} + (1.0\times10^3\ \text{kg/m}^3)(9.80\ \text{m/s}^2)(20.0\ \text{m})\right]}$$

$$= \boxed{11\ \text{min}}$$

57. Strategy Use the microscopic form of the ideal gas law, Eq. (13-13).

Solution Find the number of air molecules released.
$N = \dfrac{PV}{kT}$, and V, k, and T are constant, so

$$\Delta N = \frac{V\Delta P}{kT} = \frac{(1.0\ \text{m}^3)(15.0\ \text{atm} - 20.0\ \text{atm})(1.013\times10^5\ \text{Pa/atm})}{(1.38\times10^{-23}\ \text{J/K})(273\ \text{K})} = -1.3\times10^{26}.$$

$\boxed{1.3\times10^{26}}$ air molecules were released.

61. Strategy The total translational kinetic energy of the gas molecules is equal to the number of molecules times the average translational kinetic energy per molecule. Use Eqs. (13-13) and (13-20).

Solution Find the total translational kinetic energy of the gas molecules.

$$K_\text{total} = N\langle K_\text{tr}\rangle = N\left(\frac{3}{2}kT\right) = \frac{3}{2}PV = \frac{3}{2}(1.013\times10^5\ \text{Pa})(0.00100\ \text{m}^3) = \boxed{152\ \text{J}}$$

65. Strategy The total internal kinetic energy of the ideal gas is equal to the number of molecules times the average kinetic energy per molecule. Use Eq. (13-20).

Solution Find the total internal kinetic energy of the ideal gas.

$$K_\text{total} = N\langle K_\text{tr}\rangle = N\left\langle\frac{3}{2}kT\right\rangle = \frac{3}{2}nRT = \frac{3}{2}(1.0\ \text{mol})[8.314\ \text{J/(mol}\cdot\text{K)}](273.15\ \text{K} + 0.0\ \text{K}) = \boxed{3.4\ \text{kJ}}$$

69. Strategy Use Eq. (13-22).

Solution Find the rms speeds of the molecules.

(a) $v_{\text{rms}} = \sqrt{\dfrac{3kT}{m}} = \sqrt{\dfrac{3(1.38\times10^{-23}\text{ J/K})(273.15\text{ K}+0.0\text{ K})}{2(14.00674\text{ u})(1.66\times10^{-27}\text{ kg/u})}} = \boxed{493\text{ m/s}}$

(b) $v_{\text{rms}} = \sqrt{\dfrac{3(1.38\times10^{-23}\text{ J/K})(273.15\text{ K}+0.0\text{ K})}{2(15.9994\text{ u})(1.66\times10^{-27}\text{ kg/u})}} = \boxed{461\text{ m/s}}$

(c) $v_{\text{rms}} = \sqrt{\dfrac{3(1.38\times10^{-23}\text{ J/K})(273.15\text{ K}+0.0\text{ K})}{[12.011\text{ u}+2(15.9994\text{ u})](1.66\times10^{-27}\text{ kg/u})}} = \boxed{393\text{ m/s}}$

73. Strategy Use Eq. (13-20).

Solution Find the temperature of the ideal gas.

$\langle K_{\text{tr}}\rangle = \dfrac{3}{2}kT$, so $T = \dfrac{2\langle K_{\text{tr}}\rangle}{3k} = \dfrac{2(4.60\times10^{-20}\text{ J})}{3(1.38\times10^{-23}\text{ J/K})} = \boxed{2220\text{ K}}$.

75. Strategy and Solution The average translational energy of a molecule in an ideal gas is $\langle K_{\text{tr}}\rangle = \dfrac{1}{2}m\langle v^2\rangle$. The rms speed is $v_{\text{rms}} = \sqrt{\langle v^2\rangle}$, so $\dfrac{1}{2}mv_{\text{rms}}^2 = \langle K_{\text{tr}}\rangle$. From Eq. (13-19), we know that $\langle K_{\text{tr}}\rangle = \dfrac{3PV}{2N}$, and using the ideal gas law, $PV = nRT$, we have $\dfrac{1}{2}mv_{\text{rms}}^2 = \dfrac{3}{2}\left(\dfrac{PV}{N}\right) = \dfrac{3}{2}\left(\dfrac{nRT}{N}\right)$, so $v_{\text{rms}} = \sqrt{3\left(\dfrac{n}{mN}\right)RT} = \sqrt{\dfrac{3RT}{M}}$.

77. Strategy Form a proportion with the two reaction rates and solve for the temperature increase. Use Eq. (13-24).

Solution Find the temperature increase.

$$\frac{1.035}{1} = \frac{e^{-\frac{E_a}{kT_2}}}{e^{-\frac{E_a}{kT_1}}} = e^{\frac{E_a}{k}\left(\frac{1}{T_1}-\frac{1}{T_2}\right)}$$

$$\ln 1.035 = \frac{E_a}{k}\left(\frac{1}{T_1}-\frac{1}{T_2}\right)$$

$$\frac{k\ln 1.035}{E_a} = \frac{1}{T_1}-\frac{1}{T_2}$$

$$\frac{1}{T_2} = \frac{1}{T_1}-\frac{k\ln 1.035}{E_a}$$

$$T_2 = \left(\frac{1}{T_1}-\frac{k\ln 1.035}{E_a}\right)^{-1}$$

$$\Delta T = \left(\frac{1}{T_1}-\frac{k\ln 1.035}{E_a}\right)^{-1} - T_1$$

$$= \left[\frac{1}{273.15\text{ K}+10.00\text{ K}} - \frac{(1.38\times10^{-23}\text{ J/K})\ln 1.035}{2.81\times10^{-19}\text{ J}}\right]^{-1} - (273.15\text{ K}+10.00\text{ K}) = \boxed{0.14°\text{C}}$$

81. Strategy Use Eq. (13-26).

Solution Estimate the time it takes a sucrose molecule to move 5.00 mm in one direction.

$$x_{\text{rms}}=\sqrt{2Dt},\text{ so } t=\frac{x_{\text{rms}}^2}{2D}=\frac{(5.00\times10^{-3}\text{ m})^2}{2(5.0\times10^{-10}\text{ m}^2/\text{s})}=\boxed{2.5\times10^4\text{ s}}.$$

85. Strategy Find the temperature at which the radius of the steel sphere and the internal radius of the brass ring are the same. Use Eq. (13-5).

Solution Set the final lengths of the radii equal and solve for the final temperature.

$$L_{\text{b}0}+L_{\text{b}0}\alpha_{\text{b}}\Delta T=L_{\text{s}0}+L_{\text{s}0}\alpha_{\text{s}}\Delta T$$
$$L_{\text{b}0}\alpha_{\text{b}}(T_{\text{f}}-T_{\text{i}})=L_{\text{s}0}-L_{\text{b}0}+L_{\text{s}0}\alpha_{\text{s}}(T_{\text{f}}-T_{\text{i}})$$
$$T_{\text{f}}(L_{\text{b}0}\alpha_{\text{b}}-L_{\text{s}0}\alpha_{\text{s}})=L_{\text{s}0}-L_{\text{b}0}+T_{\text{i}}(L_{\text{b}0}\alpha_{\text{b}}-L_{\text{s}0}\alpha_{\text{s}})$$
$$T_{\text{f}}=\frac{L_{\text{s}0}-L_{\text{b}0}}{L_{\text{b}0}\alpha_{\text{b}}-L_{\text{s}0}\alpha_{\text{s}}}+T_{\text{i}}$$
$$T_{\text{f}}=\frac{1.0010\text{ cm}-1.0000\text{ cm}}{(1.0000\text{ cm})(19\times10^{-6}\text{ K}^{-1})-(1.0010\text{ cm})(12\times10^{-6}\text{ K}^{-1})}+22.0^\circ\text{C}=\boxed{165^\circ\text{C}}$$

89. Strategy Use the ideal gas law.

Solution

(a) Find the number of moles in terms of the pressure.

$$PV=nRT,\text{ so } n=\frac{PV}{RT}.$$

Find the percent change in the number of moles of air in the cabin.

$$\frac{n_{\text{f}}-n_{\text{i}}}{n_{\text{i}}}\times100\%=\frac{\frac{P_{\text{f}}V}{RT}-\frac{P_{\text{i}}V}{RT}}{\frac{P_{\text{i}}V}{RT}}\times100\%=\frac{P_{\text{f}}-P_{\text{i}}}{P_{\text{i}}}\times100\%=\frac{7.62\times10^4\text{ Pa}-1.01\times10^5\text{ Pa}}{1.01\times10^5\text{ Pa}}\times100\%=-25\%$$

The number of moles $\boxed{\text{decreases by 25\%}}$.

(b) Find the final temperature in terms of the pressure.

$$\frac{T_{\text{f}}}{T_{\text{i}}}=\frac{\frac{P_{\text{f}}V}{nR}}{\frac{P_{\text{i}}V}{nR}}=\frac{P_{\text{f}}}{P_{\text{i}}},\text{ so } T_{\text{f}}=\frac{P_{\text{f}}}{P_{\text{i}}}T_{\text{i}}=\frac{7.62\times10^4\text{ Pa}}{1.01\times10^5\text{ Pa}}(25.0\text{ K}+273.15\text{ K})=225\text{ K}=\boxed{-48^\circ\text{C}}.$$

93. (a) Strategy The slope of a graph is rise over run.

Solution Find the slope of pressure versus temperature.

$$\frac{\Delta P}{\Delta T}=\frac{8.00\text{ mm}}{20.0^\circ\text{C}}=\boxed{0.400\text{ mm Hg}/^\circ\text{C}}$$

(b) Strategy Use the ideal gas law and the result of part (a).

Solution Find the number of moles of gas present.

$(\Delta P)V=nR(\Delta T)$, so

$$n=\frac{V}{R}\frac{\Delta P}{\Delta T}=\frac{0.500\text{ L}}{8.314\text{ J/(mol}\cdot\text{K)}}(0.400\text{ mm Hg}/^\circ\text{C})\frac{10^{-3}\text{ m}^3}{1\text{ L}}(1.333\times10^2\text{ Pa/mm Hg})=\boxed{3.21\times10^{-3}\text{ mol}}.$$

95. Strategy Assume that each air molecule is at the center of a sphere (with volume V/N) of diameter d. Then the average distance between air molecules is approximately d. Use the microscopic form of the ideal gas law, Eq. (13-13).

Solution Estimate the average distance between air molecules.

$$PV = P\left(N\frac{1}{6}\pi d^3\right) = NkT, \text{ so } d = \left(\frac{6kT}{\pi P}\right)^{1/3} = \left[\frac{6(1.38\times10^{-23}\text{ J/K})(273.15\text{ K}+0.0\text{ K})}{\pi(1.00\text{ atm})(1.013\times10^5\text{ Pa/atm})}\right]^{1/3} \approx \boxed{4\text{ nm}}.$$

97. (a) Strategy Use Eq. (13-10) to find the number of air molecules per m^3; multiply by 0.21 to find the number of O_2 molecules.

Solution Find the number of oxygen molecules per cubic meter.

$$\frac{N_{O_2}}{V} = \frac{0.21\rho_{\text{air}}}{m_{\text{air}}} = \frac{0.21(1.20\times10^3\text{ g/m}^3)(6.022\times10^{23}\text{ mol}^{-1})}{0.78(2)(14.00674\text{ g/mol})+0.21(2)(15.9994\text{ g/mol})+0.01(39.948\text{ g/mol})} = \boxed{5.2\times10^{24}\text{ m}^{-3}}$$

(b) Strategy The ratio of the surface volume of air to the volume of the air at a depth of 100.0 m multiplied by 21% gives the percentage of O_2 molecules. Use the ideal gas law with N, k, and T constant and Eq. (9-3).

Solution Find the appropriate percentage of oxygen molecules in the tank.

$$\frac{V_2}{V_1}(21\%) = \frac{P_1}{P_2}(21\%) = \frac{P_{\text{atm}}}{P_{\text{atm}}+\rho gh}(21\%) = \frac{(1.013\times10^5\text{ Pa})(21\%)}{1.013\times10^5\text{ Pa}+(1025\text{ kg/m}^3)(9.80\text{ m/s}^2)(100.0\text{ m})} = \boxed{1.9\%}$$

101. Strategy The relative number of atoms of an element contained within the molecule is equal to the molecular mass of the molecule times the percentage of the molecular mass of the molecule that is the mass of the atoms of element divided by the molecular mass of the element.

Solution Find the chemical formula.

$$\frac{(63\text{ u})(0.016)}{1.00794\text{ u}} = 1.0\text{ H};\quad \frac{(63\text{ u})(0.222)}{14.00674\text{ u}} = 1.0\text{ N};\quad \frac{(63\text{ u})(0.762)}{15.9994\text{ u}} = 3.0\text{ O}$$

The chemical formula is $\boxed{HNO_3}$.

105. Strategy Use the microscopic form of the ideal gas law, Eq. (13-13).

Solution Find the number of air molecules N.

$$NkT = PV, \text{ so } N = \frac{PV}{kT} = \frac{(1.00\times10^5\text{ Pa})\frac{4}{3}\pi(0.125\times10^{-3}\text{ m})^3}{(1.38\times10^{-23}\text{ J/K})(310\text{ K})} = \boxed{1.9\times10^{14}\text{ molecules}}.$$

109. Strategy Use ideal gas law and Hooke's law.

Solution Find the pressure of the gas.

$PV = nRT$, so $P_{\text{gas}} = \dfrac{nRT}{V}$. The force with which the piston pushes on the spring is equal to

$F = (P_{\text{gas}} - P_{\text{atm}})A_{\text{piston}}$. Set this equal to $F = k\Delta x$ to find the spring constant.

$k\Delta x = (P_{\text{gas}} - P_{\text{atm}})A_{\text{piston}} = \left(\frac{nRT}{V} - P_{\text{atm}}\right)A_{\text{piston}}$, so

$$k = \left[\frac{(6.50\times10^{-2}\text{ mol})[8.314\text{ J/(mol·K)}](20.0\text{ K}+273.15\text{ K})}{(0.120\text{ m}+0.0540\text{ m})\pi(0.0800/2\text{ m})^2} - 1.013\times10^5\text{ Pa}\right]\frac{\pi(0.0800/2\text{ m})^2}{0.0540\text{ m}} = \boxed{7.4\times10^3\text{ N/m}}.$$

Chapter 14

HEAT

Problems

1. (a) Strategy The gravitational potential energy of the 1.4 kg of water is converted to internal energy in the 6.4-kg system.

Solution Compute the increase in internal energy.

$U = mgh = (1.4 \text{ kg})(9.80 \text{ m/s}^2)(2.5 \text{ m}) = \boxed{34 \text{ J}}$

(b) Strategy and Solution Yes; the increase in internal energy increases the average kinetic energy of the water molecules, thus the temperature is slightly increased.

3. Strategy The amount of internal energy generated is equal to the decrease in kinetic energy of the bullet.

Solution Compute the amount of internal energy generated.

$|\Delta K| = \frac{1}{2}mv_i^2 = \frac{1}{2}(0.0200 \text{ kg})(7.00\times10^2 \text{ m/s})^2 = \boxed{4.90 \text{ kJ}}$

5. (a) Strategy The decrease in gravitational potential energy of the child is equal to the amount of internal energy generated.

Solution Compute the amount of internal energy generated.

$U = mgh = (15 \text{ kg})(9.80 \text{ m/s}^2)(1.7 \text{ m}) = \boxed{250 \text{ J}}$

(b) Strategy and Solution Friction warms the slide and the child, and the air molecules are deflected by the child's body. The energy goes into $\boxed{\text{all three}}$.

9. Strategy The conversion factor is $1 \text{ kW}\cdot\text{h} = 3.600 \text{ MJ}$.

Solution Convert 1.00 kJ to kilowatt-hours.

$(1.00\times10^3 \text{ J})\frac{1 \text{ kW}\cdot\text{h}}{3.600\times10^6 \text{ J}} = \boxed{2.78\times10^{-4} \text{ kW}\cdot\text{h}}$

11. Strategy The heat capacity of an object is equal to its mass times its specific heat.

Solution Find the heat capacity of the 5.00-g gold ring.

$C = mc = (0.00500 \text{ kg})[0.128 \text{ kJ/(kg}\cdot\text{K)}] = \boxed{6.40\times10^{-4} \text{ kJ/K}}$

13. Strategy Use Eq. (14-4).

Solution Find the amount of heat that must flow into the water.

$Q = mc\Delta T = (2.0\times10^{-3} \text{ m}^3)(1.0\times10^3 \text{ kg/m}^3)[4186 \text{ J/(kg}\cdot\text{K)}](80.0-20.0) \text{ K} = \boxed{0.50 \text{ MJ}}$

15. Strategy The 3.3% of the energy from the food is converted to gravitational potential energy of the high jumper.

Solution Find the height the athlete could jump.

$U = mgh,$ so $h = \frac{U}{mg} = \frac{(3.00\times10^6\ \text{cal})(4.186\ \text{J/cal})(0.033)}{(60.0\ \text{kg})(9.80\ \text{m/s}^2)} = \boxed{700\ \text{m}}.$

17. Strategy The heat capacity of an object is equal to its mass times its specific heat. The mass of an object is equal to its density times its volume.

Solution Find the heat capacities.

(a) $C = mc = \rho Vc = (2702\ \text{kg/m}^3)(1.00\ \text{m}^3)[0.900\ \text{kJ/(kg}\cdot\text{K)}] = \boxed{2430\ \text{kJ/K}}$

(b) $C = mc = \rho Vc = (7860\ \text{kg/m}^3)(1.00\ \text{m}^3)[0.44\ \text{kJ/(kg}\cdot\text{K)}] = \boxed{3500\ \text{kJ/K}}$

19. Strategy Use Eq. (14-4) to find the heat required.

Solution The heat capacity of the system is $C = m_{\text{Al}}c_{\text{Al}} + m_{\text{w}}c_{\text{w}}$.

$Q = mc\Delta T = \{(0.400\ \text{kg})[0.900\ \text{kJ/(kg}\cdot\text{K)}] + (2.00\ \text{kg})[4.186\ \text{kJ/(kg}\cdot\text{K)}]\}(100.0°\text{C} - 15.0°\text{C}) = \boxed{742\ \text{kJ}}$

21. Strategy Use Eq. (14-4).

Solution Find the specific heat of lead.

$c = \frac{Q}{m\Delta T} = \frac{(210\times10^{-3}\ \text{kcal})(4.186\ \text{kJ/kcal})}{(0.35\ \text{kg})(20.0\ \text{K})} = \boxed{0.13\ \text{kJ/(kg}\cdot\text{K)}}$

25. Strategy The change in internal energy of the water is equal to the work done on the water by the mixer plus the heat that flows into the water: $Q + W = \Delta U = mc\Delta T$.

Solution Find the quantity of heat that flowed into the water.

$Q = mc\Delta T - W = (2.00\ \text{kg})[4.186\ \text{kJ/(kg}\cdot\text{K)}](4.00\ \text{K}) - 6.0\ \text{kJ} = \boxed{27.5\ \text{kJ}}.$

27. Strategy The energy required to increase the internal energy of the gas is equal to $nC_\text{V}\Delta T$ where $C_\text{V} = 20.4\ \text{J/(mol}\cdot\text{K)}$ for H_2. Use the ideal gas law to find the number of moles of H_2.

Solution Find the energy required.

$$\Delta U = nC_\text{V}\Delta T = \frac{PV}{RT}C_\text{V}\Delta T = \frac{(10.0\ \text{atm})(1.013\times10^5\ \text{Pa/atm})(250\ \text{L})(10^{-3}\ \text{m}^3/\text{L})[20.4\ \text{J/(mol}\cdot\text{K)}](25.0\ \text{K})}{[8.314\ \text{J/(mol}\cdot\text{K)}](273.15\ \text{K} + 0.0\ \text{K})}$$

$$= \boxed{57\ \text{kJ}}$$

29. Strategy The number of moles of air is given by the ideal gas law, $n = PV/(RT)$, where these are the initial quantities. The power generated by the crowd is $501P_{av} = \Delta Q/\Delta t$ and $\Delta Q = \frac{5}{2}nR\Delta T$ for a diatomic gas.

Solution Find ΔT.

$$\frac{5}{2}nR\Delta T = \frac{5}{2}\left(\frac{PV}{RT}\right)R\Delta T = \Delta Q = 501P_{av}\Delta t, \text{ so}$$

$$\Delta T = \frac{2(501)TP_{av}\Delta t}{5PV} = \frac{2(501)(273.15\text{ K} + 20.0\text{ K})(110\text{ W})(2.0\text{ h})(3600\text{ s/h})}{5(1.01\times10^5\text{ Pa})(8.00\times10^3\text{ m}^3)} = \boxed{58°\text{C}}.$$

33. Strategy The sum of the heat flows is zero. Use Eqs. (14-4) and (14-9).

Solution Find the heat of fusion of water.

$$0 = Q_{ice} + Q_w + Q_c = m_{ice}c_{ice}\Delta T_{ice1} + m_{ice}L_f + m_{ice}c_w\Delta T_{ice2} + m_w c_w \Delta T_w + m_c c_c \Delta T_c, \text{ so}$$

$$L_f = -\frac{m_{ice}c_{ice}\Delta T_{ice1} + c_w(m_{ice}\Delta T_{ice2} + m_w\Delta T_w) + m_c c_c \Delta T_c}{m_{ice}}$$

$$= -\frac{(11.6\text{ g})[0.50\text{ cal/(g}\cdot\text{K)}](5.0\text{ K}) + [1.000\text{ cal/(g}\cdot\text{K)}][(11.6\text{ g})(15.0\text{ K}) + (2.00\times10^2\text{ g})(-5.0\text{ K})] + (3.00\times10^2\text{ g})[0.090\text{ cal/(g}\cdot\text{K)}](-5.0\text{ K})}{11.6\text{ g}}$$

$$= \boxed{80\text{ cal/g}}$$

37. Strategy Find the heat required to melt the ice, and compare it to the available heat energy in the water. If the ice does not melt completely, the final temperature will be $0°\text{C}$. If the ice does melt completely, find the final temperature of the system.

Solution The heat required to melt the ice is

$$Q = m_{ice}L_f + m_{ice}c_{ice}\Delta T = (0.075\text{ kg})\{333.7\text{ kJ/kg} + [2.1\text{ kJ/(kg}\cdot\text{K)}](10.0\text{ K})\} = 27\text{ kJ}.$$

The heat available in the water is $|Q| = m_w c_w |\Delta T| = (0.500\text{ kg})[4.186\text{ kJ/(kg}\cdot\text{K)}](50.0\text{ K}) = 105\text{ kJ}.$

Since 105 kJ > 27 kJ, $\boxed{\text{the ice will melt completely}}$.

Set the sum of the heat flows equal to zero and solve for the final temperature, T.

$$0 = Q_w + Q_{ice}$$
$$0 = m_w c_w \Delta T_w + m_{ice}L_f + m_{ice}c_{ice}\Delta T_1 + m_{ice}c_w\Delta T_2$$
$$0 = m_w c_w (T - T_w) + m_{ice}L_f + m_{ice}c_{ice}(273.15\text{ K} - T_{ice}) + m_{ice}c_w(T - 273.15\text{ K})$$
$$0 = (m_w c_w + m_{ice}c_w)T + m_{ice}L_f + m_{ice}(c_{ice} - c_w)(273.15\text{ K}) - m_w c_w T_w - m_{ice}c_{ice}T_{ice}$$
$$T = \frac{m_w c_w T_w + m_{ice}c_{ice}T_{ice} - m_{ice}L_f - m_{ice}(c_{ice} - c_w)(273.15\text{ K})}{c_w(m_w + m_{ice})}$$

$$T = \frac{(0.500\text{ kg})[4.186\text{ kJ/(kg}\cdot\text{K)}](50.0 + 273.15)\text{ K} + (0.075\text{ kg})\{[2.1\text{ kJ/(kg}\cdot\text{K)}](-10.0 + 273.15)\text{ K} - 333.7\text{ kJ/kg}\} - (0.075\text{ kg})[2.1\text{ kJ/(kg}\cdot\text{K)} - 4.186\text{ kJ/(kg}\cdot\text{K)}](273.15\text{ K})}{[4.186\text{ kJ/(kg}\cdot\text{K)}](0.500\text{ kg} + 0.075\text{ kg})} = 305.6\text{ K} = \boxed{32°\text{C}}$$

39. Strategy Heat flows from the water to the ice, melting some of it. Find the mass of ice required to lower the temperature of the water to 0.0°C. Use Eqs. (14-4) and (14-9).

Solution Find the required mass of ice.

$0 = Q_{\text{ice}} + Q_{\text{w}} = m_{\text{ice}} L_{\text{f}} + m_{\text{w}} c_{\text{w}} \Delta T_{\text{w}}$, so

$$m_{\text{ice}} = -\frac{m_{\text{w}} c_{\text{w}} \Delta T_{\text{w}}}{L_{\text{f}}} = -\frac{(5.00\times10^2 \text{ mL})(1.00 \text{ g/mL})[4.186 \text{ J/(g}\cdot\text{K)}](-25.0 \text{ K})}{333.7 \text{ J/g}} = \boxed{157 \text{ g}}.$$

41. Strategy The sum of the heat flows is zero. The tea is basically water. The mass of the tea is found by multiplying the density of water by the volume of the tea. Do not neglect the temperature change of the glass. Use Eqs. (14-4) and (14-9).

Solution Find the mass of the ice required to cool the tea to 10.0°C. Let ΔT be the temperature change of the tea and the glass.

$$0 = Q_{\text{t}} + Q_{\text{ice}} + Q_{\text{g}}$$

$$0 = \rho_{\text{w}} V_{\text{t}} c_{\text{w}} \Delta T + m_{\text{ice}} L_{\text{f}} + m_{\text{ice}} c_{\text{ice}} \Delta T_1 + m_{\text{ice}} c_{\text{w}} \Delta T_2 + m_{\text{g}} c_{\text{g}} \Delta T$$

$$0 = (\rho_{\text{w}} V_{\text{t}} c_{\text{w}} + m_{\text{g}} c_{\text{g}}) \Delta T + m_{\text{ice}} (L_{\text{f}} + c_{\text{ice}} \Delta T_1 + c_{\text{w}} \Delta T_2)$$

$$m_{\text{ice}} = \frac{-(\rho_{\text{w}} V_{\text{t}} c_{\text{w}} + m_{\text{g}} c_{\text{g}}) \Delta T}{L_{\text{f}} + c_{\text{ice}} \Delta T_1 + c_{\text{w}} \Delta T_2}$$

$$= \frac{-\{(1.00\times10^3 \text{ kg/m}^3)(2.00\times10^{-4} \text{ m}^3)[4.186 \text{ kJ/(kg}\cdot\text{K)}] + (0.35 \text{ kg})[0.837 \text{ kJ/(kg}\cdot\text{K)}]\}(-85.0 \text{ K})}{333.7 \text{ kJ/kg} + [2.1 \text{ kJ/(kg}\cdot\text{K)}](10.0 \text{ K}) + [4.186 \text{ kJ/(kg}\cdot\text{K)}](10.0 \text{ K})}$$

$$= \boxed{242 \text{ g}}$$

The percentage change from the answer for Problem 40 is $\frac{242 \text{ g} - 179 \text{ g}}{179 \text{ g}} \times 100\% = \boxed{35\%}$.

45. Strategy Heat flows from the aluminum into the ice. Use Eqs. (14-4) and (14-9).

Solution Find the mass of aluminum required to melt 10.0 g of ice.

$Q_{\text{Al}} + Q_{\text{ice}} = m_{\text{Al}} c_{\text{Al}} \Delta T_{\text{Al}} + m_{\text{ice}} L_{\text{f}} = 0$, so

$$m_{\text{Al}} = -\frac{m_{\text{ice}} L_{\text{f}}}{c_{\text{Al}} \Delta T_{\text{Al}}} = -\frac{(10.0 \text{ g})(333.7 \text{ J/g})}{[0.900 \text{ J/(g}\cdot\text{K)}](0.0°\text{C} - 80.0°\text{C})} = \boxed{46.3 \text{ g}}.$$

49. Strategy The rate of heat loss is $\Delta Q/\Delta t = L_{\text{v}} \Delta m/\Delta t$ since $Q = mL_{\text{v}}$ for evaporation and where Δm represents the mass of water evaporated.

Solution Compute the rate of heat lost by the dog through panting.

$$\frac{\Delta Q}{\Delta t} = (670 \text{ min}^{-1})(0.010 \text{ g})(2256 \text{ J/g})\left(\frac{1 \text{ min}}{60 \text{ s}}\right) = \boxed{250 \text{ W}}$$

53. Strategy Use Eq. (14-12).

Solution Compute the thermal resistance for each material.

(a) $R = \frac{d}{\kappa A} = \frac{2.0\times10^{-2} \text{ m}}{[0.17 \text{ W/(m}\cdot\text{K)}](1.0 \text{ m}^2)} = \boxed{0.12 \text{ K/W}}$

(b) $R = \frac{2.0\times10^{-2} \text{ m}}{[80.2 \text{ W/(m}\cdot\text{K)}](1.0 \text{ m}^2)} = \boxed{2.5\times10^{-4} \text{ K/W}}$

(c) $R = \dfrac{2.0\times10^{-2}\text{ m}}{[401\text{ W/(m}\cdot\text{K)}](1.0\text{ m}^2)} = \boxed{5.0\times10^{-5}\text{ K/W}}$

55. Strategy From the given information, $\Delta T = \mathcal{P}_1 R_1 = \mathcal{P}_2 R_2 = \mathcal{P}(R_1 + R_2)$.

Solution Find the rate of heat flow per unit area.

$\dfrac{R_1}{R_2} = \dfrac{\mathcal{P}_2}{\mathcal{P}_1}$ and $\mathcal{P} = \dfrac{\mathcal{P}_2 R_2}{R_1 + R_2}$. Thus, $\mathcal{P} = \dfrac{\mathcal{P}_2}{\frac{R_1}{R_2}+1} = \dfrac{\mathcal{P}_2}{\frac{\mathcal{P}_2}{\mathcal{P}_1}+1} = \dfrac{20.0\text{ W/m}^2}{\frac{20.0}{10.0}+1} = \boxed{6.67\text{ W/m}^2}$.

57. Strategy Use Fourier's law of heat conduction, Eq. (14-10).

Solution

(a) $\mathcal{P} = \kappa A\dfrac{\Delta T}{d} = [401\text{ W/(m}\cdot\text{K)}](1.0\times10^{-6}\text{ m}^2)\dfrac{104°\text{C} - 24°\text{C}}{0.10\text{ m}} = \boxed{0.32\text{ W}}$

(b) $\dfrac{\Delta T}{d} = \dfrac{104°\text{C} - 24°\text{C}}{0.10\text{ m}} = \boxed{800\text{ K/m}}$

(c) The effective length has doubled.

$\mathcal{P} = [401\text{ W/(m}\cdot\text{K)}](1.0\times10^{-6}\text{ m}^2)\dfrac{104°\text{C} - 24°\text{C}}{0.20\text{ m}} = \boxed{0.16\text{ W}}$

(d) The effective area has doubled.

$\mathcal{P} = [401\text{ W/(m}\cdot\text{K)}](2.0\times10^{-6}\text{ m}^2)\dfrac{104°\text{C} - 24°\text{C}}{0.10\text{ m}} = \boxed{0.64\text{ W}}$

(e) Since the bars are identical, the temperature at the junction will be midway between the temperatures of the baths.

$\dfrac{104°\text{C} + 24°\text{C}}{2} = \boxed{64°\text{C}}$

59. Strategy Use Fourier's law of heat conduction, Eq. (14-10).

Solution Find the lowest temperature the dog can withstand without increasing its heat output.

$\mathcal{P} = \kappa A\dfrac{\Delta T}{d} = \kappa A\dfrac{T_\text{i} - T_\text{o}}{d}$, so $T_\text{o} = T_\text{i} - \dfrac{d\mathcal{P}}{\kappa A} = 38°\text{C} - \dfrac{(0.050\text{ m})(51\text{ W})}{[0.026\text{ W/(m}\cdot\text{K)}](1.31\text{ m}^2)} = \boxed{-37°\text{C}}$.

61. Strategy Use the latent heat of vaporization for water to determine the rate of heat flow. Then, use Fourier's law of heat conduction to find the temperature at the base of the pan.

Solution The temperature of the water and the top side of the bottom of the pan is $T = 100°\text{C}$. Determine the rate of heat flow.

$0.730\dfrac{Q}{\Delta t} = \dfrac{mL_\text{v}}{\Delta t}$, so $\dfrac{Q}{\Delta t} = \dfrac{mL_\text{v}}{0.730\Delta t}$

Find the temperature at the base of the pan, T_b.

$\dfrac{Q}{\Delta t} = \kappa A\dfrac{\Delta T}{d} = \kappa A\dfrac{T_\text{b} - T}{d} = \dfrac{mL_\text{v}}{0.730\Delta t}$, so

$T_\text{b} = \dfrac{dmL_\text{v}}{0.730\kappa A\Delta t} + T = \dfrac{(0.00300\text{ m})(10.0\text{ g})(2256\text{ J/g})}{0.730[237\text{ W/(m}\cdot\text{K)}](325\times10^{-4}\text{ m}^2)(1.00\text{ s})} + 100.0°\text{C} = \boxed{112.0°\text{C}}$.

63. Strategy Use Eq. (14-14).

Solution Find the rate of convective heat loss.

$\mathcal{P} = hA\Delta T = [22\ \text{W}/(\text{m}^2\cdot{}^\circ\text{C})](1.4\ \text{m}^2)(0.85)(29.0^\circ\text{C} - 35.0^\circ\text{C}) = -160\ \text{W}$

The marathon runner loses heat at a rate of $\boxed{160\ \text{W}}$ due to convection.

65. Strategy Use Eq. (14-14).

Solution Find the rate of convective heat loss per unit area.

$$\frac{\mathcal{P}}{A} = h\Delta T = [84.8\ \text{W}/(\text{m}^2\cdot{}^\circ\text{C})](5.0^\circ\text{C}) = \boxed{420\ \text{W}/\text{m}^2}$$

69. Strategy Use Stefan's law of radiation, Eq. (14-17). Intensity is power per unit area.

Solution Let $r_1 = 0.40$ m be the radius of the sphere and $r_2 = 2.0$ m be the distance from the center of the sphere where the intensity is measured. Compute the emissivity.

$\mathcal{P} = e\sigma AT^4 = e\sigma(4\pi r_1^2)T^4 = 4\pi r_2^2 I$, so

$$e = \frac{r_2^2 I}{\sigma r_1^2 T^4} = \frac{(2.0\ \text{m})^2(102\ \text{W}/\text{m}^2)}{[5.670\times10^{-8}\ \text{W}/(\text{m}^2\cdot\text{K}^4)](0.40\ \text{m})^2(250\ \text{K} + 273.15\ \text{K})^4} = \boxed{0.60}.$$

71. Strategy Use Stefan's law of radiation, Eq. (14-17).

Solution Compute the power radiated by the bulb.

$\mathcal{P} = e\sigma AT^4 = 0.32[5.670\times10^{-8}\ \text{W}/(\text{m}^2\cdot\text{K}^4)](1.00\times10^{-4}\ \text{m}^2)(3.00\times10^3\ \text{K})^4 = \boxed{150\ \text{W}}$

73. Strategy Approximate the pots as cubes of similar volume. Use Eq. (14-19).

Solution Find the net rate of radiative heat loss from the two pots.

$s^3 = V$, so $s = V^{1/3}$ and $6s^2 = A = 6V^{2/3}$.

Coffeepot:

$$\begin{aligned}\mathcal{P}_{\text{net}} &= e\sigma A(T^4 - T_s^4)\\ &= 0.12[5.670\times10^{-8}\ \text{W}/(\text{m}^2\cdot\text{K}^4)][6(1.00\ \text{L})^{2/3}(10^{-3}\ \text{m}^3/\text{L})^{2/3}][(98\ \text{K} + 273.15\ \text{K})^4 - (25\ \text{K} + 273.15\ \text{K})^4]\\ &= \boxed{4.5\ \text{W}}\end{aligned}$$

Teapot:

$$\begin{aligned}\mathcal{P}_{\text{net}} &= 0.65[5.670\times10^{-8}\ \text{W}/(\text{m}^2\cdot\text{K}^4)][6(1.00\ \text{L})^{2/3}(10^{-3}\ \text{m}^3/\text{L})^{2/3}][(98\ \text{K} + 273.15\ \text{K})^4 - (25\ \text{K} + 273.15\ \text{K})^4]\\ &= \boxed{24\ \text{W}}\end{aligned}$$

77. (a) Strategy The power absorbed by the leaf must equal that radiated away. Power is equal to intensity times area. Use Stefan's law of radiation, Eq. (14-17).

Solution Absorbed:

$$\begin{aligned}I_{\text{top}}A + e\sigma AT_s^4 &= 0.700(9.00\times10^2\ \text{W}/\text{m}^2)(5.00\times10^{-3}\ \text{m}^2) +\\ &\quad (1)[5.670\times10^{-8}\ \text{W}/(\text{m}^2\cdot\text{K}^4)](5.00\times10^{-3}\ \text{m}^2)(273.15\ \text{K} + 25.0\ \text{K})^4\\ &= 5.39\ \text{W}\end{aligned}$$

Find the temperature of the leaf. The area is now $2(5.00\times10^{-3}\text{ m}^2)=10.0\times10^{-3}\text{ m}^2$ (both sides of the leaf).

$\mathcal{P}=e\sigma AT^4$, so

$$T=\left(\frac{\mathcal{P}}{e\sigma A}\right)^{1/4}=\left[\frac{5.39\text{ W}}{(1)[5.670\times10^{-8}\text{ W/(m}^2\cdot\text{K}^4)](10.0\times10^{-3}\text{ m}^2)}\right]^{1/4}=312\text{ K}-273\text{ K}=\boxed{39°\text{C}}.$$

(b) Strategy Since the bottom of the leaf absorbs and emits at the same rate, it can be ignored.

Solution Find the power per unit area that must be lost by other methods.

$$\frac{\mathcal{P}_{\text{abs,Sun}}}{A}=\frac{\mathcal{P}_{\text{rad}}}{A}+\frac{\mathcal{P}_{\text{other}}}{A}=e\sigma T^4+\frac{\mathcal{P}_{\text{other}}}{A},\text{ so}$$

$$\frac{\mathcal{P}_{\text{other}}}{A}=\frac{\mathcal{P}_{\text{abs,Sun}}}{A}-e\sigma T^4=0.700(9.00\times10^2\text{ W/m}^2)-(1)[5.670\times10^{-8}\text{ W/(m}^2\cdot\text{K}^4)](273.15\text{ K}+25.0\text{ K})^4$$
$$=182\text{ W/m}^2$$

Thus, the power per unit area that must be lost by other methods is $\boxed{182\text{ W/m}^2}$.

81. Strategy The temperature of the ice must be raised to 0°C. Next, is must be melted. Then, the resulting liquid water must be raised to 100°C. Finally, the water must be vaporized. Use Eqs. (14-4) and (14-9).

Solution Find the heat energy required to convert the ice to steam.

$$Q=mc_{\text{ice}}\Delta T_{\text{ice}}+mL_{\text{f}}+mc_{\text{water}}\Delta T_{\text{water}}+mL_{\text{v}}=m(c_{\text{ice}}\Delta T_{\text{ice}}+L_{\text{f}}+c_{\text{water}}\Delta T_{\text{water}}+L_{\text{v}})$$
$$=5(0.0220\text{ kg})\{[2.1\text{ kJ/(kg}\cdot\text{K)}](50.0\text{ K})+333.7\text{ kJ/kg}+[4.186\text{ kJ/(kg}\cdot\text{K)}](100.0\text{ K})+2256\text{ kJ/kg}\}$$
$$=\boxed{342\text{ kJ}}$$

85. Strategy Heat flows from the tetrachloromethane to the water. Use Eq. (14-4).

Solution Find the specific heat of CCl_4, c_{t}.

$$0=Q_{\text{w}}+Q_{\text{t}}=m_{\text{w}}c_{\text{w}}\Delta T_{\text{w}}+m_{\text{t}}c_{\text{t}}\Delta T_{\text{t}},\text{ so}$$

$$c_{\text{t}}=-\frac{m_{\text{w}}c_{\text{w}}\Delta T_{\text{w}}}{m_{\text{t}}\Delta T_{\text{t}}}=-\frac{(2.00\text{ kg})[4.186\text{ kJ/(kg}\cdot\text{K)}](18.54-18.00)\text{ K}}{(2.50\times10^{-1}\text{ kg})(18.54-40.00)\text{ K}}=\boxed{0.84\text{ kJ/(kg}\cdot\text{K)}}.$$

89. Strategy The potential energy of the spring is equal to $\frac{1}{2}kx^2$. Use Eq. (14-4).

Solution Find the temperature change of the water.

$$Q=mc\Delta T=\Delta U=\frac{1}{2}kx^2,\text{ so }\Delta T=\frac{kx^2}{2mc}=\frac{(8.4\times10^3\text{ N/m})(0.10\text{ m})^2}{2(1.0\text{ kg})[4186\text{ J/(kg}\cdot\text{K)}]}=\boxed{0.010°\text{C}}.$$

93. (a) Strategy Use Eqs. (14-4) and (14-9).

Solution Find the heat given up by the steam.

$$Q=-mc_{\text{w}}\Delta T+mL_{\text{v}}=(4.0\text{ g})\{-[4.186\text{ J/(g}\cdot\text{K)}](45.0-100.0)\text{ K}+2256\text{ J/g}\}=\boxed{9.9\text{ kJ}}$$

(b) Strategy Use Eq. (14-4).

Solution Compute the mass of the tissue.

$$m=\frac{Q}{c\Delta T}=\frac{9945\text{ J}}{[3.5\text{ J/(g}\cdot\text{K)}](45.0-37.0)\text{ K}}=\boxed{360\text{ g}}$$

97. Strategy The heat loss is given by $\mathcal{P}\Delta t$ and it is equal to $Q = mL_v$.

Solution Find the mass of water required to replenish the fluid loss.

$$m = \frac{\mathcal{P}\Delta t}{L_v} = \frac{(650\text{ W})(30.0\text{ min})(60\text{ s/min})}{2430\text{ J/g}} = \boxed{480\text{ g}}$$

101. Strategy Use Eq. (14-19).

Solution

(a) Compute the net rate of heat loss through radiation.

$$\mathcal{P}_{net} = e\sigma A(T^4 - T_s^4)$$
$$= 0.97[5.670\times10^{-8}\text{ W/(m}^2\cdot\text{K}^4)](2.20\text{ m}^2)[(273.15\text{ K}+37.0\text{ K})^4 - (273.15\text{ K}+23.0\text{ K})^4] = \boxed{190\text{ W}}$$

(b) Find the skin temperature such that the net heat loss due to radiation is equal to the basal metabolic rate.

$e\sigma A(T^4 - T_s^4) = \text{BMR}$, so

$$T = \left(\frac{\text{BMR}}{e\sigma A} + T_s^4\right)^{1/4} = \left[\frac{(2167\text{ kcal/day})(4186\text{ J/kcal})\left(\frac{1\text{ day}}{86{,}400\text{ s}}\right)}{0.97[5.670\times10^{-8}\text{ W/(m}^2\cdot\text{K}^4)](2.20\text{ m}^2)} + (273.15\text{ K}+23.0\text{ K})^4\right]^{1/4} - 273.15\text{ K}$$
$$= \boxed{31°\text{C}}$$

(c) Wearing clothing slows heat loss by radiation because air layers trapped between clothing layers act as insulation and thus reduce the net radiative heat loss. ($T^4 - T_s^4$ is reduced.)

105. Strategy Set the rates of heat conduction equal. Use Fourier's law of heat conduction, Eq. (14-10).

Solution Compute the ratio of the depths.

$$\mathcal{P} = \kappa_b A\frac{\Delta T}{d_b} = \kappa_s A\frac{\Delta T}{d_s},\text{ so }\frac{d_b}{d_s} = \frac{\kappa_b}{\kappa_s} = \frac{3.1}{2.4} = 1.3,\text{ or }\boxed{d_b = 1.3d_s}.$$

107. Strategy Heat flows from the copper block to the water and iron pot. Use Eq. (14-4).

Solution Find the final temperature of the system.

$$0 = Q_w + Q_{Cu} + Q_{Fe}$$
$$Q_w = -Q_{Cu} - Q_{Fe}$$
$$m_w c_w(T_f - T_i) = -m_{Cu}c_{Cu}(T_f - T_{Cu}) - m_{Fe}c_{Fe}(T_f - T_i)$$
$$T_f(m_w c_w + m_{Cu}c_{Cu} + m_{Fe}c_{Fe}) = m_{Cu}c_{Cu}T_{Cu} + (m_{Fe}c_{Fe} + m_w c_w)T_i$$

Solve for T_f.

$$T_f = \frac{m_{Cu}c_{Cu}T_{Cu} + (m_{Fe}c_{Fe} + m_w c_w)T_i}{m_w c_w + m_{Cu}c_{Cu} + m_{Fe}c_{Fe}}$$
$$= \frac{(2.0\text{ kg})[385\text{ J/(kg}\cdot\text{K)}](100.0°\text{C}) + \{(2.0\text{ kg})[440\text{ J/(kg}\cdot\text{K)}] + (1.0\text{ kg})[4186\text{ J/(kg}\cdot\text{K)}]\}(25.0°\text{C})}{(1.0\text{ kg})[4186\text{ J/(kg}\cdot\text{K)}] + (2.0\text{ kg})[385\text{ J/(kg}\cdot\text{K)}] + (2.0\text{ kg})[440\text{ J/(kg}\cdot\text{K)}]} = \boxed{35°\text{C}}$$

109. Strategy The shaking will heat the water, although very slowly. The work done by the scientist is converted to heat in the water. The rate at which energy is supplied by the shaking is equal to the gravitational potential energy lost by the water during each fall times the frequency of the shaking. Use Eq. (14-4).

Solution Find the time it will take to heat the water.

$$\frac{mc\Delta T}{\Delta t} = mghf,\text{ so }\Delta t = \frac{c\Delta T}{ghf} = \frac{[4186\text{ J/(kg}\cdot\text{K)}](87°\text{C} - 12°\text{C})}{(9.80\text{ m/s}^2)(0.333\text{ m})(30\text{ min}^{-1})(1440\text{ min/day})} = \boxed{2\text{ days}}.$$

Chapter 15

THERMODYNAMICS

Problems

1. **Strategy** The work done by Ming is equal to the magnitude of the force of friction $f = \mu N$ times the total "rubbing" distance. Use the first law of thermodynamics.

   Solution Find the change in internal energy.
   $\Delta U = Q + W = 0 + \mu N d = 0.45(5.0\text{ N})[8(0.16\text{ m})] = \boxed{2.9\text{ J}}$

3. **Strategy** Use the first law of thermodynamics. $\Delta U > 0$ and $W > 0$.

   Solution Find the heat flow.
   $Q = \Delta U - W = 400\text{ J} - 500\text{ J} = -100\text{ J}$

   $\boxed{\text{100 J of heat flows out of the system.}}$

5. **Strategy** No work is done during the constant volume process, but work is done during the constant pressure process. Use Eq. (15-3).

   Solution Compute the total work done by the gas.
   $W = P_i\Delta V = (2.000\text{ atm})(1.013\times10^5\text{ Pa/atm})(2.000\text{ L} - 1.000\text{ L})(10^{-3}\text{ m}^3/\text{L}) = \boxed{202.6\text{ J}}$

9. **(a) Strategy** For A–C (constant temperature), $W = nRT\ln V_i/V_f$, and for C–D (constant pressure), $W = -P_i\Delta V$. Use the ideal gas law to find T.

   Solution

$$W_{\text{total}} = nRT\ln\frac{V_i}{V_f} - P\Delta V = nR\left(\frac{P_AV_A}{nR}\right)\ln\frac{V_A}{V_C} - P_i\Delta V$$
$$= \left[(2.000\text{ atm})(4.000\text{ L})\ln\frac{4.000\text{ L}}{8.000\text{ L}} - (1.000\text{ atm})(16.000\text{ L} - 8.000\text{ L})\right](1.013\times10^5\text{ Pa/atm})(10^{-3}\text{ m}^3/\text{L})$$
$$= \boxed{-1372\text{ J}}$$

   (b) Strategy For constant temperature, $\Delta U = 0$. For constant pressure,
$$\Delta U = Q + W = nC_P\Delta T - P_i\Delta V = \frac{5}{2}nR\left(\frac{P_i\Delta V}{nR}\right) - P_i\Delta V = \frac{3}{2}P_i\Delta V.$$

   Solution
$$\Delta U = \frac{3}{2}(1.000\text{ atm})(16.000\text{ L} - 8.000\text{ L})(1.013\times10^5\text{ Pa/atm})(10^{-3}\text{ m}^3/\text{L}) = \boxed{1216\text{ J}}.$$

   The total heat flow is $Q = \Delta U - W = 1216\text{ J} + 1372\text{ J} = \boxed{2588\text{ J}}$.

13. Strategy Use Eq. (15-12).

Solution Find the efficiency of the generator.

$$e=\frac{W_{\text{net}}}{Q_{\text{in}}}=\frac{1.17\text{ kW}\cdot\text{h}}{6.71\times10^6\text{ J}}\left(\frac{3.600\times10^6\text{ J}}{1\text{ kW}\cdot\text{h}}\right)=\boxed{0.628}$$

15. (a) Strategy Use Eq. (15-12).

Solution Find the heat absorbed by the engine.

$$Q_{\text{H}}=\frac{W_{\text{net}}}{e}=\frac{1.00\times10^3\text{ J}}{0.333}=\boxed{3.00\text{ kJ}}$$

(b) Strategy The net work done by an engine during one cycle is equal to the net heat flow into the engine during the cycle.

Solution Find the heat exhausted by the engine.

$W_{\text{net}}=Q_{\text{H}}-Q_{\text{C}}$, so $Q_{\text{C}}=Q_{\text{H}}-W_{\text{net}}=3.00\text{ kJ}-1.00\text{ kJ}=\boxed{2.00\text{ kJ}}$.

17. Strategy The net rate of work done by the engine is $\Delta K/\Delta t$.

Solution The net rate of work done is

$$\frac{W_{\text{net}}}{\Delta t}=\frac{\Delta K}{\Delta t}=\frac{\frac{1}{2}mv^2}{\Delta t}=\frac{mv^2}{2\Delta t}.$$

According to the definition of efficiency, the rate of heat flow in at the high temperature is

$$\frac{Q_{\text{in}}}{\Delta t}=\frac{1}{e}\frac{W_{\text{net}}}{\Delta t}=\frac{mv^2}{2e\Delta t}=\frac{(1800\text{ kg})(27\text{ m/s})^2}{2(0.27)(9.5\text{ s})}=\boxed{2.6\times10^5\text{ W}}.$$

Referring to the result of Example 15.5, the rate of heat flow out at the low temperature is

$$\frac{Q_{\text{out}}}{\Delta t}=\frac{W_{\text{net}}}{\Delta t}\left(\frac{1}{e}-1\right)=\frac{(1800\text{ kg})(27\text{ m/s})^2}{2(9.5\text{ s})}\left(\frac{1}{0.27}-1\right)=\boxed{1.9\times10^5\text{ W}}.$$

21. Strategy The work done by the engine is equal to the increase in gravitational potential energy of the crate plus the increase in kinetic energy. Use Eq. (15-12).

Solution Find the required heat input.

$$Q_{\text{in}}=\frac{W_{\text{net}}}{e}=\frac{mgh+\frac{1}{2}mv^2}{e}=\frac{m}{e}\left(gh+\frac{v^2}{2}\right)=\frac{5.00\text{ kg}}{0.300}\left[(9.80\text{ m/s}^2)(10.0\text{ m})+\frac{(4.00\text{ m/s})^2}{2}\right]=\boxed{1770\text{ J}}$$

25. (a) Strategy Use Eq. (15-12). The net work done by an engine during one cycle is equal to the net heat flow into the engine during the cycle.

Solution Compute the efficiency of the engine.

$$e=\frac{W_{\text{net}}}{Q_{\text{in}}}=\frac{Q_{\text{in}}-Q_{\text{out}}}{Q_{\text{in}}}=1-\frac{Q_{\text{out}}}{Q_{\text{in}}}=1-\frac{82}{125}=\boxed{0.34}$$

(b) Strategy Use Eq. (15-17).

Solution Compute the efficiency of an ideal engine.

$$e_{\text{r}}=1-\frac{T_{\text{C}}}{T_{\text{H}}}=1-\frac{293\text{ K}}{815\text{ K}}=\boxed{0.640}$$

27. Strategy The maximum efficiency is that of a reversible heat engine. Use Eq. (15-17).

Solution Calculate the maximum possible efficiency.

$$e_r = 1 - \frac{T_C}{T_H} = 1 - \frac{273.15\text{ K} + 4.0\text{ K}}{273.15\text{ K} + 18.0\text{ K}} = \boxed{0.0481}$$

29. Strategy The maximum rate at which the river can carry away heat is $Q_C/\Delta t = mc\Delta T/\Delta t$. Use energy conservation and the definition of efficiency of an engine.

Solution Find the maximum possible power the plant can produce.

$$\frac{W_{net}}{\Delta t} = \frac{Q_H - Q_C}{\Delta t} = \frac{W_{net}}{e\Delta t} - \frac{mc\Delta T}{\Delta t}, \text{ so}$$

$$\frac{W_{net}}{\Delta t} = \left(\frac{1}{e} - 1\right)^{-1} \frac{m}{\Delta t} c\Delta T = \left(\frac{1}{0.300} - 1\right)^{-1} (5.0\times10^6\text{ kg/s})[4186\text{ J/(kg}\cdot\text{K)}](0.50\text{ K}) = \boxed{4.5\text{ GW}}.$$

31. Strategy Assume constant rates and reversibility. Use Eq. (15-17) and conservation of energy. $P = W_{net}/\Delta t$.

Solution Compute the efficiency.

$$e_r = 1 - \frac{T_C}{T_H} = 1 - \frac{273.15\text{ K} + 2.0\text{ K}}{273.15\text{ K} + 40.0\text{ K}} = 0.1213$$

Find the power used.

$$Q_C = Q_H - W_{net} = \frac{W_{net}}{e} - W_{net} = W_{net}\left(\frac{1}{e} - 1\right), \text{ so } \frac{Q_C}{\Delta t} = \frac{W_{net}}{\Delta t}\left(\frac{1}{e} - 1\right), \text{ and } P = \frac{Q_C/\Delta t}{\frac{1}{e} - 1} = \frac{0.10\times10^3\text{ W}}{\frac{1}{0.1213} - 1} = \boxed{14\text{ W}}.$$

33. Strategy Use Eq. (15-17) and $W_{net} = Q_H - Q_C$.

Solution Find the waste heat exhausted.

Coal:

$$e_r = 1 - \frac{T_C}{T_H} = 1 - \frac{273.15\text{ K} + 27\text{ K}}{273.15\text{ K} + 727\text{ K}} = 0.700 \text{ and}$$

$$Q_C = Q_H - W_{net} = \frac{W_{net}}{e} - W_{net} = W_{net}\left(\frac{1}{e} - 1\right) = (1.00\text{ MJ})\left(\frac{1}{0.700} - 1\right) = 0.43\text{ MJ}.$$

Nuclear:

$$e_r = 1 - \frac{T_C}{T_H} = 1 - \frac{273.15\text{ K} + 27\text{ K}}{273.15\text{ K} + 527\text{ K}} = 0.625 \text{ and } Q_C = (1.00\text{ MJ})\left(\frac{1}{0.625} - 1\right) = 0.60\text{ MJ}.$$

$\boxed{\text{The coal-fired plant and the nuclear plant exhaust 0.43 MJ and 0.60 MJ of heat, respectively.}}$

35. (a) Strategy Use Eq. (15-17).

Solution Find the temperature of the hot reservoir.

$$e_r = 1 - \frac{T_C}{T_H}, \text{ so } T_H = \frac{T_C}{1 - e_r} = \frac{310.0\text{ K}}{1 - 0.300} = \boxed{443\text{ K}}.$$

(b) Strategy Use $W_{net} = Q_H - Q_C$ and the definition of efficiency of an engine.

Solution Find the amount of heat exhausted to the cold reservoir.

$$W_{net} = Q_H - Q_C = \frac{W_{net}}{e_r} - Q_C, \text{ so } Q_C = W_{net}\left(\frac{1}{e_r} - 1\right) = (0.100\times10^3\text{ J})\left(\frac{1}{0.300} - 1\right) = \boxed{233\text{ J}}.$$

37. Strategy For maximum efficiency, assume reversibility. Use Eq. (15-17).

Solution Find the percent decrease in theoretical maximum efficiency.

$$\frac{\Delta e_r}{e_r}\times 100\% = \frac{1-\frac{T_{Cf}}{T_H}-\left(1-\frac{T_{Ci}}{T_H}\right)}{1-\frac{T_{Ci}}{T_H}}\times 100\% = \frac{-\frac{T_{Cf}}{T_H}+\frac{T_{Ci}}{T_H}}{1-\frac{T_{Ci}}{T_H}}\times 100\% = \frac{T_{Ci}-T_{Cf}}{T_H-T_{Ci}}\times 100\% = \frac{27°C-47°C}{500.0°C-27°C}\times 100\%$$
$$= -4.2\%$$

The theoretical maximum efficiency would decrease by $\boxed{4.2\%}$.

41. Strategy The maximum possible efficiency occurs if the engine is reversible. Use Eq. (15-17).

Solution Find the maximum possible efficiency.

$$e_r = 1-\frac{T_C}{T_H} = 1-\frac{273.15\text{ K}+10.0\text{ K}}{273.15\text{ K}+15.0\text{ K}} = \boxed{0.0174}$$

45. Strategy The refrigerator is reversible and the electric motor does 250 J of work per second. Use Eqs. (15-12) and (15-17), and $W_{net} = Q_H - Q_C$.

Solution Calculate e_r.

$$e_r = 1-\frac{T_C}{T_H} = 1-\frac{256\text{ K}}{320\text{ K}} = 0.20$$

Calculate Q_C/Q_H for the refrigerator.

$$\frac{Q_C}{Q_H} = \frac{Q_H - W_{net}}{Q_H} = 1-\frac{W_{net}}{Q_H} = 1-e_r = 1-0.20 = 0.80$$

Since $Q_C/Q_H < 1$, more heat is added to the room than is removed. Assuming constant rates, the net rate is

$$\frac{Q_H}{\Delta t}-\frac{Q_C}{\Delta t} = \frac{Q_H}{\Delta t}\left(1-\frac{Q_C}{Q_H}\right) = \frac{W_{net}/e_r}{\Delta t}(1-0.80) = \frac{0.20}{e_r}\times\frac{W_{net}}{\Delta t} = \frac{W_{net}}{\Delta t} = \boxed{+250\text{ W}}.$$

Or, more simply, since the room is insulated, and the electric motor is ideal (250 W of electrical energy is converted to 250 W of work), by conservation of energy, the net rate of heat added to the room must be +250 W.

47. Strategy and Solution The mass is the same for each case. For equal masses, water has more entropy than ice, and warmer water has more entropy than cooler water, so the order is $\boxed{\text{(b), (a), (c), (d)}}$.

49. Strategy The potential energy of the ball is converted into kinetic energy as it falls. When it hits the ground, the kinetic energy is converted into heat. Use Eq. (15-20) and $U = mgh = Q$.

Solution Find the increase in the entropy of the universe.

$$\Delta S = \frac{Q}{T} = \frac{mgh}{T} = \frac{(0.15\text{ kg})(9.80\text{ m/s}^2)(24\text{ m})}{19\text{ K}+273.15\text{ K}} = \boxed{0.12\text{ J/K}}$$

51. Strategy The temperature is constant and the heat entering the system is $Q = mL_V$. Use Eq. (15-20).

Solution Find the change in the entropy of the water.

$$\Delta S = \frac{Q}{T} = \frac{mL_V}{T} = \frac{(1.00\text{ kg})(2256\text{ kJ/kg})}{273.15\text{ K}+100.0\text{ K}} = \boxed{+6.05\text{ kJ/K}}$$

Gas is more disordered than liquid, so the entropy increases.

53. Strategy Use Eq. (15-20).

Solution

(a) Compute the change in entropy of the block.

$$\Delta S = \frac{Q}{T_C} = \frac{1.0\text{ J}}{273.15\text{ K} + 20.0\text{ K}} = \boxed{3.4\times10^{-3}\text{ J/K}}$$

(b) Compute the change in entropy for the water.

$$\Delta S = \frac{Q}{T_H} = \frac{-1.0\text{ J}}{273.15\text{ K} + 80.0\text{ K}} = \boxed{-2.8\times10^{-3}\text{ J/K}}$$

(c) Calculate the change in entropy of the universe.

$$\Delta S = -\frac{Q}{T_H} + \frac{Q}{T_C} = 2.8\times10^{-3}\text{ J/K} + 3.4\times10^{-3}\text{ J/K} = \boxed{6.2\times10^{-3}\text{ J/K}}$$

57. Strategy Refer to Example 15.19. Use Eq. (15-20).

Solution Find the entropy change for each stage.

$1 \to 2$: isothermal expansion; $T = 1000.0$ K and $\Delta S = \frac{Q}{T_H} = \frac{25\text{ J}}{1000.0\text{ K}} = 0.025\text{ J/K} > 0.$

$2 \to 3$: adiabatic expansion; T decreases while S is constant.

$3 \to 4$: isothermal compression; $T = 300.0$ K and $\Delta S < 0$.

$4 \to 1$: adiabatic compression; T increases while S is constant.

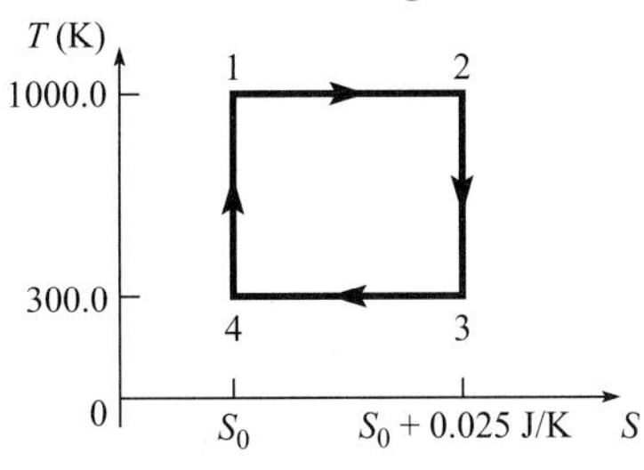

61. Strategy Identify the most likely macrostate. Then count the number of microstates and use Eq. (15-22).

Solution A sum of 7 is the most likely macrostate, and can occur 6 different ways (microstates: 1 and 6, 6 and 1, 2 and 5, 5 and 2, 3 and 4, 4 and 3). So, the maximum entropy is $S = k\ln\Omega = k\ln 6 \approx \boxed{1.79k}$.

63. Strategy A macrostate for the system of two identical dice is specified by the sum of the two numbers that come up on the dice. Use the rules of probability.

Solution

(a) The microstate is specified by the product of each die's possible outcomes, N_1 and N_2. So, the microstate is specified by $\boxed{N_1N_2}$.

(b) There are 6 numbers on each die, so $N_1N_2 = 6\cdot 6 = \boxed{36}$.

(c) The possible sums are 2 through 12, so there are $\boxed{11}$ macrostates.

(d) A sum of 7 is the most likely outcome, so $\boxed{7}$ is the most probable macrostate.

(e) The first die can have any value. Then, the second die has a 1-in-6 chance of summing to 7 with the first, so $1\cdot\frac{1}{6}=\boxed{\frac{1}{6}}$ is the probability.

(f) The probability to roll a one is $\frac{1}{6}$ for each die. The probability of rolling two ones is the product of the individual probabilities, or $\frac{1}{6}\cdot\frac{1}{6}=\boxed{\frac{1}{36}}$.

65. Strategy Use Eq. (15-22) and the definition of latent heat.

Solution $\Delta S = k\ln\Omega_f - k\ln\Omega_i = k\ln\Omega_f/\Omega_i$. so

$$\frac{\Delta S}{k}=\ln\frac{\Omega_f}{\Omega_i}$$

$$e^{\Delta S/k}=\frac{\Omega_f}{\Omega_i}$$

Since T is constant and $Q=mL_f$, $\Delta S=\frac{Q}{T}=\frac{mL_f}{T}$, and

$$\frac{\Delta S}{k}=\frac{mL_f}{kT}=\frac{(1.0\text{ g})(333.7\text{ J/g})}{(1.38\times10^{-23}\text{ J/K})(273.15\text{ K}+0.0\text{ K})}=8.9\times10^{22}.$$

So, the number of microstates has increased by a factor of $\boxed{e^{8.9\times10^{22}}}$.

69. (a) Strategy The work done per cycle is equal to the area contained within the curve.

Solution Compute the work done per cycle.

$$W=\frac{1}{2}(5.00\text{ atm}-1.00\text{ atm})(1.013\times10^5\text{ Pa/atm})(2.00\text{ m}^3-0.500\text{ m}^3)=\boxed{304\text{ kJ}}$$

(b) Strategy Use the ideal gas law to compute the temperatures at the upper left and lower right points on the curve.

Solution Compute the temperatures.

$PV=nRT$, so $\frac{P_2V_2}{P_1V_1}=\frac{T_2}{T_1}$.

$$T_{ul}=\frac{P_2V_2T_1}{P_1V_1}=\frac{(5.00)(0.500)(470.0\text{ K})}{(1.00)(0.500)}=2350\text{ K}$$

$$T_{lr}=\frac{P_2V_2T_1}{P_1V_1}=\frac{(1.00)(2.00)(470.0\text{ K})}{(1.00)(0.500)}=1880\text{ K}$$

The maximum temperature is $\boxed{2350\text{ K}}$.

(c) Strategy Use the ideal gas law at the lower-left corner of the diagram.

Solution Find the number of moles of gas used in the engine.

$$n=\frac{PV}{RT}=\frac{(1.00\text{ atm})(1.013\times10^5\text{ Pa/atm})(0.500\text{ m}^3)}{[8.314\text{ J/(mol}\cdot\text{K)}](470.0\text{ K})}=\boxed{13.0\text{ mol}}$$

73. Strategy At constant pressure, the work done by the expanding gas is $W = -P\Delta V$. $\Delta U = Q + W$, so $\Delta U = Q - P\Delta V$. According to the ideal gas law, $V_f = T_f V_i / T_i$ (n, R, and P are constant).

Solution Find the increase in internal energy of the gas.

$$\Delta U = Q - P(V_f - V_i) = Q - PV_i\left(\frac{T_f}{T_i} - 1\right) = 25 \text{ kJ} - (2.0\times10^5 \text{ Pa})(0.10 \text{ m}^3)\left(\frac{273.15 \text{ K} + 27 \text{ K}}{273.15 \text{ K} - 73 \text{ K}} - 1\right) = \boxed{15 \text{ kJ}}$$

77. Strategy Count the number of microstates for each macrostate and use Eq. (15-22).

Solution For the macrostate of half heads and half tails, there are $8!/(4!4!) = 70$ microstates. For the macrostate of all heads, there is only one microstate, since there is only one way this can happen. So, the change in entropy is $\Delta S = k \ln \Omega_f - k \ln \Omega_i = (1.381\times10^{-23} \text{ J/K})(\ln 1 - \ln 70) = \boxed{-5.867\times10^{-23} \text{ J/K}}$.

The entropy of the universe can increase in this process because the entropy required to flip the coins is greater than the decrease in entropy of the coins.

81. Strategy The energy of the mixed state, U, must equal the sum of the original (unmixed) states, U_1 and U_2. The energy for a monatomic ideal gas is related to the temperature by $U = \frac{3}{2}NkT$.

Solution Find the final temperature T of the mixture.

$$U = U_1 + U_2$$

$$\frac{3}{2}(N_1 + N_2)kT = \frac{3}{2}N_1kT_1 + \frac{3}{2}N_2kT_2$$

$$\frac{3}{2}(n_1 + n_2)RT = \frac{3}{2}n_1RT_1 + \frac{3}{2}n_2RT_2$$

$$T = \frac{n_1T_1 + n_2T_2}{n_1 + n_2} = \frac{(4.0 \text{ mol})(20.0°\text{C}) + (3.0 \text{ mol})(30.0°\text{C})}{4.0 \text{ mol} + 3.0 \text{ mol}} = \boxed{24°\text{C}}$$

85. (a) Strategy The temperature is constant and the heat entering the system is $Q = mL_f$. Use Eq. (15-20).

Solution Find the change in the entropy.

$$\Delta S = \frac{Q}{T} = \frac{mL_f}{T} = \frac{(1.00 \text{ mol})(15.9994 \text{ g/mol} + 2\times1.00794 \text{ g/mol})(333.7 \text{ J/g})}{273.15 \text{ K} + 0.0 \text{ K}} = \boxed{22.0 \text{ J/K}}$$

(b) Strategy Use Eq. (15-20).

Solution Compute the entropy change of the universe when the ice melts.

$$\Delta S = Q\left(\frac{1}{T_C} - \frac{1}{T_H}\right)$$

$$= (1.00 \text{ mol})(15.9994 \text{ g/mol} + 2\times1.00794 \text{ g/mol})(333.7 \text{ J/g})\left(\frac{1}{273.15 \text{ K} + 0.0 \text{ K}} - \frac{1}{273.15 \text{ K} + 10.0 \text{ K}}\right)$$

$$= \boxed{0.777 \text{ J/K}}$$

89. Strategy Assume the freezer is reversible. Use Eqs. (14-4), (14-9), (15-16), and (15-19).

Solution Find the minimum work input required to freeze the ice.

$$K_r = \frac{\text{heat removed}}{\text{net work input}} = \frac{Q_C}{W_{net}} = \frac{1}{T_H/T_C - 1}, \text{ so}$$

$$W_{net} = Q_C\left(\frac{T_H}{T_C} - 1\right) = (mL_f - mc_w\Delta T_w - mc_{ice}\Delta T_{ice})\left(\frac{T_H}{T_C} - 1\right)$$

$$= (1.20 \text{ kg})\{333{,}700 \text{ J/kg} - [4186 \text{ J/(kg}\cdot\text{K)}](-20.0 \text{ K}) - [2100 \text{ J/(kg}\cdot\text{K)}](-20.0 \text{ K})\}\left(\frac{273.15 \text{ K} + 20.0 \text{ K}}{273.15 \text{ K} - 20.0 \text{ K}} - 1\right)$$

$$= \boxed{87.1 \text{ kJ}}$$

93. (a) Strategy Use Eq. (15-17).

Solution Compute the efficiency.

$$e_r = 1 - \frac{T_C}{T_H} = 1 - \frac{273.15 \text{ K} + 22 \text{ K} - 15 \text{ K}}{273.15 \text{ K} + 22 \text{ K}} = \boxed{0.051}$$

(b) Strategy The power supplied to the town is equal to the efficiency times the rate at which heat is supplied by the lake. Use Eq. (14-4) and the relationship between mass, density, and volume.

Solution Find the volume of water used each second.

$$P = e\frac{\Delta Q}{\Delta t} = \frac{emc\Delta T}{\Delta t} = \frac{e\rho Vc\Delta T}{\Delta t}, \text{ so}$$

$$V = \frac{P\Delta t}{e\rho c\Delta T} = \frac{(1.0\times10^8 \text{ W})(1.0 \text{ s})}{0.051(1.00\times10^3 \text{ kg/m}^3)[4186 \text{ J/(kg}\cdot\text{K)}](15 \text{ K})} = \boxed{31 \text{ m}^3}.$$

(c) Strategy The incident power of the Sun must be greater than the power required to run the engine. Power is equal to intensity times area.

Solution Compare the power supplied to the power required by the town.

$$P_{Sun} = IA = (200 \text{ W/m}^2)(8.0\times10^7 \text{ m}^2) = 1.6\times10^{10} \text{ W}$$

$$P_{engine} = \frac{P_{town}}{e} = \frac{1.0\times10^8 \text{ W}}{0.051} = 2.0\times10^9 \text{ W}$$

$1.6\times10^{10} \text{ W} > 2.0\times10^9 \text{ W}$, so $P_{sun} > P_{engine}$, and $\boxed{\text{yes}}$, the lake can supply enough heat to meet the town's needs.

REVIEW AND SYNTHESIS: CHAPTERS 13–15

Review Exercises

1. Strategy Assume no heat is lost to the air. The potential energy of the water is converted into heating of the water. The internal energy of the water increases by an amount equal to the initial potential energy.

Solution Find the change in internal energy.

$$\Delta U = mgh = (1.00\text{ m}^3)(1.00\times10^3\text{ kg/m}^3)(9.80\text{ m/s}^2)(11.0\text{ m}) = \boxed{108\text{ kJ}}$$

5. Strategy Use the ideal gas law.

Solution Find the number of moles of air when the balloon is at 40.0°C.

$$PV = nRT,\text{ so } n_2 = \frac{PV}{RT_2} = \frac{(1.00\text{ atm})(1.013\times10^5\text{ Pa/atm})(12.0\text{ m}^3)}{[8.314\text{ J/(mol}\cdot\text{K)}](273.15\text{ K}+40.0\text{ K})} = \boxed{467\text{ mol}}.$$

9. (a) Strategy Use the ideal gas law.

Solution Find the pressure at point A, which is the same as the pressure at point D.

$$P_D = P_A = \frac{nRT_A}{V_A} = \frac{(2.00\text{ mol})[8.314\text{ J/(mol}\cdot\text{K)}](800.0\text{ K})}{1.50\text{ m}^3} = \boxed{8.87\text{ kPa}}$$

Find the temperature at point D.

$$T_D = \frac{P_DV_D}{nR} = \frac{(8.87\times10^3\text{ Pa})(2.25\text{ m}^3)}{(2.00\text{ mol})[8.314\text{ J/(mol}\cdot\text{K)}]} = \boxed{1200\text{ K}}$$

(b) Strategy The net work done on the gas is equal to the area inside the graph.

Solution Find the net work done on the gas as it is taken though four cycles.

$$W = 4(8.87\text{ kPa} - 1.30\text{ kPa})(2.25\text{ m}^3 - 1.50\text{ m}^3) = \boxed{23\text{ kJ}}$$

(c) Strategy The internal energy of an ideal monatomic gas is given by $U = \frac{3}{2}nRT$.

Solution Compute the internal energy of the gas at point A.

$$U = \frac{3}{2}nRT = \frac{3}{2}(2.00\text{ mol})[8.314\text{ J/(mol}\cdot\text{K)}](800.0\text{ K}) = \boxed{20.0\text{ kJ}}$$

(d) Strategy and Solution The total change in internal energy in four complete cycles is $\boxed{0}$, since the change in temperature is zero.

13. Strategy Use Fourier's law of heat conduction for the copper rod and the heat of fusion for the ice.

Solution Find the rate of melting.

$$\frac{Q}{\Delta t} = \frac{mL_\text{f}}{\Delta t} = \kappa A\frac{\Delta T}{d},\text{ so } \frac{m}{\Delta t} = \frac{\kappa A\Delta T}{L_\text{f}d} = \frac{[401\text{ W/(m}\cdot\text{K)}]\pi(0.0100\text{ m})^2(100.0\text{ K})}{(333.7\text{ J/g})(1.00\text{ m})}\left(\frac{3600\text{ s}}{1\text{ h}}\right) = \boxed{136\text{ g/h}}.$$

17. (a) Strategy The maximum possible efficiency is that of a reversible engine.

Solution Compute the maximum possible efficiency.

$$e_r = 1 - \frac{T_C}{T_H} = 1 - \frac{323\text{ K}}{535\text{ K}} = 0.396 \text{ or } \boxed{39.6\%}$$

(b) Strategy Use energy conservation and the definition of efficiency of an engine.

Solution Find the rate at which heat must be removed by means of a cooling tower.

$$\frac{W_{net}}{\Delta t} = \frac{Q_H}{\Delta t} - \frac{Q_C}{\Delta t} = \frac{W_{net}/e}{\Delta t} - \frac{Q_C}{\Delta t}, \text{ so}$$

$$\frac{Q_C}{\Delta t} = \frac{W_{net}}{e\Delta t} - \frac{W_{net}}{\Delta t} = \frac{W_{net}}{\Delta t}\left(\frac{1}{0.500e_r} - 1\right) = (1.23\times10^8\text{ W})\left(\frac{1}{0.500(0.396)} - 1\right) = \boxed{4.98\times10^8\text{ W}}.$$

MCAT Review

1. Strategy and Solution According to the second law of thermodynamics, heat never flows spontaneously from a colder body to a hotter body, therefore, heat will not flow from bar A to bar B. The correct answer is $\boxed{\text{C}}$.

2. Strategy Assume that the specific heat capacity of seawater is approximately the same at 0°C and 5°C.

Solution Find the approximate temperature T.

$0 = Q_0 + Q_5 = mc(T - 0°\text{C}) + mc(T - 5°\text{C})$, so $2T = 5°\text{C}$ or $T = 2.50°\text{C}$.

The correct answer is $\boxed{\text{B}}$.

3. Strategy Use the latent heat of fusion for water.

Solution The heat gained by the ice when melting is $Q = mL_f = (0.0180\text{ kg})(333.7\text{ kJ/kg}) = 6.01\text{ kJ}$.

The correct answer is $\boxed{\text{C}}$.

4. Strategy and Solution Since $e = 1 - Q_C/Q_H = 1 - T_C/T_H$, decreasing the exhaust temperature will increase the steam engine's efficiency. The correct answer is $\boxed{\text{B}}$.

5. Strategy and Solution Since refrigerators remove heat by transferring it to a liquid that vaporizes, refrigerators are primarily dependent upon the heat of vaporization of the refrigerant liquid. The correct answer is $\boxed{\text{A}}$.

6. Strategy and Solution Steam is generally at a higher temperature than water and the specific heat of steam is lower than that of water, so water would be more effective than steam for changing steam to water. Circulating water brings more mass of water in contact with the condenser than stationary water, so it can carry away heat at a faster rate, therefore, it would be more effective for changing steam to water. The correct answer is $\boxed{\text{D}}$.

7. Strategy and Solution Since it is not possible to convert all of the input heat into output work, the amount of useful work that can be generated from a source of heat can only be less than the amount of heat. The correct answer is $\boxed{\text{A}}$.

8. Strategy and Solution The internal energy of the steam is converted into mechanical energy as it expands and moves the piston of the steam engine to the right, therefore, the correct answer is $\boxed{\text{C}}$.

9. Strategy and Solution The refrigerant must be able to vaporize (boil) at temperatures lower than the freezing point of water so that it can carry away heat (as a gas) from the contents of the refrigerator (which contain water) to cool and possibly freeze the contents. The correct answer is $\boxed{\text{B}}$.

10. Strategy The heat transferred to the water by the heaters was $Q_\text{w} = m_\text{w} c_\text{w} \Delta T_\text{w}$. The heat required for the oil is $Q_\text{o} = m_\text{o} c_\text{o} \Delta T_\text{o}$.

Solution Form a proportion and use the temperature changes of the oil and water and the specific heat and the specific gravity of the oil to obtain a ratio of heat required for the oil to that transferred to the water.

$$\frac{Q_\text{o}}{Q_\text{w}} = \frac{m_\text{o} c_\text{o} \Delta T_\text{o}}{m_\text{w} c_\text{w} \Delta T_\text{w}} = \frac{(0.7 m_\text{w})(0.60 c_\text{w})(60-20)}{m_\text{w} c_\text{w}(100-20)} = 0.21$$

So, 21% of the amount of heat transferred to the water is required to heat the oil to 60°C. Assuming the heaters work at the same rate for both the water and the oil, the time required to raise the temperature of the oil from 20°C to 60°C is $0.21(15\text{ h}) = 3.2\text{ h}$. The correct answer is $\boxed{\text{A}}$.

11. Strategy and Solution The high pressure would increase the pressure on the plug, making it more difficult to lift. The pressure difference between the air in the tank and the air outside of the tank would increase the fluid velocity when the tank is drained, thus, decreasing the time required to drain the tank. The time required to heat the oil would be the least likely affected, since the oil is fairly incompressible. The correct answer is $\boxed{\text{A}}$.

Chapter 16

ELECTRIC FORCES AND FIELDS

Problems

1. **Strategy** There are 10 protons in each water molecule. Multiply the elementary charge by Avogadro's number and the number of protons per molecule.

   Solution Find the total positive charge.

   $$10(1.0\text{ mol})(6.022\times10^{23}\text{ mol}^{-1})(1.602\times10^{-19}\text{ C}) = \boxed{9.6\times10^{5}\text{ C}}$$

3. **(a) Strategy and Solution** Since electrons have negative charge, and since the balloon acquired a negative net charge, electrons were $\boxed{\text{added}}$ to the balloon.

   (b) Strategy Divide the net charge by the charge of an electron.

   Solution Compute the number of electrons transferred.

   $$\frac{-0.60\times10^{-9}\text{ C}}{-1.602\times10^{-19}\text{ C}} = \boxed{3.7\times10^{9}}$$

5. **Strategy and Solution**

   (a) When the rod is brought near sphere A, negative charge flows from sphere B to sphere A. The spheres are then moved apart and the rod is removed, so A is left with a net $\boxed{\text{negative charge}}$.

   (b) Sphere B has $\boxed{\text{an equal magnitude of positive charge}}$, since the two spheres were initially uncharged.

7. **Strategy** Each time a pair of spheres makes contact, their net charge is shared equally, with the exception of the time when C is grounded.

   Solution After A and B make contact and are separated, each sphere has a charge of $Q/2$. After B and C make contact and are separated, each sphere has zero charge because sphere C was grounded. After A and C make contact and are separated, each sphere has a charge of $(Q/2+0)/2 = Q/4$. The charges on spheres A and C are $\boxed{Q/4}$. The charge on sphere B is $\boxed{0}$.

9. **Strategy** Use Coulomb's law, Eq. (16-2).

   Solution Find the distance between the charges.

   $$F = \frac{k|q_1||q_2|}{r^2},\text{ so } r = \sqrt{\frac{k|q_1||q_2|}{F}} = \sqrt{\frac{(8.988\times10^{9}\text{ N}\cdot\text{m}^2/\text{C}^2)(1\text{ C})^2}{10\text{ N}}} = \boxed{30\text{ km}}.$$

13. Strategy Divide the magnitude of the Coulomb force by the magnitude of the gravitational force.

Solution Compute the ratio.

$$\frac{F_q}{F_g} = \frac{\frac{kq^2}{r^2}}{\frac{Gm_pm_e}{r^2}} = \frac{kq^2}{Gm_pm_e} = \frac{(8.988\times10^9\ \text{N}\cdot\text{m}^2/\text{C}^2)(1.602\times10^{-19}\ \text{C})^2}{(6.674\times10^{-11}\ \text{N}\cdot\text{m}^2/\text{kg}^2)(1.673\times10^{-27}\ \text{kg})(9.109\times10^{-31}\ \text{kg})} = \boxed{2.268\times10^{39}}$$

15. Strategy Use Coulomb's law, Eq. (16-2). The force on the 1.0-μC charge due to the −0.60-μC charge is to the left and that due to the 0.80-μC charge is along the line between the charges and away from the 0.80-μC charge.

Solution Calculate the components of the force.

$$F_x = -\frac{(8.988\times10^9\ \text{N}\cdot\text{m}^2/\text{C}^2)(0.60\times10^{-6}\ \text{C})(1.0\times10^{-6}\ \text{C})}{\left(\sqrt{(0.100\ \text{m})^2-(0.080\ \text{m})^2}\right)^2}$$

$$+\frac{(8.988\times10^9\ \text{N}\cdot\text{m}^2/\text{C}^2)(0.80\times10^{-6}\ \text{C})(1.0\times10^{-6}\ \text{C})}{(0.100\ \text{m})^2}\left(\frac{\sqrt{(0.100\ \text{m})^2-(0.080\ \text{m})^2}}{0.100\ \text{m}}\right) = -1.1\ \text{N}$$

$$F_y = -\frac{(8.988\times10^9\ \text{N}\cdot\text{m}^2/\text{C}^2)(0.80\times10^{-6}\ \text{C})(1.0\times10^{-6}\ \text{C})}{(0.100\ \text{m})^2}\left(\frac{0.080\ \text{m}}{0.100\ \text{m}}\right) = -0.58\ \text{N}$$

Calculate the magnitude of the force.

$$F = \sqrt{F_x^2+F_y^2} = \sqrt{(-1.067\ \text{m})^2+(-0.575\ \text{m})^2} = 1.2\ \text{N}$$

Calculate the direction.

$$\theta = \tan^{-1}\frac{F_y}{F_x} = \tan^{-1}\frac{-0.575}{-1.067} = 28°$$

So, $\vec{\mathbf{F}} = \boxed{1.2\ \text{N at } 28° \text{ below the negative } x\text{-axis}}$.

17. Strategy The force is attractive. Use Coulomb's law, Eq. (16-2).

Solution

(a) Find the electric force on the positive charge.

$$F = -\frac{k|q_1||q_2|}{r^2} = -\frac{(8.988\times10^9\ \text{N}\cdot\text{m}^2/\text{C}^2)(2.0\times10^{-9}\ \text{C})(3.0\times10^{-9}\ \text{C})}{(0.030\ \text{m})^2} = -6.0\times10^{-5}\ \text{N}$$

So, $\vec{\mathbf{F}} = \boxed{6.0\times10^{-5}\ \text{N toward the } -3.0\text{-nC charge}}$.

(b) The force is equal in magnitude and opposite in direction to that found in part (a).

So, $\vec{\mathbf{F}} = \boxed{6.0\times10^{-5}\ \text{N toward the 2.0-nC charge}}$.

21. Strategy The force is attractive. Use Coulomb's law, Eq. (16-2).

Solution Find the electric force on the potassium ion.

$$F = -\frac{k|q_1||q_2|}{r^2} = -\frac{ke^2}{r^2} = -\frac{(8.988\times10^9\ \text{N}\cdot\text{m}^2/\text{C}^2)(1.602\times10^{-19}\ \text{C})^2}{(9.0\times10^{-9}\ \text{m})^2} = -2.8\times10^{-12}\ \text{N}$$

So, $\vec{\mathbf{F}} = \boxed{2.8\times10^{-12}\ \text{N toward the Cl}^-\ \text{ion}}$.

25. Strategy The net force on the charge at the origin is the vector sum of the forces due to each of the other two charges. Use Coulomb's law, Eq. (16-2).

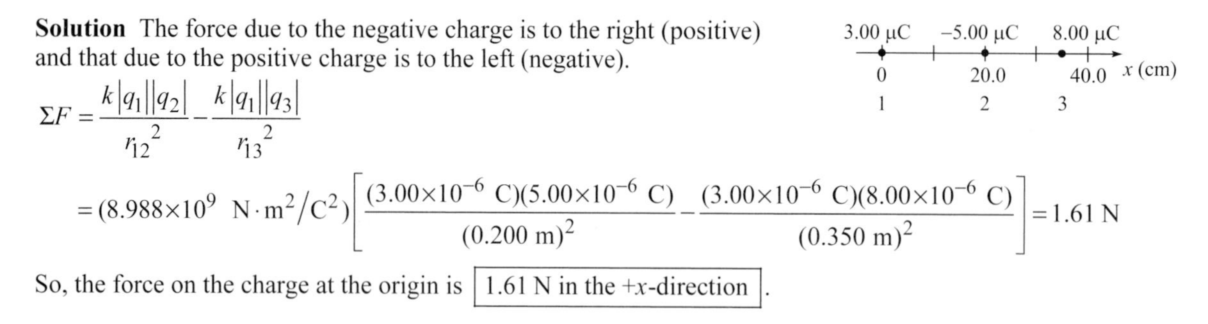

Solution The force due to the negative charge is to the right (positive) and that due to the positive charge is to the left (negative).

$$\Sigma F = \frac{k|q_1||q_2|}{r_{12}^2} - \frac{k|q_1||q_3|}{r_{13}^2}$$

$$= (8.988\times10^9\ \text{N}\cdot\text{m}^2/\text{C}^2)\left[\frac{(3.00\times10^{-6}\ \text{C})(5.00\times10^{-6}\ \text{C})}{(0.200\ \text{m})^2} - \frac{(3.00\times10^{-6}\ \text{C})(8.00\times10^{-6}\ \text{C})}{(0.350\ \text{m})^2}\right] = 1.61\ \text{N}$$

So, the force on the charge at the origin is $\boxed{1.61\ \text{N in the } +x\text{-direction}}$.

29. (a) Strategy and Solution Since positive charges move along the direction of electric field lines, the sodium ions flow $\boxed{\text{into the cell}}$.

(b) Strategy Use Eq. (16-4b).

Solution Compute the magnitude of the electric force on the sodium ion.

$$F = eE = (1.602\times10^{-19}\ \text{C})(1.0\times10^7\ \text{N/C}) = \boxed{1.6\times10^{-12}\ \text{N}}$$

31. Strategy Electric field lines begin on positive charges and end on negative charges. The magnitude of the negative charge is twice that of the positive charges (which have equal magnitude). The same number of field lines begins on each of the positive charges and all end on the negative charge. Field lines never cross. Use the principles of superposition and symmetry.

Solution The electric field lines for the system of three charges:

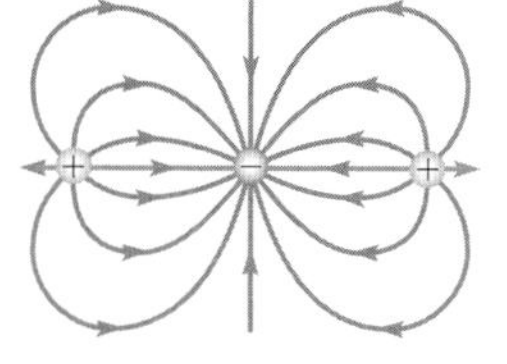

33. Strategy Use Newton's second law and Eq. (16-4b).

Solution $\vec{\mathbf{F}} = q\vec{\mathbf{E}} = e\vec{\mathbf{E}}$ for a proton. Find the acceleration.

$\vec{\mathbf{E}}\uparrow\ \oplus_\text{p}\ \uparrow\vec{\mathbf{a}}$

$$m\vec{\mathbf{a}} = e\vec{\mathbf{E}},\ \text{so}\ \vec{\mathbf{a}} = \frac{e\vec{\mathbf{E}}}{m} = \frac{(1.602\times10^{-19}\ \text{C})(33\times10^3\ \text{N/C up})}{1.673\times10^{-27}\ \text{kg}} = \boxed{3.2\times10^{12}\ \text{m/s}^2\ \text{up}}.$$

35. Strategy Use Eq. (16-5) and the principles of superposition and symmetry.

Solution The field is directed along the x-axis.

$$E = E_q + E_{2q} = \frac{kq}{d^2} - \frac{k(2q)}{(2d)^2} = \frac{kq}{d^2}\left(1 - \frac{1}{2}\right) = \frac{kq}{2d^2},\ \text{so}\ \vec{\mathbf{E}} = \boxed{\frac{kq}{2d^2}\ \text{in the } +x\text{-direction}}.$$

37. Strategy Use Eq. (16-5) and the principles of superposition and symmetry. If r is the distance to $(x, y) = (0.50 \text{ m}, 0.50 \text{ m})$ from each charge, then $\cos\theta = x/r$ and $\sin\theta = y/r$.

Solution Find the magnitude of the electric field.

$$E = \sqrt{E_x^2 + E_y^2} = \sqrt{\left[\frac{kq_1}{r^2}\left(\frac{x}{r}\right) - \frac{kq_2}{r^2}\left(\frac{x}{r}\right)\right]^2 + \left[\frac{kq_1}{r^2}\left(\frac{y}{r}\right) + \frac{kq_2}{r^2}\left(\frac{y}{r}\right)\right]^2}$$

$$= \frac{k}{r^3}\sqrt{[(q_1 - q_2)x]^2 + [(q_1 + q_2)y]^2}$$

y (m), 0.50, $\vec{\mathbf{E}}_2$, $\vec{\mathbf{E}}_1$, 20.0 nC, 10.0 nC, 0, 0.50, 1.00, x (m), 1, 2

$$= \frac{8.988\times10^9 \text{ N}\cdot\text{m}^2/\text{C}^2}{\left(\sqrt{(0.50 \text{ m})^2 + (0.50 \text{ m})^2}\right)^3}\sqrt{\begin{matrix}[(20.0\times10^{-9} \text{ C} - 10.0\times10^{-9} \text{ C})(0.50 \text{ m})]^2 \\ +[(20.0\times10^{-9} \text{ C} + 10.0\times10^{-9} \text{ C})(0.50 \text{ m})]^2\end{matrix}} = \boxed{400 \text{ N/C}}$$

41. Strategy Electric field lines begin on positive charges and end on negative charges. The same number of field lines begins on the plate and ends on the negative point charge. Field lines never cross. Use the principles of superposition and symmetry.

Solution The electric field lines for the system of the point charge and metal plate:

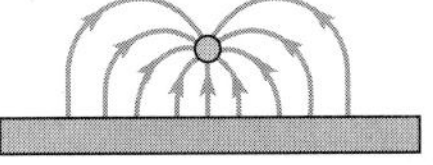

43. (a) Strategy Use Eq. (16-4b).

Solution Find the force on the electron.

$$\vec{\mathbf{F}} = -e\vec{\mathbf{E}} = -(1.602\times10^{-19} \text{ C})(500.0 \text{ N/C up}) = \boxed{8.010\times10^{-17} \text{ N down}}$$

(b) Strategy Use the work-kinetic energy theorem. The work done on the electron is equal to the force on the electron times the deflection.

Solution Find the increase in the kinetic energy of the electron.

$$\Delta K = W = Fd = (8.010\times10^{-17} \text{ N})(0.00300 \text{ m}) = \boxed{2.40\times10^{-19} \text{ J}}$$

45. (a) Strategy Compare the electrical and gravitational forces.

Solution The gravitational force is $mg = (0.00230 \text{ kg})(9.80 \text{ m/s}^2) = 2.25\times10^{-2} \text{ N}$. The electrical force is $qE = (10.0\times10^{-6} \text{ C})(6.50\times10^3 \text{ N/C}) = 6.50\times10^{-2} \text{ N}$.

The gravitational force is about 1/3 of the electrical force, so the gravitational force can't be neglected.

(b) Strategy Add the forces and find the total acceleration using Newton's second law. Then, use the formula for the range of a projectile.

v_i, 55.0°, $\Delta x = 1.78$ m

Solution Find the downward acceleration.

$$a = \frac{F_g + F_e}{m}$$

Find Δx.

$$\Delta x = R = \frac{v_i^2 \sin 2\theta}{a} = \frac{mv_i^2 \sin 2\theta}{F_g + F_e} = \frac{(0.00230 \text{ kg})(8.50 \text{ m/s})^2 \sin[2(55.0°)]}{2.25\times10^{-2} \text{ N} + 6.50\times10^{-2} \text{ N}} = \boxed{1.78 \text{ m}}$$

49. (a) Strategy and Solution Electrons have negative charge, so the field must be oriented $\boxed{\text{vertically downward}}$ for them to be deflected upward.

(b) Strategy Use Eq. (4-9), $\Delta x = v\Delta t$, and Newton's second law.

Solution Find the deflection d in terms of the time.

$$\Delta y = \frac{1}{2}a_y(\Delta t)^2 = d$$

Find the time Δt.

$$\Delta t = \frac{\Delta x}{v_\text{i}}$$

Find a_y.

$$a_y = \frac{2d}{(\Delta t)^2} = \frac{2dv_\text{i}^2}{(\Delta x)^2}$$

$\Sigma F_y = eE = ma_y$, so $E = ma_y/e$. Calculate E.

$$E = \frac{m}{e}\left[\frac{2dv_\text{i}^2}{(\Delta x)^2}\right] = \frac{2(9.109\times10^{-31}\text{ kg})(0.0020\text{ m})(8.4\times10^6\text{ m/s})^2}{(1.602\times10^{-19}\text{ C})(0.0250\text{ m})^2} = \boxed{2600\text{ N/C}}$$

(c) Strategy Use Eq. (4-9) and $\Delta x = v\Delta t$.

Solution Find the deflection of the electrons due to the gravitational force.

$$d = \frac{1}{2}g(\Delta t)^2 = \frac{1}{2}g\left(\frac{\Delta x}{v_\text{i}}\right)^2 = \frac{(9.80\text{ m/s}^2)(0.0250\text{ m})^2}{2(8.4\times10^6\text{ m/s})^2} = \boxed{4.3\times10^{-17}\text{ m}}$$

53. Strategy The charge on the inner surface is induced by the net charge contained within the shell. The charge on the outer surface is equal in magnitude and opposite in sign to the charge on the inner surface plus the net charge.

Solution

(a) The 6 μC of charge within the shell induces a $\boxed{-6\ \mu\text{C}}$ charge on the inner surface of the shell.

(b) The shell has a net charge of 6 μC, so the charge on the outer surface is $6\ \mu\text{C} + 6\ \mu\text{C} = \boxed{12\ \mu\text{C}}$.

57. Strategy Since the charge is located at the center of the cube, the electric flux through one side of the cube is one-sixth of the total flux. Use Gauss's law, Eq. (16-8).

Solution Find the flux through one side of the cube.

$$\Phi_\text{E} = 4\pi kq = \frac{4\pi(8.988\times10^9\text{ N}\cdot\text{m}^2/\text{C}^2)(0.890\times10^{-6}\text{ C})}{6} = \boxed{1.68\times10^4\text{ N}\cdot\text{m}^2/\text{C}}$$

61. Strategy Use the properties of electric fields and the rules for sketching field lines, and Gauss's law, Eq. (16-8).

Solution

(a) The electric field lines due to the (finite) sheet:

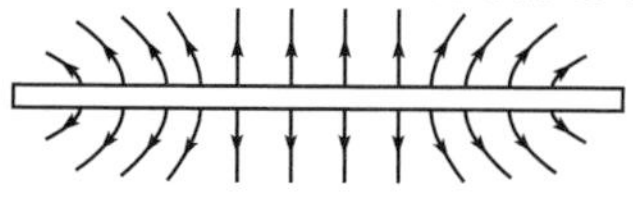

(b) The electric field lines due to an infinitely large sheet:

(c) The electric field lines in (b) are uniform, so

the field strength is independent of the distance from the sheet.

(d) The electric field lines in (a) are nearly uniform close to the sheet and far from the edges, so the answer is

yes.

(e) The Gaussian surface is a "pill box." It is a cylinder with its top and bottom circular surfaces parallel to the surface of the sheet, which bisects the cylinder. The electric field lines are approximately parallel to the side of the cylinder, so $\Phi_{\text{E side}} = E_{\perp}A_{\text{side}} = 0$, or $E_{\perp} = 0$.

$$\Phi_{\text{E net}} = E_{\text{top}}A_{\text{top}} + E_{\text{bottom}}A_{\text{bottom}} = \frac{q}{\epsilon_0}$$

$\vec{\mathbf{E}}_{\text{top}} = -\vec{\mathbf{E}}_{\text{bottom}}$, and the outward normal of A_{top} is opposite to that for A_{bottom} and the areas are equal. Find the magnitude.

$$EA + (-E)(-A) = 2EA = \frac{q}{\epsilon_0}, \text{ so } E = \frac{1}{2\epsilon_0}\left(\frac{q}{A}\right) = \frac{\sigma}{2\epsilon_0}.$$

65. Strategy Use the results of Problem 64.

Solution

(a) The electric field magnitude due to a solid sphere of radius R with a uniform charge q spread throughout is

$$\boxed{E(r \geq R) = \frac{kq}{r^2}}.$$

(b) At some point $r \leq R$, find the total charge enclosed within a spherical Gaussian surface.

$q = \frac{4}{3}\pi R^3 \rho$, where ρ is the uniform volume charge density.

Therefore, $q_{\text{enc}} = \frac{4}{3}\pi r^3 \rho = \frac{4}{3}\pi r^3 \times \frac{q}{\frac{4}{3}\pi R^3} = q\left(\frac{r}{R}\right)^3$. By Gauss's law, $\Phi_{\text{E}} = EA = 4\pi k q_{\text{enc}} = 4\pi kq\left(\frac{r}{R}\right)^3$.

The area is $A = 4\pi r^2$, so $E(4\pi r^2) = 4\pi kq\frac{r^3}{R^3}$ or $\boxed{E(r \leq R) = \frac{kq}{R^3}r}$. ($E$ varies linearly with r.)

(c) The maximum magnitude of the electric field is $E_{\text{max}} = E(R) = kq/R^2$. Sketch the graph.

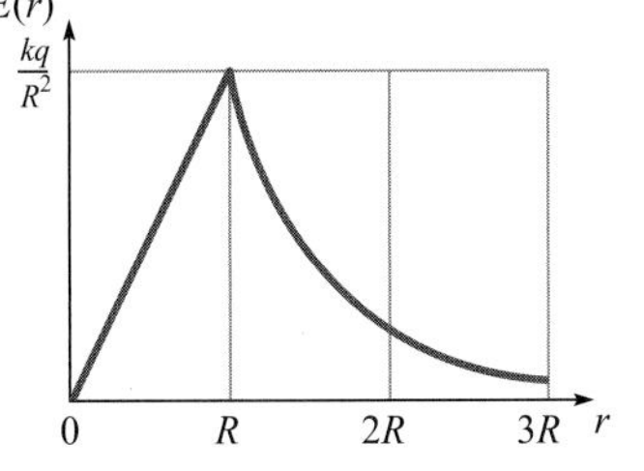

69. Strategy The net electric field at P is the vector sum of the electric fields at that location due to both of the charges. Let the left-hand charge by 1 and the right-hand charge be 2. Also, let $a = 0.0340\text{ m}$ and $b = 0.0140\text{ m}$. Use Eq. (16-5) and the principle of superposition.

Solution Find the components of the electric field.

$$E_x = E_{1x} + E_{2x} = 0 + \frac{k|q_2|}{a^2+b^2}\left(\frac{a}{\sqrt{a^2+b^2}}\right) = \frac{k|q_2|a}{(a^2+b^2)^{3/2}}$$

$$= \frac{(8.988\times10^9\text{ N}\cdot\text{m}^2/\text{C}^2)(47.0\times10^{-9}\text{ C})(0.0340\text{ m})}{[(0.0340\text{ m})^2+(0.0140\text{ m})^2]^{3/2}}$$

$$= \boxed{2.89\times10^5\text{ N/C}}$$

$$E_y = E_{1y} + E_{2y} = \frac{k|q_1|}{b^2} - \frac{k|q_2|}{a^2+b^2}\left(\frac{b}{\sqrt{a^2+b^2}}\right) = k\left[\frac{|q_1|}{b^2} - \frac{|q_2|b}{(a^2+b^2)^{3/2}}\right]$$

$$= (8.988\times10^9\text{ N}\cdot\text{m}^2/\text{C}^2)\left\{\frac{63.0\times10^{-9}\text{ C}}{(0.0140\text{ m})^2} - \frac{(47.0\times10^{-9}\text{ C})(0.0140\text{ m})}{[(0.0340\text{ m})^2+(0.0140\text{ m})^2]^{3/2}}\right\}$$

$$= \boxed{2.77\times10^6\text{ N/C}}$$

73. Strategy Use the properties of electric fields and the rules for sketching field lines.

Solution Since the semicircle is positively charged, the field lines point toward the center of curvature. Let the semicircle be oriented such that its ends are on the x-axis and its midpoint is on the negative y-axis. The x-components of $\vec{\mathbf{E}}$ all cancel due to symmetry, and the y-components all add and point in the positive y-direction. So, the electric field at the center points away from the midpoint of the semicircle.

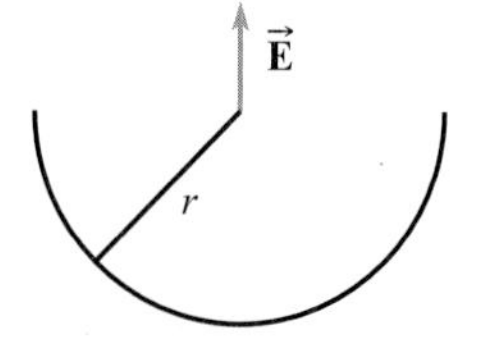

77. Strategy Use Coulomb's law, Eq. (16-2). Set the sum of the forces due to q_2 and q_3 equal to zero.

Solution Find the location of the third point charge.

$$0 = -F_{12} + F_{13} = -\frac{k|q_1||q_2|}{x_2^2} + \frac{k|q_1||q_3|}{x_3^2}, \text{ so } \frac{|q_2|}{x_2^2} = \frac{|q_3|}{x_3^2} \text{ and}$$

$$x_3 = \pm\sqrt{\frac{|q_3|}{|q_2|}}x_2 = \pm\sqrt{\frac{8.0\ \mu\text{C}}{3.0\ \mu\text{C}}}(-20.0\text{ cm}) = \pm33\text{ cm}.$$

-33 cm is extraneous (the force on $q_1 \neq 0$), so q_3 must be placed at $\boxed{x = 33\text{ cm}}$.

81. (a) Strategy Set the magnitude of the electric force equal to the magnitude of the gravitational force.

Solution Find the required magnitudes of the net charges on the Earth and the Sun.

$$\frac{k|q_S||q_E|}{r^2} = \frac{Gm_Sm_E}{r^2}, \text{ so } |q_S||q_E| = \frac{G}{k}m_Sm_E.$$

Let $\frac{|q_E|}{m_E} = \frac{|q_S|}{m_S}$. Then, $|q_E| = \frac{m_E}{m_S}|q_S|$. Find $|q_S|$.

$$|q_S| = \frac{G}{k}m_S m_E\left(\frac{1}{|q_E|}\right) = \frac{G}{k}m_S m_E\left(\frac{m_S}{m_E|q_S|}\right)$$

$$q_S^2 = \frac{Gm_S^2}{k}$$

$$|q_S| = m_S\sqrt{\frac{G}{k}} = (1.987\times10^{30}\text{ kg})\sqrt{\frac{6.674\times10^{-11}\text{ N}\cdot\text{m}^2/\text{kg}^2}{8.988\times10^9\text{ N}\cdot\text{m}^2/\text{C}^2}} = \boxed{1.712\times10^{20}\text{ C}}$$

Find $|q_E|$.

$$|q_E| = \frac{G}{k}m_S m_E\left(\frac{1}{|q_S|}\right) = \frac{G}{k}m_S m_E\left(\frac{m_E}{m_S|q_E|}\right)$$

$$q_E^2 = \frac{Gm_E^2}{k}$$

$$|q_E| = m_E\sqrt{\frac{G}{k}} = (5.974\times10^{24}\text{ kg})\sqrt{\frac{6.674\times10^{-11}\text{ N}\cdot\text{m}^2/\text{kg}^2}{8.988\times10^9\text{ N}\cdot\text{m}^2/\text{C}^2}} = \boxed{5.148\times10^{14}\text{ C}}$$

(b) Strategy and Solution If the magnitude of the charges of the proton and electron were not exactly equal, astronomical bodies would have net charges with the same sign, so the force between them would be repulsive. The force responsible for the Earth's orbit is attractive, so this charge imbalance could not possibly be the explanation for Earth's orbit. The answer is $\boxed{\text{no}}$.

85. Strategy Use Coulomb's law and the principle of superposition.

Solution Find q.

$$0 = E_x(1.0\text{ m}, 0) = \frac{kq_0}{d^2} + \frac{kq}{(d/2)^2} = q_0 + 4q,\text{ so}$$

$$q = -\frac{1}{4}q_0 = -\frac{1}{4}(6.0\text{ nC}) = \boxed{-1.5\text{ nC}}.$$

89. Strategy Use Coulomb's law, Eq. (16-2).

Solution

(a) Approximate the force.

$$F = \frac{k|q_1||q_2|}{r^2} = \frac{(8.988\times10^9\text{ N}\cdot\text{m}^2/\text{C}^2)(1.0\times10^{-6}\text{ C})(0.2\times10^{-6}\text{ C})}{(1.00\text{ m})^2} = \boxed{2\text{ mN}}$$

(b) Coulomb's law is only valid for point charges or when the sizes of the charge distributions are much smaller than their separation.

(c) Since the positive charges on each sphere will move toward the outer (opposite) sides of the spheres due to the repulsive force between them, the average distance separating charges will be larger than 12 cm, so the actual force would be $\boxed{\text{smaller}}$.

Chapter 17

ELECTRIC POTENTIAL

Problems

1. Strategy Use Eq. (17-1).

Solution Compute the electric potential energy.

$$U_E = k\frac{q_1 q_2}{r} = \frac{(8.988\times10^9\ \text{N}\cdot\text{m}^2/\text{C}^2)(5.0\times10^{-6}\ \text{C})(-2.0\times10^{-6}\ \text{C})}{5.0\ \text{m}} = \boxed{-18\ \text{mJ}}$$

5. Strategy The work done on the charges is equal to their potential energy. Let the upper charge by 1, the lower left-hand charge be 2, and the right-hand charge be 3. Also, let $a = 0.16$ m and $b = 0.12$ m. Use Eq. (17-2).

Solution Compute the work done on the charges.

$$W = U_E = k\left(\frac{q_1q_2}{b} + \frac{q_1q_3}{\sqrt{a^2+b^2}} + \frac{q_2q_3}{a}\right)$$

$$= (8.988\times10^9\ \text{N}\cdot\text{m}^2/\text{C}^2)\left[\frac{(5.5\times10^{-6}\ \text{C})(-6.5\times10^{-6}\ \text{C})}{0.12\ \text{m}} + \frac{(5.5\times10^{-6}\ \text{C})(2.5\times10^{-6}\ \text{C})}{\sqrt{(0.16\ \text{m})^2+(0.12\ \text{m})^2}}\right.$$

$$\left. + \frac{(-6.5\times10^{-6}\ \text{C})(2.5\times10^{-6}\ \text{C})}{0.16\ \text{m}}\right] = \boxed{-3.0\ \text{J}}$$

7. Strategy Use Eq. (17-2).

Solution Find the electric potential energy.

$$U_E = k\left(\frac{q_1q_2}{r_{12}} + \frac{q_1q_3}{r_{13}} + \frac{q_2q_3}{r_{23}}\right)$$

$$= (8.988\times10^9\ \text{N}\cdot\text{m}^2/\text{C}^2)\left[(4.0\times10^{-6}\ \text{C})\left(\frac{3.0\times10^{-6}\ \text{C}}{\sqrt{(4.0\ \text{m})^2+(3.0\ \text{m})^2}} + \frac{-1.0\times10^{-6}\ \text{C}}{3.0\ \text{m}}\right)\right.$$

$$\left. + \frac{(3.0\times10^{-6}\ \text{C})(-1.0\times10^{-6}\ \text{C})}{4.0\ \text{m}}\right] = \boxed{2.8\ \text{mJ}}$$

9. Strategy Use the principle of superposition and Eq. (17-9).

Solution Sum the electric fields at the center due to each charge.

$$\vec{\mathbf{E}} = \vec{\mathbf{E}}_a + \vec{\mathbf{E}}_b + \vec{\mathbf{E}}_c + \vec{\mathbf{E}}_d = \vec{\mathbf{E}}_a + \vec{\mathbf{E}}_b - \vec{\mathbf{E}}_a - \vec{\mathbf{E}}_b = \boxed{0}$$

Do the same for the potential at the center.

$$V = \Sigma\frac{kQ_i}{r_i} = \frac{4kQ}{r} = \frac{4(8.988\times10^9\ \text{N}\cdot\text{m}^2/\text{C}^2)(9.0\times10^{-6}\ \text{C})}{\frac{\sqrt{(0.020\ \text{m})^2+(0.020\ \text{m})^2}}{2}} = \boxed{2.3\times10^7\ \text{V}}$$

13. Strategy and Solution

(a) Since V is positive, q is $\boxed{\text{positive}}$.

(b) $V \propto \frac{1}{r}$, so since the potential is doubled, the distance is halved or $\boxed{10.0\text{ cm}}$.

17. Strategy Use Eq. (17-9).

Solution Find the electric potential at the third corner, B.

$$V = \Sigma\frac{kQ_i}{r_i} = \frac{k}{r}(Q_A + Q_B) = \frac{8.988\times10^9\text{ N}\cdot\text{m}^2/\text{C}^2}{1.0\text{ m}}(2.0\times10^{-9}\text{ C} - 1.0\times10^{-9}\text{ C}) = \boxed{9.0\text{ V}}$$

21. Strategy Rewrite each unit in terms of kg, m, s, and C.

Solution Show that $1\text{ N/C} = 1\text{ V/m}$.

$$1\text{ N/C} = \frac{\text{kg}\cdot\text{m/s}^2}{\text{C}}\text{ and }1\text{ V/m} = \frac{\text{J}}{\text{m}\cdot\text{C}} = \frac{\text{kg}\cdot\text{m}^2/\text{s}^2}{\text{m}\cdot\text{C}} = \frac{\text{kg}\cdot\text{m/s}^2}{\text{C}},\text{ therefore }1\text{ N/C} = 1\text{ V/m}.$$

25. Strategy Equipotential surfaces are perpendicular to electric field lines at all points. For equipotential surfaces drawn such that the potential difference between adjacent surfaces is constant, the surfaces are closer together where the field is stronger. The electric field always points in the direction of maximum potential decrease.

Solution Outside the sphere, $\vec{\mathbf{E}}$ is radially directed (toward the sphere), and $V \propto r^{-1}$. The equipotential surfaces are perpendicular to $\vec{\mathbf{E}}$ at any point, so they are $\boxed{\text{spheres}}$. Inside the sphere, $\vec{\mathbf{E}} = 0$ and V is constant.

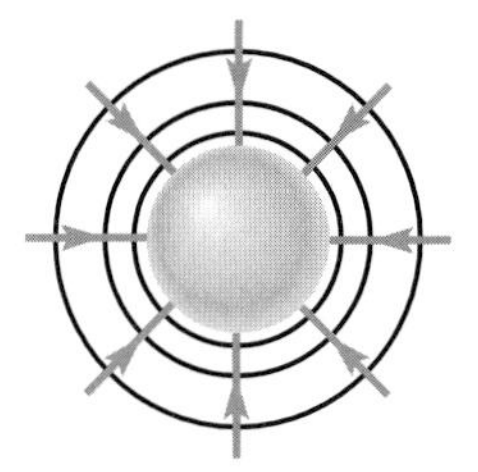

29. Strategy Use Eq. (16-4b).

Solution Find the magnitude of the charge on the drop in terms of the elementary charge e.

$$F = qE,\text{ so }q = \frac{F}{E} = \frac{Fe}{Ee} = \frac{9.6\times10^{-16}\text{ N}}{(3000\text{ V/m})(1.602\times10^{-19}\text{ C})}e = \boxed{2e}.$$

33. (a) Strategy The electric field always points in the direction of maximum potential decrease. Electrons, being negatively charged, move in the direction opposite the direction of the electric field; that is, in the direction of potential increase.

Solution Since the speed of the electron decreased, it must have traveled in the direction of the electric field, so it moved in the direction of potential decrease; that is, $\boxed{\text{to a lower potential}}$.

(b) Strategy The kinetic energy of the electron decreased, so its potential energy increased. Use conservation of energy and Eq. (17-7).

Solution Compute the potential difference the electron moved through.

$$\Delta V = \frac{\Delta U}{q} = \frac{-\Delta K}{-e} = \frac{m(v_f^2 - v_i^2)}{2e} = \frac{(9.109\times10^{-31}\text{ kg})[(2.50\times10^6\text{ m/s})^2 - (8.50\times10^6\text{ m/s})^2]}{2(1.602\times10^{-19}\text{ C})} = \boxed{-188\text{ V}}$$

35. Strategy and Solution Since positive charges move through decreases in potential, and since the potential and potential energy are greatest at A, the proton will spontaneously travel from point A to point E. So, $K_A = \boxed{0}$.

37. Strategy and Solution

(a) Electrons travel opposite the direction of the electric field, so $\vec{\mathbf{E}}$ is directed $\boxed{\text{upward}}$.

(b) For a uniform electric field, $\Sigma F_y = eE = \dfrac{e\Delta V}{d} = ma_y$, so $a_y = \dfrac{e\Delta V}{md}$. Thus, $\Delta t = \dfrac{v_y}{a_y} = \boxed{\dfrac{v_y md}{e\Delta V}}$.

(c) Since the electron gains kinetic energy, its potential energy $\boxed{\text{decreases}}$.

41. Strategy and Solution

(a) Since $\vec{\mathbf{E}}$ does not depend upon the separation of the plates $(E = \sigma/\epsilon_0)$, it $\boxed{\text{stays the same}}$.

(b) Since $\Delta V \propto d$, ΔV $\boxed{\text{increases}}$ if d increases.

45. Strategy Use the definition of capacitance, Eq. (17-14).

Solution Find the capacitance of the spheres.

$$Q = C\Delta V,\text{ so } C = \frac{Q}{\Delta V} = \frac{3.2\times10^{-14}\text{ C}}{0.0040\text{ V}} = \boxed{8.0\text{ pF}}.$$

49. (a) Strategy Use Eq. (16-6).

Solution The electric field between the plates is

$$E = \frac{Q}{\epsilon_0 A} = \frac{4.0\times10^{-11}\text{ C}}{[8.854\times10^{-12}\text{ C}^2/(\text{N}\cdot\text{m}^2)](0.062\text{ m})(0.022\text{ m})} = \boxed{3.3\times10^3\text{ V/m}}.$$

(b) Strategy Use the definition of the dielectric constant, Eq. (17-17).

Solution Find the electric field between the plates of the capacitor with the dielectric.

$$\kappa = \frac{E_0}{E},\text{ so } E = \frac{E_0}{\kappa} = \frac{3.3\times10^3\text{ V/m}}{5.5} = \boxed{6.0\times10^2\text{ V/m}}.$$

53. Strategy The spark flies between the spheres when the electric field between them exceeds the dielectric strength. The magnitude of the electric field is given by $\Delta V/d$, where d is the distance between the spheres.

Solution Find d.

$$E=\frac{\Delta V}{d},\text{ so } d=\frac{\Delta V}{E}=\frac{900\text{ V}}{3.0\times10^6\text{ V/m}}=\boxed{0.30\text{ mm}}.$$

55. Strategy Use Eq. (17-16).

Solution Compute the capacitance of the capacitor.

$$C=\kappa\frac{\epsilon_0 A}{d}=\frac{2.5[8.854\times10^{-12}\text{ C}^2/(\text{N}\cdot\text{m}^2)](0.30\text{ m})(0.40\text{ m})}{0.030\times10^{-3}\text{ m}}=\boxed{89\text{ nF}}$$

57. (a) Strategy Use Eq. (17-18c).

Solution Compute the capacitance.

$$U=\frac{Q^2}{2C},\text{ so } C=\frac{Q^2}{2U}=\frac{(8.0\times10^{-2}\text{ C})^2}{2(450\text{ J})}=\boxed{7.1\ \mu\text{F}}.$$

(b) Strategy Use Eq. (17-18a).

Solution Compute the potential difference.

$$U=\frac{1}{2}Q\Delta V,\text{ so } \Delta V=\frac{2U}{Q}=\frac{2(450\text{ J})}{8.0\times10^{-2}\text{ C}}=\boxed{1.1\times10^4\text{ V}}.$$

61. (a) Strategy Use Eq. (17-15).

Solution Find the capacitance for the thundercloud.

$$C=\frac{\epsilon_0 A}{d}=\frac{[8.854\times10^{-12}\text{ C}^2/(\text{N}\cdot\text{m}^2)](4500\text{ m})(2500\text{ m})}{550\text{ m}}=\boxed{0.18\ \mu\text{F}}$$

(b) Strategy Use Eq. (17-18c).

Solution Find the energy stored in the capacitor.

$$U=\frac{Q^2}{2C}=\frac{(18\text{ C})^2}{2(0.1811\times10^{-6}\text{ F})}=\boxed{8.9\times10^8\text{ J}}$$

63. (a) Strategy Use the definition of capacitance, Eq. (17-14), and Eq. (17-15).

Solution Find the charge on the capacitor.

$$Q=C\Delta V=\frac{\epsilon_0 A}{d}\Delta V=\frac{[8.854\times10^{-12}\text{ C}^2/(\text{N}\cdot\text{m}^2)](0.100\text{ m})^2(150\text{ V})}{0.75\times10^{-3}\text{ m}}=\boxed{18\text{ nC}}$$

(b) Strategy Use Eq. (17-18a).

Solution Compute the energy stored in the capacitor.

$$U=\frac{1}{2}Q\Delta V=\frac{1}{2}(17.7\times10^{-9}\text{ C})(150\text{ V})=\boxed{1.3\ \mu\text{J}}$$

65. (a) Strategy $U = P\Delta t$ where $P = 10.0$ kW and $\Delta t = 2.0$ ms. Use Eq. (17-18b).

Solution Find the initial potential difference.

$$U = \frac{1}{2}C(\Delta V)^2, \text{ so } \Delta V = \sqrt{\frac{2U}{C}} = \sqrt{\frac{2(10.0\text{ kW})(2.0\text{ ms})}{100.0\times10^{-6}\text{ F}}} = \boxed{630\text{ V}}.$$

(b) Strategy Use Eq. (17-18c).

Solution Find the initial charge.

$$U = \frac{Q^2}{2C}, \text{ so } Q = \sqrt{2CU} = \sqrt{2(100.0\times10^{-6}\text{ F})(10.0\text{ kW})(2.0\text{ ms})} = \boxed{0.063\text{ C}}.$$

69. (a) Strategy Use the definition of capacitance, Eq. (17-14).

Solution Compute the charge that passes through the body tissues.

$$Q = C\Delta V = (15\times10^{-6}\text{ F})(9.0\times10^3\text{ V}) = \boxed{0.14\text{ C}}$$

(b) Strategy Use Eq. (17-18b) and the definition of average power.

Solution Find the average power delivered to the tissues.

$$P_{\text{av}} = \frac{\Delta E}{\Delta t} = \frac{U}{\Delta t} = \frac{C(\Delta V)^2}{2\Delta t} = \frac{(15\times10^{-6}\text{ F})(9.0\times10^3\text{ V})^2}{2(2.0\times10^{-3}\text{ s})} = \boxed{0.30\text{ MW}}$$

73. (a) Strategy Let $q_{\text{L}} = -q_{\text{R}} = 10.0$ nC. Use Eqs. (17-5) and (17-9).

Solution Find the potential energy of the point charge at each location.

$$U_a = qV_a = q\left(\frac{kq_{\text{L}}}{r_{\text{L}}} + \frac{kq_{\text{R}}}{r_{\text{R}}}\right) = kqq_{\text{L}}\left(\frac{1}{r_{\text{L}}} - \frac{1}{r_{\text{R}}}\right)$$

$$= (8.988\times10^9\text{ N}\cdot\text{m}^2/\text{C}^2)(-4.2\times10^{-9}\text{ C})(10.0\times10^{-9}\text{ C})\left(\frac{1}{0.0400\text{ m}} - \frac{1}{0.1200\text{ m}}\right) = \boxed{-6.3\ \mu\text{J}}$$

$$U_b = qV_b = (8.988\times10^9\text{ N}\cdot\text{m}^2/\text{C}^2)(-4.2\times10^{-9}\text{ C})(10.0\times10^{-9}\text{ C})\left(\frac{1}{0.0400\text{ m}} - \frac{1}{0.0400\text{ m}}\right) = \boxed{0}$$

$$U_c = qV_c = (8.988\times10^9\text{ N}\cdot\text{m}^2/\text{C}^2)(-4.2\times10^{-9}\text{ C})(10.0\times10^{-9}\text{ C})\left(\frac{1}{0.0800\text{ m}} - \frac{1}{0.0800\text{ m}}\right) = \boxed{0}$$

(b) Strategy The work done by the external force is negative the work done by the field. Use Eq. (6-8).

Solution Find the work required to move the point charge.

$$W = -W_{\text{field}} = \Delta U = U_a - U_b = -6.3\ \mu\text{J} - 0 = \boxed{-6.3\ \mu\text{J}}$$

77. (a) Strategy Electric field lines begin on positive charges and end on negative charges. The same number of field lines begins on the plate and ends on the negative point charge. Field lines never cross. Use the principles of superposition and symmetry.

Solution The electric field lines for the cylinder and sheet:

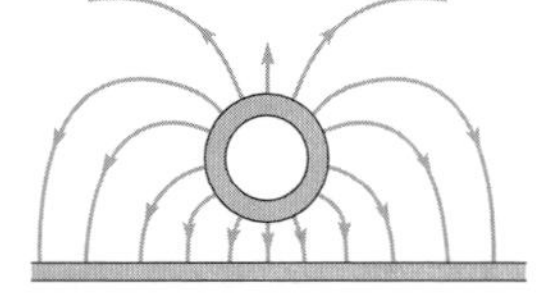

(b) Strategy Equipotential surfaces are perpendicular to electric field lines at all points. For equipotential surfaces drawn such that the potential difference between adjacent surfaces is constant, the surfaces are closer together where the field is stronger. The electric field always points in the direction of maximum potential decrease.

Solution The equipotential surfaces for the cylinder and sheet:

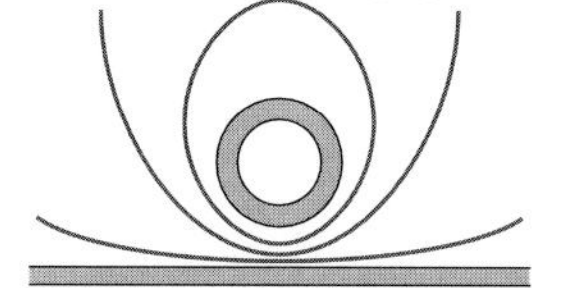

81. Strategy Assume the field is uniform. Use Eq. (17-10).

Solution Compute the magnitude of the electric field in the membrane.

$$E = \frac{\Delta V}{d} = \frac{90\times10^{-3}\text{ V}}{10\times10^{-9}\text{ m}} = \boxed{9\times10^{6}\text{ V/m}}.$$

85. Strategy Find the charge on the capacitor. Use Eqs. (17-14) and (17-15).

Solution The charge on the capacitor is $Q = Ne$, where N is the number of excess electrons.

$$Q = Ne = C\Delta V = \frac{\epsilon_0 A}{d}\Delta V,\text{ so } N = \frac{\epsilon_0 A\Delta V}{de} = \frac{[8.854\times10^{-12}\text{ C}^2/(\text{N}\cdot\text{m}^2)](0.0100\text{ m})^2(3.00\text{ V})}{(0.00200\text{ m})(1.602\times10^{-19}\text{ C})} = \boxed{8.29\times10^{6}}.$$

89. Strategy Use Eq. (17-16).

Solution Find the capacitance of the axon.

$$C = \kappa\frac{\epsilon_0 A}{d} = \frac{5[8.854\times10^{-12}\text{ C}^2/(\text{N}\cdot\text{m}^2)](5\times10^{-12}\text{ m}^2)}{4.4\times10^{-9}\text{ m}} = \boxed{5\times10^{-14}\text{ F}}$$

93. Strategy Use conservation of energy, $W_{\text{field}} = -\Delta U$, and the fact that the field is uniform.

Solution Find the kinetic energy of each electron when it leaves the space between the plates.

$$\Delta K = K_\text{f} - K_\text{i} = -\Delta U = W_{\text{field}} = eE\Delta y = e\left(\frac{\Delta V}{d}\right)\Delta y,\text{ so}$$

$$K_\text{f} = K_\text{i} + e\left(\frac{\Delta V}{d}\right)\Delta y = 2.0\times10^{-15}\text{ J} + \frac{(1.602\times10^{-19}\text{ C})(100.0\times10^{3}\text{ V})(0.0030\text{ m})}{0.0120\text{ m}} = \boxed{6.0\times10^{-15}\text{ J}}.$$

97. Strategy Use Eq. (17-18b) and form a proportion. ΔV is constant.

Solution Find the energy stored in the capacitor after the dielectric is inserted.

$$\frac{U}{U_0} = \frac{\frac{1}{2}C(\Delta V)^2}{\frac{1}{2}C_0(\Delta V)^2} = \frac{C}{C_0} = \frac{\kappa C_0}{C_0} = \kappa,\text{ so } U = \kappa U_0 = \boxed{3.0U_0}.$$

Chapter 18

ELECTRIC CURRENT AND CIRCUITS

Problems

1. **Strategy** Use the definition of electric current.

   Solution Compute the total charge.

   $I=\dfrac{\Delta q}{\Delta t}$, so $\Delta q = I\Delta t = (3.0\text{ A})(4.0\text{ h})(3600\text{ s/h}) = \boxed{4.3\times 10^4\text{ C}}$.

3. **(a) Strategy and Solution** The electrons flow from the filament to the anode; since they are negatively charged, the current flows $\boxed{\text{from the anode to the filament}}$.

   (b) Strategy Use the definition of electric current.

   Solution Compute the current in the tube.

   $I=\dfrac{\Delta q}{\Delta t} = \Delta q \times f = (1.602\times 10^{-19}\text{ C})(6.0\times 10^{12}\text{ s}^{-1}) = \boxed{0.96\ \mu\text{A}}$

5. **Strategy** Use the definition of electric current and the elementary charge of an electron.

   Solution Find the number of electrons per second that hit the screen.

   $I=\dfrac{\Delta q}{\Delta t}=\dfrac{Ne}{\Delta t}$, so $\dfrac{N}{\Delta t}=\dfrac{I}{e}=\dfrac{320\times 10^{-6}\text{ A}}{1.602\times 10^{-19}\ \frac{\text{C}}{\text{electron}}} = \boxed{2.0\times 10^{15}\text{ electrons/s}}$.

7. **Strategy** Since the oppositely charged ions move in opposite directions, they both contribute to the current in the same direction. Use the definition of electric current.

   Solution Compute the current in the solution.

   $I=\dfrac{\Delta q}{\Delta t}=\dfrac{Ne}{\Delta t}=[2(3.8\times 10^{16})+6.2\times 10^{16}](1.602\times 10^{-19}\text{ C/s}) = \boxed{22.1\text{ mA}}$

9. **Strategy** The total energy stored in a battery is equal to the total work the battery is able to do. Use Eq. (18-2).

   Solution Compute the energy stored in the battery.

   $W=\mathscr{E}q=(1.20\text{ V})(675\text{ C}) = \boxed{810\text{ J}}$

11. **(a) Strategy** Use the definition of electric current.

    Solution Compute the amount of charge pumped by the battery.

    $\Delta q = I\Delta t = (220.0\text{ A})(1.20\text{ s}) = \boxed{264\text{ C}}$

    (b) Strategy The electrical energy supplied is equal to the work done by the battery. Use Eq. (18-2).

    Solution Compute the amount of electrical energy supplied by the battery.

    $W=\mathscr{E}q=(12.0\text{ V})(264\text{ C}) = \boxed{3.17\text{ kJ}}$

13. Strategy Use Eq. (18-3).

Solution Form a proportion.

$$\frac{I_1}{I_2}=1=\frac{neA_1v_1}{neA_2v_2}=\frac{ne\left(\frac{1}{4}\pi d_1^2\right)v_1}{ne\left(\frac{1}{4}\pi d_2^2\right)v_2}=\frac{d_1^2v_1}{d_2^2v_2},\text{ so } v_1=\left(\frac{d_2}{d_1}\right)^2 v_2=\left(\frac{2}{1}\right)^2 v_2=4v_2.$$

The relationship between the drift speeds is $\boxed{v_1=4v_2}$.

15. Strategy Use Eq. (18-3) and $\Delta x=v_D\Delta t$.

Solution Find the drift speed of the conduction electrons in the wire.

$$I=neAv_D,\text{ so } v_D=\frac{I}{neA}=\frac{I}{ne(\pi r^2)}=\frac{I}{\pi ner^2}.$$

Find the time to travel 1.00 m along the wire.

$$\Delta t=\frac{\Delta x}{v_D}=\frac{\Delta x}{\frac{I}{\pi ner^2}}=\frac{\pi ner^2\Delta x}{I}=\frac{\pi(8.47\times10^{28}\text{ m}^{-3})(1.602\times10^{-19}\text{ C})(0.00100\text{ m}/2)^2(1.00\text{ m})}{10.0\text{ A}}\left(\frac{1\text{ min}}{60\text{ s}}\right)$$

$$=\boxed{17.8\text{ min}}$$

17. Strategy Let h be the thickness of the strip so that the cross-sectional area is $A=hw$, where w is the width. Use Eq. (18-3).

Solution Find the thickness of the strip.

$$I=neAv_D=ne(hw)v_D,\text{ so } h=\frac{I}{newv_D}=\frac{130\times10^{-6}\text{ A}}{(8.8\times10^{22}\text{ m}^{-3})(1.602\times10^{-19}\text{ C})(260\times10^{-6}\text{ m})(0.44\text{ m/s})}=\boxed{81\ \mu\text{m}}.$$

19. Strategy Use Eq. (18-3) and $n=1.3\rho N_A/M$, the number of electrons per unit volume.

Solution Find the drift speed of the conduction electrons.

$$v_D=\frac{I}{neA}=\frac{IM}{1.3\rho N_A eA}=\frac{(2.0\text{ A})(64\text{ g/mol})}{1.3(9.0\times10^6\text{ g/m}^3)(6.022\times10^{23}\text{ mol}^{-1})(1.602\times10^{-19}\text{ C})(1.00\times10^{-6}\text{ m}^2)}$$

$$=\boxed{0.11\text{ mm/s}}$$

21. Strategy Use the definition of resistance.

Solution Compute the current through the resistor.

$$R=\frac{\Delta V}{I},\text{ so } I=\frac{\Delta V}{R}=\frac{16\text{ V}}{12\ \Omega}=\boxed{1.3\text{ A}}.$$

25. (a) Strategy Use the definition of resistance.

Solution Compute the required potential difference between the electrician's hands.

$\Delta V=IR=(50\text{ mA})(1\text{ k}\Omega)=\boxed{50\text{ V}}$

(b) Strategy and Solution An electrician working on a "live" circuit keeps one hand behind his or her back to avoid becoming part of the circuit.

29. Strategy As found in Example 18.4, $R/R_0 = 1 + \alpha\Delta T$. Find T using this and the definition of resistance.

Solution Estimate the temperature of the tungsten filament.

$$1 + \alpha\Delta T = \frac{R}{R_0}, \text{ so } T = \frac{1}{\alpha}\left(\frac{R}{R_0} - 1\right) + T_0 = \frac{1}{4.50\times10^{-3}\ ^\circ\text{C}^{-1}}\left(\frac{\frac{2.90\text{ V}}{0.300\text{ A}}}{1.10\ \Omega} - 1\right) + 20.0^\circ\text{C} = \boxed{1750^\circ\text{C}}.$$

33. Strategy Use the definition of resistance, the relationship between voltage and uniform electric field, and Eq. (18-8).

Solution $V = IR = EL$ and $R = \rho L/A$. Find E.

$$V = EL = IR = I\rho\frac{L}{A}, \text{ so } \boxed{E = \rho\frac{I}{A}, \text{ where } \rho \text{ is the resistivity}}.$$

37. (a) Strategy Sum the individual emfs with those with their left terminal at the higher potential being positive.

Solution Compute the equivalent emf.

$$\mathscr{E}_{\text{eq}} = 3.0\text{ V} + 3.0\text{ V} + 2.5\text{ V} - 1.5\text{ V} = \boxed{7.0\text{ V}}$$

(b) Strategy Use the definition of resistance.

Solution Find the value of the resistor.

$$R = \frac{\Delta V}{I} = \frac{\mathscr{E}_{\text{eq}}}{I} = \frac{7.0\text{ V}}{0.40\text{ A}} = \boxed{18\ \Omega}$$

39. (a) Strategy $C_{\text{eq}} = \Sigma C_i$ for capacitors in parallel.

Solution Compute the equivalent capacitance.

$$C_{\text{eq}} = 4.0\ \mu\text{F} + 2.0\ \mu\text{F} + 3.0\ \mu\text{F} + 9.0\ \mu\text{F} + 5.0\ \mu\text{F} = \boxed{23.0\ \mu\text{F}}$$

(b) Strategy Use Eq. (17-14).

Solution Compute the charge on the equivalent capacitor.

$$Q = C\Delta V = C_{\text{eq}}\mathscr{E} = (23.0\times10^{-6}\text{ F})(16.0\text{ V}) = \boxed{368\ \mu\text{C}}.$$

(c) Strategy Use Eq. (17-14).

Solution Compute the charge on the capacitor.

$$Q = C\Delta V = C\mathscr{E} = (3.0\times10^{-6}\text{ F})(16.0\text{ V}) = \boxed{48\ \mu\text{C}}.$$

41. (a) Strategy Use Eqs. (18-13) and (18-17).

Solution Compute the resistance between points A and B.

$$R_{\text{eq}} = \left(\frac{1}{2.0\ \Omega} + \frac{1}{1.0\ \Omega + 1.0\ \Omega}\right)^{-1} + 4.0\ \Omega = \boxed{5.0\ \Omega}$$

(b) Strategy Label the currents on a diagram. Use Kirchhoff's rules.

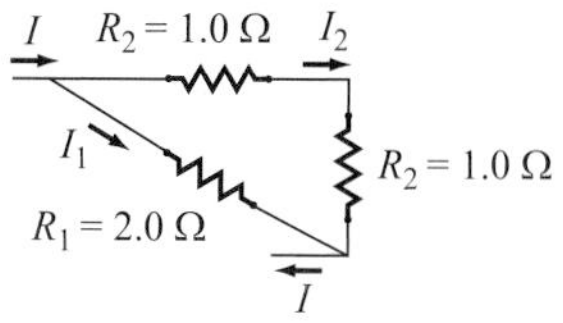

Solution The current through the emf is $I = \mathcal{E}/R_{\text{eq}} = I_1 + I_2$, where the currents labeled 1 and 2 are shown in the diagram. Applying the loop rule, we have $I_2(2R_2) - I_1R_1 = 0$, so $I_2 = \dfrac{R_1}{2R_2}I_1 = \dfrac{2.0\ \Omega}{2(1.0\ \Omega)}I_1 = 1.0I_1$.

Solve for the current through the 2.0-Ω resistor, I_1.

$$I = I_1 + I_2 = I_1 + 1.0I_1 = 2.0I_1,\ \text{so}\ I_1 = \frac{I}{2.0} = \frac{\mathcal{E}}{2.0R_{\text{eq}}} = \frac{20\ \text{V}}{2.0(5.0\ \Omega)} = \boxed{2.0\ \text{A}}.$$

45. Strategy Use the concept of equivalent resistance. The equivalent resistance of two identical resistances R in parallel is half or $R/2$.

Solution

(a) The two 2.0-Ω resistors are in series, so their equivalent resistance is $2.0\ \Omega + 2.0\ \Omega = 4.0\ \Omega$. These two resistors are in parallel with the rightmost 4.0-Ω resistor. Because the resistances of each branch of this parallel circuit are equal, the current is split evenly. Let the current through each branch be called I_3. We must determine I_3. Now, the equivalent resistance of this parallel circuit is 2.0 Ω, and this is in series with the rightmost 3.0-Ω resistor and the rightmost 1.0-Ω, so the equivalent series resistance is 6.0 Ω. This resistance is in parallel with the 6.0-Ω resistor, so the current is again split evenly. Let it be called I_2; then, $I_3 = I_2/2$. The equivalent resistance of this parallel circuit is 3.0-Ω, and this is in series with the middle 1.0-Ω resistor, so the equivalent series resistance is 4.0 Ω. This equivalent resistance is in parallel with the leftmost 4.0-Ω resistor, so the current is again split evenly. Let it be called I_1; then, $I_3 = I_2/2 = I_1/4$. The equivalent resistance of this parallel circuit is 2.0-Ω, and this is in series with the leftmost 1.0-Ω and 3.0-Ω resistors, so the equivalent resistance of the entire circuit is 6.0 Ω. If the current through the emf is I; then, $I_3 = I_2/2 = I_1/4 = I/8$. The current though the emf is given by $I = \mathcal{E}/R_{\text{eq}}$. Compute the current through one of the 2.0-Ω resistors.

$$I_3 = \frac{I}{8} = \frac{\mathcal{E}}{8R_{\text{eq}}} = \frac{24\ \text{V}}{8(6.0\ \Omega)} = \boxed{0.50\ \text{A}}$$

(b) The current through the 6.0-Ω resistor is I_2, which is one-fourth of the current through the emf.

$$I_2 = \frac{I}{4} = \frac{\mathcal{E}}{4R_{\text{eq}}} = \frac{24\ \text{V}}{4(6.0\ \Omega)} = \boxed{1.0\ \text{A}}$$

(c) The current through the leftmost 4.0-Ω resistor is I_1, which is half of the current through the emf.

$$I_1 = \frac{I}{2} = \frac{\mathcal{E}}{2R_{\text{eq}}} = \frac{24\ \text{V}}{2(6.0\ \Omega)} = \boxed{2.0\ \text{A}}$$

49. (a) Strategy Use Eqs. (18-15) and (18-18).

Solution Find the equivalent capacitance.

$$C_{eq} = \left(\frac{1}{12\ \mu F} + \frac{1}{12\ \mu F + 12\ \mu F}\right)^{-1} = \boxed{8.0\ \mu F}$$

(b) Strategy Since the capacitor at the left side of the diagram (1) is in series with the parallel combination of the other two capacitors (2), the charge Q on the capacitor 1 is the same as that on capacitor 2. (Think of the parallel combination as one capacitor with capacitance $C_2 = 12\ \mu F + 12\ \mu F = 24\ \mu F$.) Use the definition of capacitance.

Solution Find the potential difference across C_1. Let this potential difference be V_1 and the potential difference across C_2 be V_2. Then, $\mathcal{E} = V_1 + V_2$. Form a proportion.

$$\frac{V_2}{V_1} = \frac{\mathcal{E} - V_1}{V_1} = \frac{\mathcal{E}}{V_1} - 1 = \frac{Q/C_2}{Q/C_1} = \frac{C_1}{C_2}, \text{ so } V_1 = \frac{\mathcal{E}}{1 + \frac{C_1}{C_2}} = \frac{25\ V}{1 + \frac{12\ \mu F}{24\ \mu F}} = \boxed{17\ V}.$$

(c) Strategy The charge on the capacitor at the far right of the circuit (1) is half of the charge on the capacitor at the left of the circuit (2).

Solution Find the charge on the capacitor.

$$Q_2 = C_2V_2 = 2Q_1, \text{ so } Q_1 = \frac{1}{2}C_2V_2 = \frac{1}{2}(12\times 10^{-6}\ F)(17\ V) = \boxed{1.0\times 10^{-4}\ C}.$$

53. Strategy Use Kirchhoff's rules. Let I_1 be the top branch, I_2 be the middle branch, and I_3 be the bottom branch. Assume that each current flows right to left.

Solution Find the current in each branch of the circuit.

(1) $I_1 = -I_2 - I_3$ (2) $0 = 25.00\ V + (5.6\ \Omega)I_2 - (122\ \Omega)I_1$

(3) $0 = 25.00\ V + 5.00\ V + (75\ \Omega)I_3 - (122\ \Omega)I_1$ (4) $0 = 5.00\ V + (75\ \Omega)I_3 - (5.6\ \Omega)I_2$

Substitute (1) into (2).

(5) $0 = 25.00\ V + (122\ \Omega + 5.6\ \Omega)I_2 + (122\ \Omega)I_3$

Multiply (4) by 5 and subtract from (5).

$$0 = [122\ \Omega + 5.6\ \Omega + 5(5.6\ \Omega)]I_2 + [122\ \Omega - 5(75\ \Omega)]I_3, \text{ so } I_2 = \frac{5(75\ \Omega) - 122\ \Omega}{122\ \Omega + 5.6\ \Omega + 5(5.6\ \Omega)}I_3 = 1.6I_3\ (1.626I_3).$$

Substitute the result above into (4).

$$0 = 5.00\ V + (75\ \Omega)I_3 - (5.6\ \Omega)(1.626I_3), \text{ so } I_3 = \frac{5.00\ V}{1.626(5.6\ \Omega) - 75\ \Omega} = -0.076\ A.$$

So, $I_2 = 1.626(-0.076\ A) = -0.12\ A$ and $I_1 = -(-0.12\ A) - (-0.076\ A) = 0.20\ A$.

Branch	I (A)	Direction
AB	0.20	right to left
FC	0.12	left to right
ED	0.076	left to right

55. Strategy Use Kirchhoff's rules. Let the current on the left be I, the one in the middle be I_1, and the one on the right be I_2. I_1 flows downward.

Solution Find the unknown emf and the unknown resistor.
$I_1 = I + I_2 = 1.00\text{ A} + 10.00\text{ A} = 11.00\text{ A}$
Loop $ABCFA$:
$0 = -\mathcal{E} - (6.00\ \Omega)(1.00\text{ A}) - (4.00\ \Omega)(11.00\text{ A}) + 125\text{ V}$, so $\mathcal{E} = \boxed{75\text{ V}}$.
Loop $ABCDEFA$:
$$0 = -\mathcal{E} - (6.00\ \Omega)(1.00\text{ A}) + (10.00\text{ A})R = -75\text{ V} - 6.00\text{ V} + (10.00\text{ A})R,\ \text{so } R = \frac{81\text{ V}}{10.00\text{ A}} = \boxed{8.1\ \Omega}.$$

57. Strategy Use Eq. (18-20).

Solution Compute the power dissipated by the resistor.
$P = \mathcal{E}I = (2.00\text{ V})(2.0\text{ A}) = \boxed{4.0\text{ W}}$

61. Strategy and Solution $\boxed{\text{Yes}}$; the power rating can be determined by $P = IV = (5.0\text{ A})(120\text{ V}) = \boxed{600\text{ W}}$.

65. (a) Strategy Use Eqs. (8-13) and (8-17) to find the equivalent resistance. Then draw the diagram.

Solution Compute the equivalent resistance.
$$R_{\text{eq}} = 20.0\ \Omega + 50.0\ \Omega + \left(\frac{1}{70.0\ \Omega + 20.0\ \Omega} + \frac{1}{40.0\ \Omega + 20.0\ \Omega}\right)^{-1} = 106.0\ \Omega$$
The simplest equivalent circuit contains the emf and one 106.0-Ω resistor.

106.0 Ω

120 V

(b) Strategy Use the definition of resistance.

Solution Compute the current that flows from the battery.
$$I = \frac{\mathcal{E}}{R_{\text{eq}}} = \frac{120\text{ V}}{106.0\ \Omega} = \boxed{1.1\text{ A}}$$

(c) Strategy Compute the resistance between A and B. Then use the definition of resistance to find the potential difference.

Solution Compute the resistance. It is in series with the two resistors not between A and B.
$106.0\ \Omega - 20.0\ \Omega - 50.0\ \Omega = 36.0\ \Omega$
Compute the potential difference.
$V = IR = (1.13\text{ A})(36.0\ \Omega) = \boxed{41\text{ V}}$

(d) Strategy The current that flows through the battery is shared by each branch between points A and B. Draw a diagram with equivalent resistances. Use Kirchhoff's rules.

Solution The diagram is shown.

60.0 Ω, I, I_1, A, I_2, B, 90.0 Ω

$I = I_1 + I_2$ and $I_1(60.0\ \Omega) - I_2(90.0\ \Omega) = 0$, so $I_2 = \dfrac{60.0}{90.0} I_1$ and

$I = I_1 + \dfrac{60.0}{90.0} I_1 = \left(1 + \dfrac{60.0}{90.0}\right) I_1.$

So, the current through the upper branch is $I_1 = \left(1 + \dfrac{60.0}{90.0}\right)^{-1} I = \left(1 + \dfrac{60.0}{90.0}\right)^{-1} (1.13\text{ A}) = \boxed{0.68\text{ A}}$, and the current through the lower branch is $I_2 = \dfrac{60.0}{90.0} I_1 = \dfrac{60.0}{90.0}(0.68\text{ A}) = \boxed{0.45\text{ A}}$.

(e) Strategy Use $P = I^2 R$.

Solution Determine the power dissipated in the resistors.

$P_{50} = (1.13\text{ A})^2(50.0\ \Omega) = \boxed{64\text{ W}}$, $P_{70} = (0.45\text{ A})^2(70.0\ \Omega) = \boxed{14\text{ W}}$, and

$P_{40} = (0.68\text{ A})^2(40.0\ \Omega) = \boxed{18\text{ W}}$.

69. (a) Strategy Use Eq. (18-21b).

Solution Find the resistance of the heater when it is turned on.

$$P = \frac{V^2}{R},\ \text{so } R = \frac{V^2}{P} = \frac{(120\text{ V})^2}{2200\text{ W}} = \boxed{6.5\ \Omega}.$$

(b) Strategy Use Eq. (18-19).

Solution Find the current in the heater.

$$P = IV,\ \text{so } I = \frac{P}{V} = \frac{2200\text{ W}}{120\text{ V}} = \boxed{18\text{ A}}.$$

(c) Strategy Use Eqs. (18-8) and (18-9). The cross-sectional area of the wire is $\frac{1}{4}\pi d^2$.

Solution Find the diameter of the wire when it is hot.

$$R = \rho\frac{L}{A} = \rho_0(1 + \alpha\Delta T)\frac{L}{\frac{1}{4}\pi d^2},\ \text{so}$$

$$d = \sqrt{\frac{4\rho_0(1+\alpha\Delta T)L}{\pi R}} = \sqrt{\frac{4(108\times10^{-8}\ \Omega\cdot\text{m})[1 + (0.00040\text{ K}^{-1})(400\text{ K})](3.0\text{ m})}{\pi(6.545\ \Omega)}} = \boxed{0.86\text{ mm}}.$$

(d) Strategy From Example 18.4, $R/R_0 = 1 + \alpha\Delta T$. Use the definition of resistance.

Solution Find the resistance of the wire when the heater is first turned on.

$$R = R_0(1 + \alpha\Delta T) = R_0[1 + (0.00040\text{ K}^{-1})(400\text{ K})] = 1.16R_0,\ \text{so } R_0 = \frac{R}{1.16}.$$

Compute the current.

$$I = \frac{V}{R_0} = \frac{1.16V}{R} = \frac{1.16V}{V^2/P} = \frac{1.16P}{V} = \frac{1.16(2200\text{ W})}{120\text{ V}} = \boxed{21\text{ A}}$$

73. (a) Strategy To measure the current in a circuit, an ammeter must be in series in the circuit. When 12.0 A pass through the ammeter, the meter should deflect full scale. Therefore, 10.0 A should pass through the ammeter and 2.0 A should pass through the resistor.

Solution Since the current must be split between the ammeter and the resistor, the resistor must be placed $\boxed{\text{in parallel}}$ with the ammeter. The voltages across the ammeter and the resistor are the same. Find the size of the resistor, R.

$$V = I_{\text{Ammeter}}R_{\text{Ammeter}} = IR, \text{ so } R = \frac{I_{\text{Ammeter}}R_{\text{Ammeter}}}{I} = \frac{(10.0 \text{ A})(24 \text{ }\Omega)}{2.0 \text{ A}} = \boxed{120 \text{ }\Omega}.$$

(b) Strategy and Solution When 12.0 A is measured, 10.0 A is the reading on the meter. $12.0/10.0 = 1.20$, so $\boxed{\text{the meter readings should be multiplied by 1.20 to get the correct current values}}$.

75. Strategy The resistances are in series, so $V = IR_{\text{eq}}$.

Solution Find the required resistance of the series resistor.

$$(0.120\times10^{-3} \text{ A})(R_S + 34.0 \text{ }\Omega) = 100.0 \text{ V}, \text{ so } R_S = \frac{100.0 \text{ V}}{0.120\times10^{-3} \text{ A}} - 34.0 \text{ }\Omega = \boxed{833 \text{ k}\Omega}.$$

77. Strategy The resistances are in series, so $V = IR_{\text{eq}}$.

Solution Find the required resistances of the series resistors.

(a) $(2.0\times10^{-3} \text{ A})(R_S + 75 \text{ }\Omega) = 50.0 \text{ V}, \text{ so } R_S = \dfrac{50.0 \text{ V}}{2.0\times10^{-3} \text{ A}} - 75 \text{ }\Omega = \boxed{25 \text{ k}\Omega}.$

(b) $R_S = \dfrac{500.0 \text{ V}}{2.0\times10^{-3} \text{ A}} - 75 \text{ }\Omega = \boxed{250 \text{ k}\Omega}$

79. Strategy Label the currents on circuit diagrams. Use Kirchhoff's rules.

Solution

(a) Find the current through the 1.40-kΩ resistor.

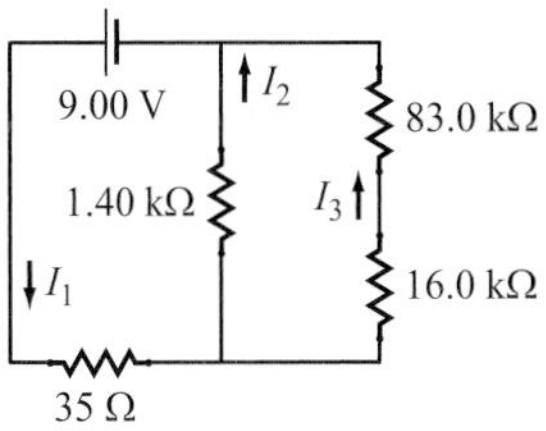

9.00 V; I_2; 83.0 kΩ; 1.40 kΩ; I_3; I_1; 16.0 kΩ; 35 Ω

(1) $I_1 = I_2 + I_3$

(2) $0 = 9.00 \text{ V} - I_1(35 \text{ }\Omega) - I_2(1.40\times10^3 \text{ }\Omega)$

$$0 = I_2(1.40\times10^3 \text{ }\Omega) - I_3(16.0\times10^3 \text{ }\Omega + 83.0\times10^3 \text{ }\Omega)$$

$$I_2 = \frac{99.0\times10^3 \text{ }\Omega}{1.40\times10^3 \text{ }\Omega}I_3$$

(3) $I_2 = 70.7I_3$

Substitute (3) into (1).

$$I_1 = I_2 + \frac{1}{70.7}I_2, \text{ so } I_1 = \frac{71.7}{70.7}I_2 \text{ (4)}.$$

Substitute (4) into (2).

$$0 = 9.00 \text{ V} - \frac{71.7}{70.7}I_2(35 \text{ }\Omega) - I_2(1.40\times10^3 \text{ }\Omega), \text{ so } I_2 = \frac{9.00 \text{ V}}{\frac{71.7}{70.7}(35 \text{ }\Omega) + 1.40\times10^3 \text{ }\Omega} = \boxed{6.27 \text{ mA}}.$$

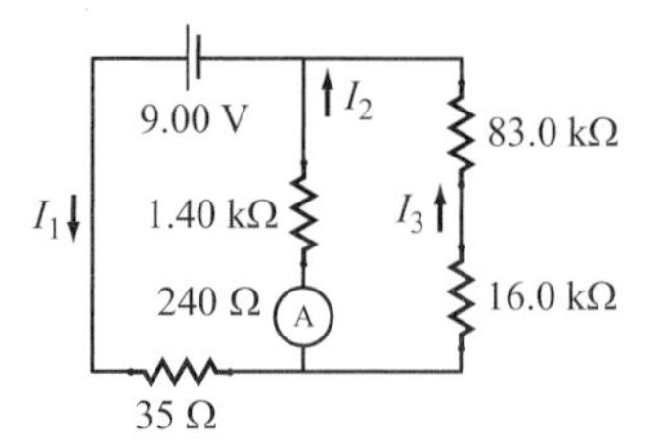

(b) The ammeter is connected in series with the 1.40-kΩ resistor. Find the new current through the resistor.

(1) $I_1 = I_2 + I_3$

(2) $0 = 9.00\text{ V} - I_1(35\ \Omega) - I_2(1.40\times10^3\ \Omega + 240\ \Omega)$

$0 = I_2(1.40\times10^3\ \Omega + 240\ \Omega) - I_3(99.0\times10^3\ \Omega)$

$$I_3 = \frac{1.40\times10^3\ \Omega + 240\ \Omega}{99.0\times10^3\ \Omega} I_2$$

(3) $I_3 = 0.0166 I_2$

Substitute (3) into (1).

$I_1 = I_2 + 0.0166 I_2$

(4) $I_1 = 1.0166 I_2$

Substitute (4) into (2).

$0 = 9.00\text{ V} - 1.0166 I_2(35\ \Omega) - I_2(1.40\times10^3\ \Omega + 240\ \Omega)$, so

$$I_2 = \frac{9.00\text{ V}}{1.0166(35\ \Omega) + 1.40\times10^3\ \Omega + 240\ \Omega} = \boxed{5.37\text{ mA}}.$$

81. Strategy Solve Eq. (18-23) for *t*.

Solution Find the time it takes for the voltage across the capacitor to be 15.0 V.

$$V_C(t) = \mathscr{E}(1 - e^{-t/\tau})$$

$$\ln e^{-t/(RC)} = \ln\left(1 - \frac{V_C(t)}{\mathscr{E}}\right)$$

$$t = -RC\ln\left(1 - \frac{V_C(t)}{\mathscr{E}}\right) = -(1.00\ \text{M}\Omega)(2.00\ \mu\text{F})\ln\left(1 - \frac{15.0}{20.0}\right) = \boxed{2.77\text{ s}}$$

83. (a) Strategy The energy dissipated by the resistor is equal to the energy initially stored in the capacitor. Use Eqs. (18-24), (18-25), and (17-18b), and the definition of resistance.

Solution Find the time constant.

$I(t = \tau) = I_0 e^{-1} \approx 0.368 I_0 = 0.368(100.0\text{ mA}) = 36.8\text{ mA}$, so $\tau \approx 12.8\text{ ms} = RC$.

$$R = \frac{V_0}{I_0} = \frac{9.0\text{ V}}{100.0\times10^{-3}\text{ A}} = \boxed{90\ \Omega};\quad C = \frac{\tau}{R} = \frac{0.0128\text{ s}}{90\ \Omega} = \boxed{140\ \mu\text{F}};$$

$$U = \frac{1}{2}CV_0^2 = \frac{1}{2}(142\times10^{-6}\text{ F})(9.0\text{ V})^2 = \boxed{5.8\text{ mJ}}$$

(b) Strategy The energy is directly proportional to the voltage across the capacitor squared. Solve Eq. (18-26) for *t*.

Solution The energy is half its initial value when

$$V_C^2 = \frac{1}{2}V_0^2 = \left(\frac{V_0}{\sqrt{2}}\right)^2 \text{ or } V_C = \frac{V_0}{\sqrt{2}}.$$

$$V_C = V_0 e^{-t/\tau} = \frac{V_0}{\sqrt{2}},\text{ so } -\frac{t}{\tau} = \ln\frac{1}{\sqrt{2}} \text{ or } t = \frac{1}{2}\tau\ln 2 = \frac{1}{2}(12.8\text{ ms})(0.693) = \boxed{4.4\text{ ms}}.$$

(c) Strategy Use Eq. (18-26).

Solution Substitute numerical values and graph the voltage across the capacitor.

$V_C(t) = V_0 e^{-t/\tau} = (9.0\text{ V})e^{-t/(12.8\text{ ms})}$

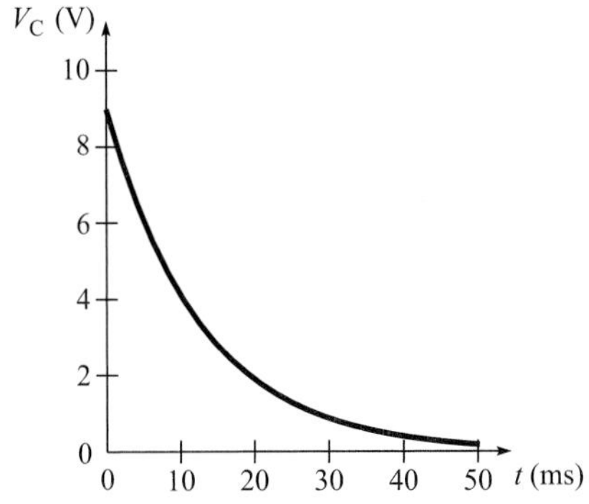

85. (a) Strategy Use Eq. (17-18b).

Solution Find the required initial potential difference.

$$U = \frac{1}{2}C(\Delta V)^2, \text{ so } \Delta V = \sqrt{\frac{2U}{C}} = \sqrt{\frac{2(20.0\text{ J})}{100.0\times10^{-6}\text{ F}}} = \boxed{632\text{ V}}.$$

(b) Strategy Use the definition of capacitance.

Solution Find the initial charge.

$$Q = C(\Delta V) = (100.0\times10^{-6}\text{ F})(632\text{ V}) = \boxed{63.2\text{ mC}}$$

(c) Strategy Solve for R using $I = I_0 e^{-t/\tau}$ where $\tau = RC$.

Solution Find the resistance of the lamp.

$$0.050 I_0 = I_0 e^{-t/\tau}$$

$$\ln 0.050 = -\frac{t}{RC}$$

$$R = -\frac{t}{C\ln 0.050} = -\frac{0.0020\text{ s}}{(100.0\times10^{-6}\text{ F})\ln 0.050} = \boxed{6.7\ \Omega}$$

89. Strategy Use Eqs. (18-24) and (18-25) and the definition of resistance.

Solution

(a) Initially ($t = 0$), the capacitor has nearly zero resistance. Find the currents and the voltages.

$$I_1 = I_2 = \frac{V}{R} = \frac{12\text{ V}}{40.0\times10^3\ \Omega} = \boxed{0.30\text{ mA}} \text{ and } V_1 = V_2 = \boxed{12\text{ V}}.$$

(b) Calculate the time constant.

$\tau = (40.0\times10^3\ \Omega)(5.0\times10^{-8}\ \text{F}) = 2.0\ \text{ms}$

Find the currents.

$I_1 = I_2 = I_0 e^{-t/\tau} = (3.0\times10^{-4}\ \text{A})e^{-(1.0\ \text{ms})/(2.0\ \text{ms})} = \boxed{0.18\ \text{mA}}$

Find V_1 and V_2.

$V_1 = \boxed{12\ \text{V}}$ and $V_2 = I_2 R = (1.82\times10^{-4}\ \text{A})(40.0\times10^3\ \Omega) = \boxed{7.3\ \text{V}}$.

(c) Find the currents and the voltages.

$I_1 = I_2 = (3.0\times10^{-4}\ \text{A})e^{-(5.0\ \text{ms})/(2.0\ \text{ms})} = \boxed{25\ \mu\text{A}}$, $V_1 = \boxed{12\ \text{V}}$, and

$V_2 = I_2 R = (2.463\times10^{-5}\ \text{A})(40.0\times10^3\ \Omega) = \boxed{0.99\ \text{V}}$.

93. (a) Strategy According to the figure, $I_0 \approx 0.070$ A. Use Eq. (18-25).

Solution Compute the current at $t = \tau$.

$I(t=\tau) = I_0 e^{-1} = (0.070\ \text{A})e^{-1} = 0.026\ \text{A}$

So, according to the figure, $\tau \approx 0.060$ s. The final charge is $Q = I_0 \Delta t = I_0 \tau \approx (0.070\ \text{A})(0.060\ \text{s}) = \boxed{4.2\ \text{mC}}$.

(b) Strategy Use the definition of capacitance.

Solution Find the capacitance.

$$C = \frac{Q}{V} \approx \frac{0.0042\ \text{C}}{9.0\ \text{V}} = \boxed{470\ \mu\text{F}}$$

(c) Strategy Use Eq. (18-24).

Solution Find the total resistance in the circuit.

$$R = \frac{\tau}{C} \approx \frac{0.060\ \text{s}}{470\times10^{-6}\ \text{F}} = \boxed{130\ \Omega}$$

(d) Strategy Use Eqs. (17-18b) and (18-23) and the fact that $U = U_0/2$.

Solution Solve for t.

$$U = \frac{1}{2}CV^2 = \frac{1}{2}CV_0^2(1-e^{-t/\tau})^2 = U_0(1-e^{-t/\tau})^2$$

$$\pm\sqrt{\frac{U}{U_0}} = 1 - e^{-t/\tau}$$

$$e^{-t/\tau} = 1 \pm \sqrt{\frac{U}{U_0}}$$

$$-\frac{t}{\tau} = \ln\left(1 \pm \sqrt{\frac{U}{U_0}}\right)$$

$$t = -\tau \ln\left(1 \pm \sqrt{\frac{U}{U_0}}\right) \approx -(0.060\ \text{s})\ln\left(1 \pm \sqrt{\frac{1}{2}}\right) = -32\ \text{ms or } 74\ \text{ms}$$

t cannot be negative, so the answer is $\boxed{74\ \text{ms}}$.

95. Strategy Use Eq. (18-19).

Solution The maximum current that can be supplied by the batteries is

$$I_{\text{max}} = \frac{P_{\text{max}}}{V} = \frac{5.0\text{ W}}{100.0\text{ V}} = 0.050\text{ A}.$$

$$I = \frac{V}{R} = \frac{100.0\text{ V}}{1.0\times10^3\ \Omega} = 0.10\text{ A} > I_{\text{max}},$$ so the current that passes through him is $\boxed{50\text{ mA}}$.

97. Strategy and Solution

(a) If the person receives a shock, $\boxed{\text{the microwave is not grounded.}}$

(b) If the cord begins to smoke, $\boxed{\text{the wires are too small to handle the current}}$ and thus overheat.

(c) If a fuse blows out, $\boxed{\text{too much current is drawn, and the appliance has a short circuit}}$.

(d) An electrical fire breaking out inside the kitchen wall is likely the result of $\boxed{\text{poor household wiring}}$.

101. Strategy Use the definition of resistance and Eqs. (18-13) and (18-17).

Solution

(a) The current through A_1 is the same as that through the emf.

$$I = \frac{V}{R_{\text{eq}}} = \frac{10.0\text{ V}}{2.00\ \Omega + \left(\frac{1}{2.00\ \Omega} + \frac{1}{3.00\ \Omega} + \frac{1}{6.00\ \Omega}\right)^{-1} + 2.00\ \Omega} = \boxed{2.00\text{ A}}$$

(b) Since $\left(\frac{1}{3.00\ \Omega} + \frac{1}{6.00\ \Omega}\right)^{-1} = 2.00\ \Omega,$ the current is split evenly at the first junction to the right of A_1. So,

$$I = \frac{2.00\text{ A}}{2} = \boxed{1.00\text{ A}}.$$

105. (a) Strategy It is okay to treat the Earth-ionosphere system as a parallel plate capacitor, since $\frac{d}{R} = \frac{5.0\times10^4\text{ m}}{6.371\times10^6\text{ m}} \approx 10^{-2}$; locally, the Earth is flat when compared with the distance between the "plates". Use Eq. (17-15).

Solution Find the capacitance.

$$C = \frac{\epsilon_0 A}{d} = \frac{[8.854\times10^{-12}\ \text{C}^2/(\text{N}\cdot\text{m}^2)]4\pi(6.371\times10^6\text{ m})^2}{5.0\times10^4\text{ m}} = \boxed{0.090\text{ F}}$$

(b) Strategy Use Eqs. (17-10) and (17-18b).

Solution Find the energy stored in the capacitor.

$$U = \frac{1}{2}CV^2 = \frac{1}{2}C(Ed)^2 = \frac{1}{2}(0.090\text{ F})(150\text{ V/m})^2(5.0\times10^4\text{ m})^2 = \boxed{2.5\text{ TJ}}$$

(c) Strategy Use the definition of resistance and Eqs. (18-8) and (17-10).

Solution Compute the resistance.

$$R = \rho\frac{L}{A} = (3.0\times10^{14}\ \Omega\cdot\text{m})\frac{5.0\times10^{4}\ \text{m}}{4\pi(6.371\times10^{6}\ \text{m})^2} = \boxed{29\ \text{k}\Omega}$$

Compute the current.

$$I = \frac{V}{R} = \frac{Ed}{R} = \frac{\left(150\ \frac{\text{V}}{\text{m}}\right)(5.0\times10^{4}\ \text{m})}{29{,}408\ \Omega} = \boxed{260\ \text{A}}.$$

(d) Strategy The system can be modeled by an *RC* circuit. The voltage across the capacitor while it is discharging is given by $V = V_0e^{-t/(RC)}$. Since $Q = CV$, $Q = Q_0e^{-t/(RC)}$ assuming C doesn't change.

Solution Solve for t.

$$Q = Q_0e^{-t/(RC)}$$

$$e^{t/(RC)} = \frac{Q_0}{Q}$$

$$\frac{t}{RC} = \ln\frac{Q_0}{Q}$$

$$t = RC\ln\frac{Q_0}{Q} = \left[(29\times10^{3}\ \Omega)(0.090\ \text{F})\ln\frac{1}{0.01}\right]\left(\frac{1\ \text{min}}{60\ \text{s}}\right) = \boxed{200\ \text{min}}$$

109. Strategy Use Eq. (18-19).

Solution Compute the current drawn by the motor.

$$I = \frac{P}{V} = \frac{1.5\ \text{hp}}{120\ \text{V}}(745.7\ \text{W/hp}) = \boxed{9.3\ \text{A}}$$

113. Strategy For the axon: r = radius; t = thickness of the membrane; L = length. Use Eq. (18-8).

Solution

(a) $$R = \rho\frac{L}{A} = \rho\frac{L}{\pi r^2} = (2.0\ \Omega\cdot\text{m})\frac{0.010\ \text{m}}{\pi(5.0\times10^{-6}\ \text{m})^2} = \boxed{250\ \text{M}\Omega}$$

(b) $$R = \rho\frac{t}{A} = \rho\frac{t}{2\pi rL} = (2.5\times10^{7}\ \Omega\cdot\text{m})\frac{8.0\times10^{-9}\ \text{m}}{2\pi(5.0\times10^{-6}\ \text{m})(0.010\ \text{m})} = \boxed{640\ \text{k}\Omega}$$

(c) Set $R_\text{a} = R_\text{b}$ and solve for L.

$$R_\text{a} = \rho_\text{a}\frac{L}{\pi r^2} = R_\text{b} = \rho_\text{b}\frac{t}{2\pi rL},\ \text{so}\ L = \sqrt{\frac{\rho_\text{b}tr}{2\rho_\text{a}}} = \sqrt{\frac{(2.5\times10^{7}\ \Omega\cdot\text{m})(8.0\times10^{-9}\ \text{m})(5.0\times10^{-6}\ \text{m})}{2(2.0\ \Omega\cdot\text{m})}} = \boxed{0.50\ \text{mm}}.$$

117. Strategy Use Eqs. (18-17), (18-21a), and (18-21b).

Solution

(a) $P = V^2/R_{eq}$, so we need to design the circuit such that V is maximized and R_{eq} is minimized. If the batteries are placed in series, $V = 2\mathcal{E}$. If the light bulbs are connected in parallel,

$$R_{eq} = \left(\frac{1}{R} + \frac{1}{R}\right)^{-1} = \frac{R}{2},$$

which is the smallest possible value. The potential across each is $V = 2\mathcal{E}$, which is the largest possible value.

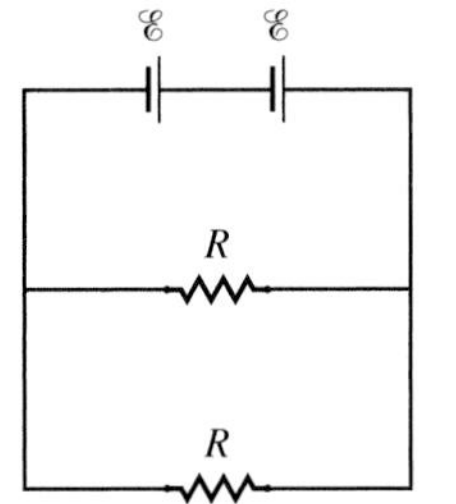

(b) The power dissipated by each bulb is the same.

$$P = \frac{V^2}{R} = \frac{(2\mathcal{E})^2}{R} = \boxed{\frac{4\mathcal{E}^2}{R}}$$

(c) The power through each bulb is given by $P = I^2R$. The circuit must be designed so that the current through the brighter bulb is larger than that through the dimmer bulb. In the circuit below, the maximum current passes through the bulb on the right, whereas only a fraction of that current passes through the bulb on the left, so $\boxed{\text{the bulb on the right is brighter.}}$

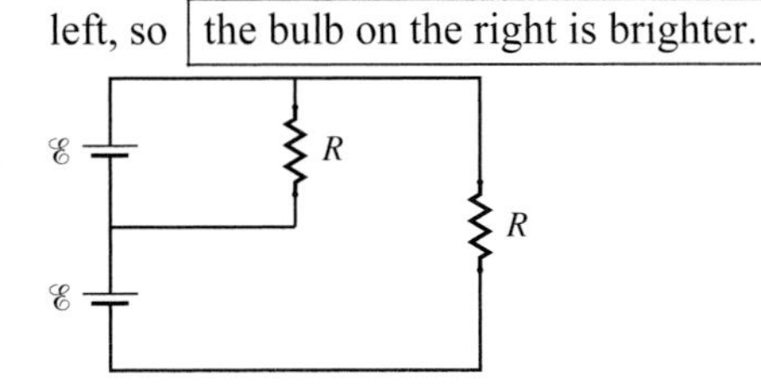

121. Strategy Use Eq. (18-21b).

Solution

(a) Form a proportion.

$$\frac{P_f}{P_i} = \frac{V_f^2/R}{V_i^2/R} = \left(\frac{V_f}{V_i}\right)^2 = \left(\frac{110}{120}\right)^2 = 0.84$$

$1 - 0.84 = 0.16$, so the heat output decreases by $\boxed{16\%}$.

(b) Form a proportion.

$$\frac{P_f}{P_i} = \frac{V_f^2/R_f}{V_i^2/R_i} = \left(\frac{V_f}{V_i}\right)^2 \frac{R_i}{R_f} = 0.84\frac{R_i}{R_f}$$

At lower temperature (at the lower voltage), the resistance is lower, so $R_f < R_i$, or $R_i/R_f > 1$. Therefore, $0.84(R_i/R_f) > 0.84$ and the actual drop in heat output is $\boxed{\text{smaller}}$ than that calculated in part (a).

125. (a) Strategy Use the definition of capacitance and Eq. (17-15).

Solution Find the charge on the upper plate.

$$Q = CV = \frac{\epsilon_0 A}{d}V = \frac{\epsilon_0 L^2}{d}V = \frac{[8.854\times10^{-12}\ \text{C}^2/(\text{N}\cdot\text{m}^2)](0.10\ \text{m})^2(10.0\ \text{V})}{89\times10^{-6}\ \text{m}} = \boxed{9.9\ \text{nC}}$$

(b) Strategy Use the definition of capacitance and Eqs. (17-15), (18-24), and (18-25).

Solution Find the time constant.

$$\tau = RC = R\frac{\epsilon_0 A}{d} = \frac{(0.100\times10^6\ \Omega)[8.854\times10^{-12}\ \text{C}^2/(\text{N}\cdot\text{m}^2)](0.10\ \text{m})^2}{89\times10^{-6}\ \text{m}} = 9.9\times10^{-5}\ \text{s}$$

Compute the initial current.

$$\frac{\mathscr{E}}{R} = \frac{10.0\ \text{V}}{0.100\times10^6\ \Omega} = 100\ \mu\text{A}$$

The current is given by $I = (100\ \mu\text{A})e^{-t/(9.9\times10^{-5}\ \text{s})}$.

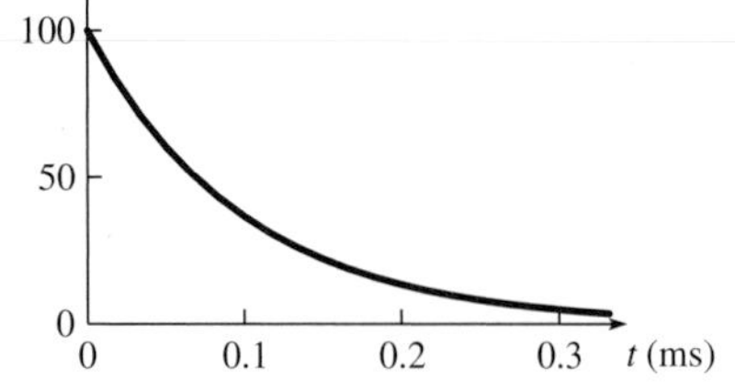

(c) Strategy The energy dissipated in R is that stored in the capacitor, U. Use Eqs. (17-15) and (17-18b).

Solution Compute the energy dissipated over the whole discharging process.

$$U = \frac{1}{2}CV^2 = \frac{1}{2}\left(\frac{\epsilon_0 A}{d}\right)\mathscr{E}^2 = \frac{[8.854\times10^{-12}\ \text{C}^2/(\text{N}\cdot\text{m}^2)](0.10\ \text{m})^2(10.0\ \text{V})^2}{2(89\times10^{-6}\ \text{m})} = \boxed{50\ \text{nJ}}$$

REVIEW AND SYNTHESIS: CHAPTERS 16–18

Review Exercises

1. Strategy and Solution Since the spheres are identical, the charge will be shared evenly by the two spheres, so the spheres will have $(18.0\ \mu\text{C} + 6.0\ \mu\text{C})/2 = \boxed{12.0\ \mu\text{C}}$ of charge each.

5. Strategy The force on charge *A* due to charge *B* is equal and opposite to the horizontal component of the tension. The sign of charge *A* is negative, since the sign of charge *B* is positive and the force is attractive. Use Newton's second law and Eq. (16-2).

Solution

(a) Find the tension.

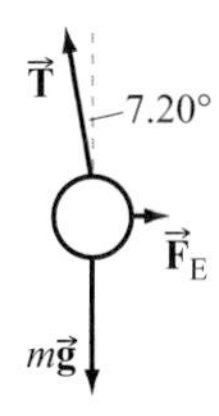

$$\Sigma F_y = T\cos 7.20° - mg = 0, \text{ so } T = \frac{mg}{\cos 7.20°}.$$

The horizontal component of the tension is $T_x = \dfrac{mg}{\cos 7.20°}\sin 7.20° = mg\tan 7.20°.$

Solve for the second charge.

$$\frac{k|q_A||q_B|}{r^2} = mg\tan 7.20°, \text{ so}$$

$$|q_A| = \frac{r^2 mg\tan 7.20°}{k|q_B|} = \frac{(0.0500\text{ m})^2(0.0900\text{ kg})(9.80\text{ m/s}^2)\tan 7.20°}{(8.988\times10^9\text{ N}\cdot\text{m}^2/\text{C}^2)(130\times10^{-9}\text{ C})} = 238\text{ nC}.$$

Thus, the charge on *A* is $\boxed{-238\text{ nC}}$.

(b) The tension in the thread is $T = \dfrac{mg}{\cos 7.20°} = \dfrac{(0.0900\text{ kg})(9.80\text{ m/s}^2)}{\cos 7.20°} = \boxed{0.889\text{ N}}$.

9. (a) Strategy Use Eqs. (18-13) and (18-17).

Solution Find the equivalent resistance.

$$R_{\text{eq}} = 15.0\ \Omega + \left(\frac{1}{40.0\ \Omega} + \frac{1}{20.0\ \Omega} + \frac{1}{40.0\ \Omega}\right)^{-1} + 10.0\ \Omega = \boxed{35.0\ \Omega}$$

(b) Strategy The current that flows through resistor R_1 is the current that flows through the emf.

Solution Find the current.

$$I = \frac{V}{R_{\text{eq}}} = \frac{24.0\text{ V}}{35.0\ \Omega} = \boxed{0.686\text{ A}}$$

(c) Strategy Use Eq. (18-21b).

Solution Find the power dissipated in the circuit.

$$P = \frac{V^2}{R_{\text{eq}}} = \frac{(24.0\text{ V})^2}{35.0\ \Omega} = \boxed{16.5\text{ W}}$$

(d) Strategy R_2, R_3, and R_4 are in parallel, so the potential difference across each is the same. Use Kirchhoff's loop rule.

Solution Find the potential difference across R_3, V_3.

$$V - IR_1 - V_3 - IR_5 = 0, \text{ so } V_3 = V - IR_1 - IR_5 = 24.0 \text{ V} - (0.686 \text{ A})(15.0\ \Omega + 10.0\ \Omega) = \boxed{6.9 \text{ V}}.$$

(e) Strategy Use the definition of resistance.

Solution Find the current through R_3, I_3.

$$I_3 = \frac{V_3}{R_3} = \frac{6.86 \text{ V}}{20.0\ \Omega} = \boxed{0.34 \text{ A}}$$

(f) Strategy Use Eq. (18-21b).

Solution Find the power dissipated in R_3.

$$P = \frac{V_3^2}{R_3} = \frac{(6.9 \text{ V})^2}{20.0\ \Omega} = \boxed{2.4 \text{ W}}$$

13. (a) Strategy After the switch has been closed for a long time, the capacitor is fully charged and acts like a resistor with infinite resistance. Thus, all of the current passes through the 12-Ω resistor, thereby bypassing the capacitor. Therefore, the current through the 12-Ω resistor is equal to the current through the emf.

Solution The resistors are in series, so the equivalent resistance of the circuit is 27 Ω. Compute the current.

$$I = \frac{V}{R_{\text{eq}}} = \frac{12 \text{ V}}{27\ \Omega} = \boxed{0.44 \text{ A}}$$

(b) Strategy The capacitor and 12-Ω resistor are in parallel, so the voltage across each is the same.

Solution Find the voltage across the capacitor.

$$V_{\text{cap}} = V_R = IR = \frac{V}{R_{\text{eq}}}R = \frac{(12\text{V})(12\ \Omega)}{27\ \Omega} = \boxed{5.3 \text{ V}}$$

17. Strategy Use the definition of resistance.

Solution No current flows through the upper branch of the circuit, since V_x is open. The voltage across R is $\mathcal{E} = 45.0$ V. So, the current is $I = \mathcal{E}/R$. The voltage across R_x is V_x, and the current is I. Find R_x.

$$R_x = \frac{V_x}{I} = \frac{V_x}{\mathcal{E}}R = \frac{30.0 \text{ V}}{45.0 \text{ V}}(100.0\ \Omega) = \boxed{66.7\ \Omega}$$

21. (a) Strategy Use Eq. (17-18b).

Solution Find the capacitance.

$$U = \frac{1}{2}C(\Delta V)^2, \text{ so } C = \frac{2U}{(\Delta V)^2} = \frac{2(32 \text{ J})}{(300 \text{ V})^2} = \boxed{710\ \mu\text{F}}.$$

(b) Strategy Use Eq. (17-16).

Solution Find the dielectric constant.

$$C=\frac{\kappa A}{4\pi kd},\text{ so }\kappa=\frac{4\pi kdC}{A}=\frac{4\pi(8.988\times10^9\text{ N}\cdot\text{m}^2/\text{C}^2)(1.1\times10^{-6}\text{ m})(710\times10^{-6}\text{ F})}{9.0\text{ m}^2}=\boxed{9.8}.$$

(c) Strategy The average power produced is equal to the energy stored in the capacitor divided by the time it takes to discharge it.

Solution Compute the average power.

$$P_{\text{av}}=\frac{U}{\Delta t}=\frac{32\text{ J}}{4.0\times10^{-3}\text{ s}}=\boxed{8.0\text{ kW}}$$

(d) Strategy The capacitance of a parallel plate capacitor is inversely proportional to the plate separation, and the energy stored in a capacitor is directly proportional to its capacitance. Thus, the energy stored in a capacitor is inversely proportional to the plate separation.

Solution Form a proportion to find the new energy capacity of the capacitor.

$$\frac{U_2}{U_1}=\frac{d_1}{d_2},\text{ so }U_2=\frac{d_1}{d_2}U_1=\frac{d_1}{\frac{1}{2}d_1}U_1=2U_1=2(32\text{ J})=\boxed{64\text{ J}}.$$

MCAT Review

1. **Strategy and Solution** At a given temperature, the resistance R of a wire to direct current is given by $R=\rho L/A$, where ρ is the resistivity, L is the length, and A is the cross-sectional area. Therefore, the correct answer is $\boxed{\text{D}}$.

2. **Strategy** There are ten electric immersion heaters that each use 5 kW of power. Use Eq. (18-19).

   Solution The total power requirement to run all ten heaters is 50 kW. Find the current.

   $$P=I\Delta V,\text{ so }I=\frac{P}{\Delta V}=\frac{50\times10^3\text{ W}}{600\text{ V}}=83\text{ A}.$$

   The correct answer is $\boxed{\text{C}}$.

3. **Strategy** Each heater draws 20 A, so five heaters draw 100 A. Use Eq. (18-19).

   Solution Find the total power usage of the heaters.
   $P=I\Delta V=(100\text{ A})(800\text{ V})=80\text{ kW}$

   The correct answer is $\boxed{\text{C}}$.

4. **Strategy** Use Kirchhoff's rules and the definition of resistance.

   Solution Find the current flowing through R_L. According to the loop rule,

   $I_\text{L}R_\text{L}-I_\text{S}R_\text{S}=0,\text{ so }I_\text{S}=\frac{R_\text{L}}{R_\text{S}}I_\text{L}=\frac{1.0\ \Omega}{2.0\ \Omega}I_\text{L}=0.50I_\text{L}$. According to the junction rule,

   $I=I_\text{L}+I_\text{S}=I_\text{L}+0.50I_\text{L}=1.50I_\text{L},\text{ so }I_\text{L}=\frac{I}{1.50}$. Thus, the voltage drop across R_L is

   $V_\text{L}=I_\text{L}R_\text{L}=\frac{IR_\text{L}}{1.50}=\frac{(0.5\text{ A})(1.0\ \Omega)}{1.50}=0.33\text{ V}$. The correct answer is $\boxed{\text{B}}$.

5. Strategy Use Kirchhoff's rules and Eq. (18-21a).

Solution Find the current flowing through R_S. According to the loop rule,

$I_L R_L - I_S R_S = 0$, so $I_L = \frac{R_S}{R_L} I_S = \frac{3.0\ \Omega}{1.0\ \Omega} I_S = 3.0 I_S$. According to the junction rule,

$I = I_L + I_S = 3.0 I_S + I_S = 4.0 I_S$, so $I_S = 0.25 I$. Thus, the power dissipated in R_S is

$P_S = I_S{}^2 R_S = (0.25I)^2 R_S = [0.25(1.2\ \text{A})]^2 (3.0\ \Omega) = 0.27\ \text{W}$. The correct answer is $\boxed{\text{A}}$.

6. Strategy and Solution As current flows through R_L, power is dissipated at a constant rate by R_L as heat that enters the water, increasing its energy and raising its temperature (and the temperature of the system). Thus, the entropy of the system increases, as well. The correct answer is $\boxed{\text{D}}$.

7. Strategy and Solution Energy is stored in the battery as chemical energy. This energy is converted into electrical energy when the current flows. The electrical energy is dissipated as heat by the resistor. The correct answer is $\boxed{\text{A}}$.

8. Strategy and Solution As R_L increases with time, the amount of the current I passing through it decreases and the amount passing through R_S increases. The correct answer is $\boxed{\text{C}}$.

9. Strategy Use Eq. (14-4).

Solution The water is heated at a rate of $Q/\Delta t = mc\Delta T/\Delta t = 1.0\ \text{W}$. So, the time it takes for the temperature of the water to increase 1.0°C is $\Delta t = \frac{mc\Delta T}{1.0\ \text{W}} = \frac{(1.0\ \text{kg})[4.2\times10^3\ \text{J/(kg}\cdot{}^\circ\text{C)}](1.0^\circ\text{C})}{1.0\ \text{W}} = 4200\ \text{s}.$

The correct answer is $\boxed{\text{D}}$.

10. Strategy Use Eq. (18-19).

Solution Compute the current required.

$P = I\Delta V$, so $I = \frac{P}{\Delta V} = \frac{1.2\times10^4\ \text{W}}{120\ \text{V}} = 100\ \text{A}$. The correct answer is $\boxed{\text{C}}$.

11. Strategy Refer to the table to compute the initial and final resistances.

Solution Compute the resistances.

$R_i = (10^5\ \text{m})\frac{3.4\times10^{-1}\ \Omega}{10^3\ \text{m}} = 34\ \Omega$ and $R_f = (10^5\ \text{m})\frac{3.8\times10^{-1}\ \Omega}{10^3\ \text{m}} = 38\ \Omega.$

The change in resistance is $R_f - R_i = 38\ \Omega - 34\ \Omega = 4\ \Omega$. The correct answer is $\boxed{\text{C}}$.

12. Strategy Use Eq. (18-21a).

Solution Compute the power lost as heat.

$P = I^2 R = (2\ \text{A})^2 (3\ \Omega) = 12\ \text{W}$, so the correct answer is $\boxed{\text{C}}$.

13. Strategy and Solution The ten residences require $10\times10^4\ \text{W} = 10^5\ \text{W}$ of power and $5\times10^3\ \text{W}$ of power is lost as heat, so the total power requirement is $10^5\ \text{W} + 5\times10^3\ \text{W} = 1.05\times10^5\ \text{W}$. The correct answer is $\boxed{\text{C}}$.

Chapter 19

MAGNETIC FORCES AND FIELDS

Problems

1. Strategy The magnetic field is strong where field lines are close together and weak where they are far apart.

Solution

(a) The field lines are farthest apart (lowest density) at point $\boxed{F}$, so the magnetic field strength is smallest there.

(b) The field lines are closest together (highest density) at point $\boxed{A}$, so the magnetic field strength is largest there.

3. Strategy A bar magnet is a magnetic dipole. Field lines emerge from a bar magnet at its north pole and enter at its south pole. Magnetic field lines are closed loops. Use symmetry.

Solution

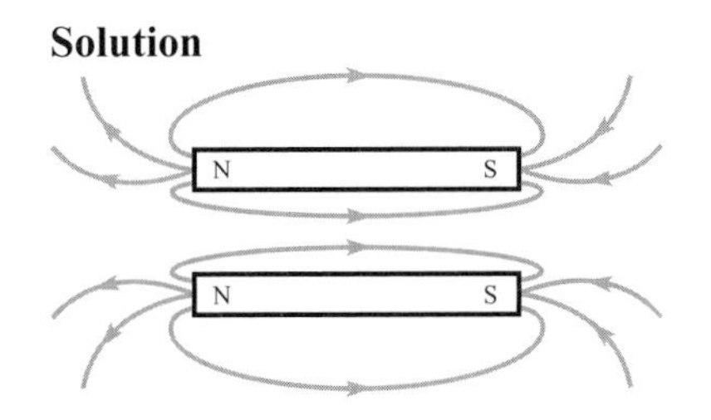

5. Strategy A bar magnet is a magnetic dipole. Field lines emerge from a bar magnet at its north pole and enter at its south pole. Magnetic field lines are closed loops. Use symmetry.

Solution

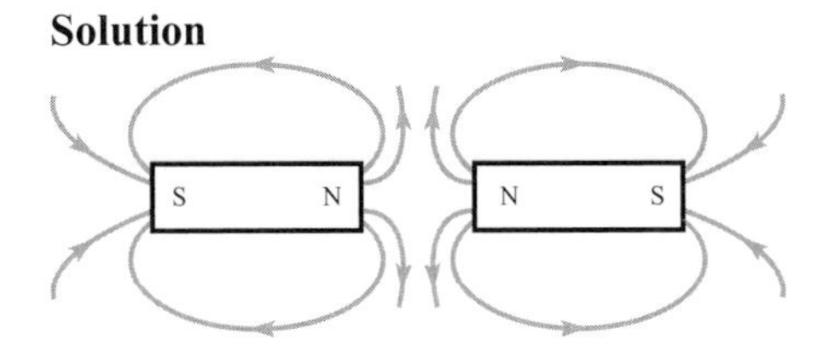

9. Strategy Determine the speed of the proton using its kinetic energy. Then determine the magnetic force on it using Eq. (19-5).

Solution Find the speed.

p $\vec{\mathbf{B}}$ $\vec{\mathbf{v}}$ $\overrightarrow{\text{N}}$

$K=\dfrac{1}{2}mv^2$, so $v=\sqrt{\dfrac{2K}{m}}$.

Calculate the force.

$$\vec{\mathbf{F}}=e\vec{\mathbf{v}}\times\vec{\mathbf{B}}=(1.602\times10^{-19}\text{ C})\left\{\left[\sqrt{\frac{2(8.0\times10^{-13}\text{ J})}{1.673\times10^{-27}\text{ kg}}}\text{ down}\right]\times(1.5\text{ T north})\right\}=\boxed{7.4\times10^{-12}\text{ N east}}$$

13. Strategy The magnetic force on a moving point charge is given by Eq. (19-5). The direction of the force is determined by the right-hand rule and the sign of the charge. Use Newton's second law.

Solution Find the acceleration.

$m\vec{\mathbf{a}} = -e\vec{\mathbf{v}} \times \vec{\mathbf{B}} = -e[(v \text{ right}) \times (B \text{ up})] = evB$ into the page, so $\vec{\mathbf{a}} = \dfrac{evB}{m}$ into the page.

Let v_d be the deflection speed and $\Delta t_d = d/v$ be the time of deflection where $d = 0.024$ m.

$$v_d = a\Delta t_d = a\left(\frac{d}{v}\right) = \frac{eBd}{m}$$

Calculate the deflection with $L = 0.25$ m.

$$\text{deflection} = v_d\Delta t = \frac{eBd}{m}\left(\frac{L}{v}\right) = \frac{(1.602\times10^{-19}\text{ C})(0.20\times10^{-2}\text{ T})(0.024\text{ m})(0.25\text{ m})}{(9.109\times10^{-31}\text{ kg})(1.8\times10^{7}\text{ m/s})} = 12\text{ cm}$$

The deflection is $\boxed{\text{12 cm into the page}}$.

15. Strategy Since the (negatively charged) electron is moving due west and the magnetic force is upward, the component of the magnetic field perpendicular to the motion of the electron points north. Use Eq. (19-1).

Solution Find the angle.

$$evB\sin\theta = F, \text{ so } \theta = \sin^{-1}\frac{F}{evB} = \sin^{-1}\frac{3.2\times10^{-14}\text{ N}}{(1.602\times10^{-19}\text{ C})(2.0\times10^{5}\text{ m/s})(1.2\text{ T})} = 56°.$$

$\vec{\mathbf{B}}$ $\vec{\mathbf{B}}$ N $\vec{\mathbf{v}}$ $\vec{\mathbf{F}}_B$

$\boxed{\text{There are two possibilities: 56° N of W and 56° N of E.}}$

17. Strategy Use Eqs. (19-1) and (19-5) to find the magnitude and direction of the force, respectively.

Solution The charge of the muon is negative. According to the RHR and $\vec{\mathbf{F}}_B = q\vec{\mathbf{v}} \times \vec{\mathbf{B}}$, the magnetic force on the muon is into the page, or to the west.

$\vec{\mathbf{B}}$ 55° $\vec{\mathbf{v}}$ N

Compute the magnitude.

$$F_B = |q|(v\sin\theta)B$$
$$= (1.602\times10^{-19}\text{ C})(4.5\times10^{7}\text{ m/s})\sin(55° + 90°)(5.0\times10^{-5}\text{ T}) = 2.1\times10^{-16}\text{ N}$$

Thus, the force on the muon is $\boxed{2.1\times10^{-16}\text{ N to the west}}$.

21. Strategy Use conservation of energy and Equation (19-7).

Solution Find the speed of the ion.

$$\Delta K = \frac{1}{2}mv^2 = -\Delta U = eV, \text{ so } v = \sqrt{\frac{2eV}{m}}.$$

Solve for the mass of the ion.

$$\frac{v^2}{r} = \frac{evB}{m}$$

$$m = \frac{erB}{v} = erB\sqrt{\frac{m}{2eV}}$$

$$\sqrt{m} = rB\sqrt{\frac{e}{2V}}$$

$$m = \frac{er^2B^2}{2V} = \frac{(1.602\times10^{-19}\text{ C})(0.125\text{ m})^2(1.2\text{ T})^2}{2(7.0\times10^{3}\text{ V})} = \boxed{2.6\times10^{-25}\text{ kg}}$$

23. Strategy Use conservation of energy and Equation (19-7).

Solution Find the speed of the ion.

$\Delta K = \frac{1}{2}mv^2 = -\Delta U = eV$, so $v = \sqrt{\frac{2eV}{m}}$.

Solve for the magnitude of the magnetic field.

$\frac{evB}{m} = \frac{v^2}{r}$, so

$$B = \frac{mv}{er} = \frac{m}{er}\sqrt{\frac{2eV}{m}} = \frac{1}{r}\sqrt{\frac{2mV}{e}} = \frac{1}{0.21\text{ m}}\sqrt{\frac{2(12.0\text{ u})(1.6605\times10^{-27}\text{ kg/u})(5.0\times10^3\text{ V})}{1.602\times10^{-19}\text{ C}}} = \boxed{0.17\text{ T}}.$$

25. Strategy Since the ions are accelerated through the same potential difference and they have the same charge, they have the same kinetic energy. Let m_u be the mass of the unknown element. Use Eq. (19-7) and refer to the periodic table.

Solution

(a) Relate the speeds to the masses.

$\frac{1}{2}m_\text{S}v_\text{S}^2 = \frac{1}{2}m_\text{u}v_\text{u}^2$, so $\frac{v_\text{S}}{v_\text{u}} = \sqrt{\frac{m_\text{u}}{m_\text{S}}}$.

$\frac{v^2}{r} = \frac{|q|vB}{m}$, so $r = \frac{mv}{|q|B}$. Form a proportion.

$$\frac{r_\text{S}}{r_\text{u}} = \frac{m_\text{S}v_\text{S}}{m_\text{u}v_\text{u}} = \frac{m_\text{S}}{m_\text{u}}\sqrt{\frac{m_\text{u}}{m_\text{S}}} = \sqrt{\frac{m_\text{S}}{m_\text{u}}}, \text{ or } m_\text{u} = \left(\frac{r_\text{u}}{r_\text{S}}\right)^2 m_\text{S} = \left(\frac{r_\text{S}+1.07\text{ cm}}{r_\text{S}}\right)^2 m_\text{S} = \left(1+\frac{1.07\text{ cm}}{r_\text{S}}\right)^2 m_\text{S}.$$

Similarly, we have $\frac{r_\text{Mn}}{r_\text{S}} = \sqrt{\frac{m_\text{Mn}}{m_\text{S}}}$, so $r_\text{Mn} = r_\text{S}\sqrt{\frac{m_\text{Mn}}{m_\text{S}}}$. Find r_S.

$$r_\text{Mn} - r_\text{S} = r_\text{S}\left(\sqrt{\frac{m_\text{Mn}}{m_\text{S}}} - 1\right), \text{ so } r_\text{S} = \frac{r_\text{Mn} - r_\text{S}}{\sqrt{\frac{m_\text{Mn}}{m_\text{S}}} - 1} = \frac{3.20\text{ cm}}{\sqrt{\frac{55\text{ u}}{32\text{ u}}} - 1} = 10.3\text{ cm}.$$

Therefore, the mass of the unknown element is $m_\text{u} = \left(1+\frac{1.07\text{ cm}}{10.3\text{ cm}}\right)^2(32\text{ u}) = \boxed{39\text{ u}}$.

(b) According to the periodic table, the unknown element is $\boxed{\text{potassium}}$.

29. Strategy The magnetic force on a moving point charge is given by Eq. (19-5). The direction of the force is determined by the right-hand rule and the sign of the charge.

Solution

(a) By the RHR and $\vec{\mathbf{F}} = q\vec{\mathbf{v}}\times\vec{\mathbf{B}}$, charge carriers will be deflected toward the top or bottom edges of the strip, depending on their sign. So, the voltmeter should be connected across the width of the strip.

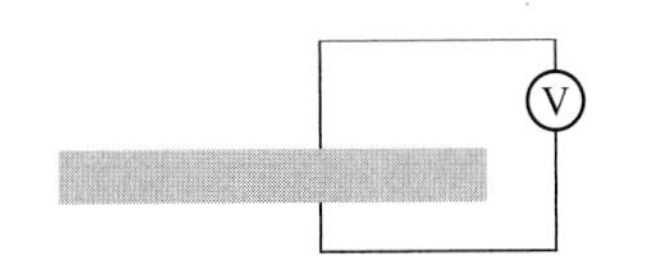

(b) Since the current flows to the right, the electrons flow to the left, so the magnetic force is upward, according to $\vec{\mathbf{F}} = -e\vec{\mathbf{v}} \times \vec{\mathbf{B}}$. Thus, the top edge of the strip has an excess of electrons, and thus, is negatively charged. The bottom edge is positively charged. So, the bottom edge is at the higher potential.

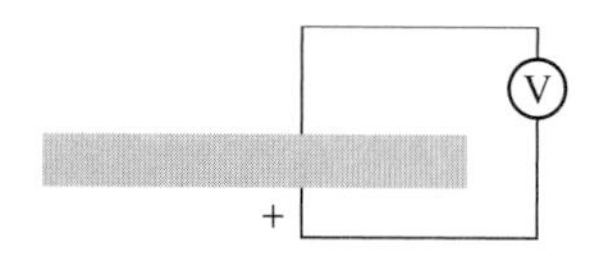

31. Strategy As found in Example 19.7, the Hall voltage is given by $V_H = BI/(net)$.

Solution Find the density of the carriers.

$$n = \frac{BI}{etV_H} = \frac{(0.43\text{ T})(54\text{ A})}{(1.602\times10^{-19}\text{ C})(0.00024\text{ m})(7.2\times10^{-6}\text{ V})} = \boxed{8.4\times10^{28}\text{ m}^{-3}}$$

33. Strategy The magnetic force on a moving point charge is given by Eq. (19-5). The direction of the force is determined by the right-hand rule and the sign of the charge. Use Eqs. (17-10) and (19-10).

Solution

(a) According to the RHR and $\vec{\mathbf{F}} = q\vec{\mathbf{v}} \times \vec{\mathbf{B}}$, the (negatively charged) electrons experience an $\boxed{\text{upward}}$ magnetic force.

(b) The drift velocity is given by $v_d = \frac{E_H}{B}$ where $E_H = \frac{V_H}{w}$. So, $v_d = \frac{V_H}{wB} = \frac{20.0\times10^{-6}\text{ V}}{(0.020\text{ m})(5.0\text{ T})} = \boxed{0.20\text{ mm/s}}$.

35. Strategy The blood speed is equal to the Hall drift speed $v_D = E_H/B = V_H/(wB)$, where the width w is the diameter of the artery. The flow rate is equal to the drift speed times the cross-sectional area of the artery. The magnetic force on a moving charge is given by Eq. (19-5). The direction of the force is determined by the right-hand rule and the sign of the charge.

Solution

(a) Compute the drift speed.

$$v_D = \frac{V_H}{wB} = \frac{0.35\times10^{-3}\text{ V}}{(0.0040\text{ m})(0.25\text{ T})} = \boxed{0.35\text{ m/s}}$$

$\vec{\mathbf{B}}\downarrow$ $\leftarrow\vec{\mathbf{v}}$ $\overrightarrow{\text{N}}$

(b) Compute the flow rate.

$$\text{flow rate} = v_D A = (0.35\text{ m/s})\pi(0.0020\text{ m})^2 = \boxed{4.4\times10^{-6}\text{ m}^3/\text{s}}$$

(c) The positive ions are deflected east (south $\times$ down), so $\boxed{\text{the east lead}}$ is at the higher potential.

37. Strategy The maximum force occurs when $\vec{\mathbf{L}}$ and $\vec{\mathbf{B}}$ are perpendicular. Use Eq. (19-12b).

Solution

(a) $F_{max} = ILB\sin 90° = (18.0\text{ A})(0.60\text{ m})(0.20\text{ T})(1) = \boxed{2.2\text{ N}}$

(b) Only the maximum possible force can be calculated, since only the magnitudes, and not the directions of $\vec{\mathbf{B}}$ and $\vec{\mathbf{L}}$ are given.

41. Strategy The net force in the *y*-direction must be zero. Use Newton's second law and Eqs. (19-12a) and (19-12b).

Solution Find the current.

$$\Sigma F_y = ILB\sin\theta - mg = 0,\ \text{so}\ I = \frac{mg}{LB\sin\theta} = \frac{(0.025\ \text{kg})(9.80\ \text{N/kg})}{(1.00\ \text{m})(0.75\ \text{T})\sin 90^\circ} = 0.33\ \text{A}.$$

The magnetic force must be upward to oppose gravity, so according to the RHR and $\vec{\mathbf{F}} = I\vec{\mathbf{L}}\times\vec{\mathbf{B}}$, the current is $\boxed{0.33\ \text{A to the left}}$.

43. Strategy Use Eq. (19-12a).

Solution

(a) Calculate the force on each wire segment.

$\vec{\mathbf{F}}_{\text{top}} = I\vec{\mathbf{L}}_{\text{top}}\times\vec{\mathbf{B}} = (1.0\ \text{A})(0.300\ \text{m right})\times(2.5\ \text{T left}) = \boxed{0}$

$\vec{\mathbf{F}}_{\text{bottom}} = I\vec{\mathbf{L}}_{\text{bottom}}\times\vec{\mathbf{B}} = -I\vec{\mathbf{L}}_{\text{top}}\times\vec{\mathbf{B}} = \boxed{0}$

$\vec{\mathbf{F}}_{\text{left}} = I\vec{\mathbf{L}}_{\text{left}}\times\vec{\mathbf{B}} = (1.0\ \text{A})(0.200\ \text{m up})\times(2.5\ \text{T left}) = \boxed{0.50\ \text{N out of the page}}$

$\vec{\mathbf{F}}_{\text{right}} = I\vec{\mathbf{L}}_{\text{right}}\times\vec{\mathbf{B}} = -I\vec{\mathbf{L}}_{\text{left}}\times\vec{\mathbf{B}} = \boxed{0.50\ \text{N into the page}}$

(b) Compute the net force.

$\vec{\mathbf{F}}_{\text{net}} = \vec{\mathbf{F}}_{\text{top}} + \vec{\mathbf{F}}_{\text{bottom}} + \vec{\mathbf{F}}_{\text{left}} + \vec{\mathbf{F}}_{\text{right}} = 0 + 0 + 0.50\ \text{N out of the page} + 0.50\ \text{N into the page} = \boxed{0}$

45. (a) Strategy The torque on a current loop is $\tau = NIAB\sin\theta$. In this case (maximum torque) $\sin\theta = \sin 90^\circ = 1$.

Solution Find *B*.

$$\tau = NIAB,\ \text{so}\ B = \frac{\tau}{NIA} = \frac{0.0020\ \text{N}\cdot\text{m}}{100(0.075\ \text{A})\pi(0.020\ \text{m})^2} = \boxed{0.21\ \text{T}}.$$

(b) Strategy Use Eq. (19-12a) and symmetry.

Solution The field points from the north pole to the south. Due to symmetry, $\vec{\mathbf{L}}_{\text{av}}$ is parallel to the faces of the poles and into the page for the right half of the loop; $\vec{\mathbf{L}}_{\text{av}}$ is perpendicular to $\vec{\mathbf{B}}$. So, $\vec{\mathbf{L}}_{\text{av}}\times\vec{\mathbf{B}}$ is directed downward. By similar reasoning, the force on the right half of the loop is equal and opposite. Thus, the torque is $\boxed{\text{clockwise}}$.

49. Strategy Assume θ is small. Use Hooke's law and Eq. (19-13a). Set the sum of the torques equal to zero.

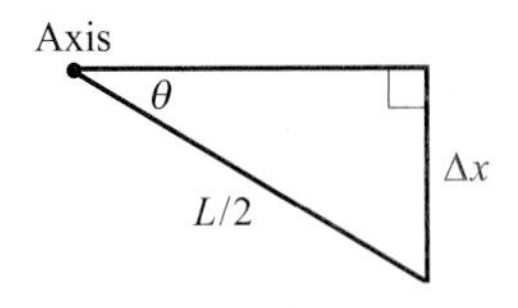

Solution The torque on the loop of wire due to the magnetic field is $\tau_{\text{B}} = IAB = IL^2B$, where *L* is the side-length of the loop. The torque on the loop due to the spring is $\tau_{\text{s}} = k\Delta x(L/2)$. Find Δx.

$$\Sigma\tau = -IL^2B + \frac{kL\Delta x}{2} = 0,\ \text{so}\ \Delta x = \frac{2ILB}{k}.\ \text{Find}\ \theta.$$

$$\theta = \tan^{-1}\frac{\Delta x}{L/2} = \tan^{-1}\frac{2ILB/k}{L/2} = \tan^{-1}\frac{4IB}{k} = \tan^{-1}\frac{4(9.0\ \text{A})(1.3\ \text{T})}{550\ \text{N/m}} = \boxed{4.9^\circ}$$

53. Strategy RHR 2 gives the direction of $\vec{\mathbf{B}}$.

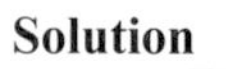

Solution

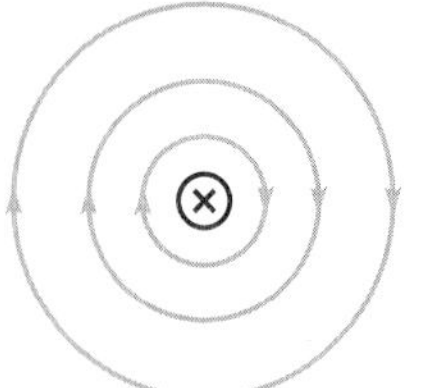

55. Strategy Use the principle of superposition and the field due to a long straight current-carrying wire, Eq. (19-14).

Solution

(a) According to RHR 2, the field due to the bottom wire is out of the page and that due to the top wire is into the page. Since the bottom wire is closer to P, the net field is out of the page.

$$B=\frac{\mu_0 I}{2\pi}\left(\frac{1}{r_{\text{bottom}}}-\frac{1}{r_{\text{top}}}\right)=\frac{(4\pi\times10^{-7}\text{ T}\cdot\text{m/A})(10.0\text{ A})}{2\pi}\left(\frac{1}{0.25\text{ m}}-\frac{1}{0.25\text{ m}+0.0030\text{ m}}\right)=9\times10^{-8}\text{ T}$$

So, $\vec{\mathbf{B}}(P)=\boxed{9\times10^{-8}\text{ T out of the page}}$.

(b) Both fields are now out of the page at P.

$$B=\frac{\mu_0 I}{2\pi}\left(\frac{1}{r_{\text{bottom}}}+\frac{1}{r_{\text{top}}}\right)=\frac{(4\pi\times10^{-7}\text{ T}\cdot\text{m/A})(10.0\text{ A})}{2\pi}\left(\frac{1}{0.25\text{ m}}+\frac{1}{0.25\text{ m}+0.0030\text{ m}}\right)=1.6\times10^{-5}\text{ T}$$

So, $\vec{\mathbf{B}}(P)=\boxed{1.6\times10^{-5}\text{ T out of the page}}$.

57. Strategy Use Eqs. (19-5) and (19-14).

Solution $\vec{\mathbf{B}}$ is into the page at the electron. $\vec{\mathbf{v}}\times\vec{\mathbf{B}}$ is to the left, so $-\vec{\mathbf{v}}\times\vec{\mathbf{B}}$ is to the right (parallel to the current).

$$\vec{\mathbf{F}}=q\vec{\mathbf{v}}\times\vec{\mathbf{B}}=ev\left(\frac{\mu_0 I}{2\pi r}\right)\text{ parallel to the current}$$
$$=\frac{(1.602\times10^{-19}\text{ C})(1.0\times10^{7}\text{ m/s})(4\pi\times10^{-7}\text{ T}\cdot\text{m/A})(50.0\text{ A})}{2\pi(0.050\text{ m})}\text{ parallel to the current}$$
$$=\boxed{3.2\times10^{-16}\text{ N parallel to the current}}$$

61. Strategy Use the field due to a long straight current-carrying wire, Eq. (19-14). Let $r=2.0$ cm.

Solution

$B_1=\dfrac{\mu_0 I}{2\pi r}$ and $B_2=\dfrac{\mu_0 I}{2\pi(3r)}=\dfrac{\mu_0 I}{6\pi r}$.

$B_2<B_1$, so the direction of the field at point P is the same as the direction of the field due to wire 1. According to RHR 2, wire 1's current must be $\boxed{\text{into the page}}$. So, wire 2's current is out of the page.

$$\vec{\mathbf{B}}_{\text{net}}=\frac{\mu_0 I}{2\pi r}\text{ down}+\frac{\mu_0 I}{6\pi r}\text{ down}=\frac{2\mu_0 I}{3\pi r}\text{ down}$$

Compute the current.

$$I=\frac{3\pi r B_{\text{net}}}{2\mu_0}=\frac{3\pi(0.020\text{ m})(1.0\times10^{-2}\text{ T})}{2(4\pi\times10^{-7}\text{ T}\cdot\text{m/A})}=\boxed{750\text{ A}}$$

65. Strategy Use Ampère's law. Currents out of the page are positive and into the page are negative.

Solution Sum the currents enclosed within each loop.

(a) $I_{net} = 14I + (-6I) + (-3I) = 5I$

So, the net current is $\boxed{5I \text{ out of the page}}$.

(b) $I_{net} = 14I + (-16I) = -2I$

So, the net current is $\boxed{2I \text{ into the page}}$.

69. Strategy and Solution Magnets must have both north and south poles. So, the new polarities are $\boxed{\text{(c) S and (d) N}}$.

73. Strategy and Solution

(a) The best permanent magnet is obtained when $B_{final} / B_{0,\,max}$ is greatest. Figure (a) shows that B is a significant fraction of $B_{0,\,max}$ when B_0 is zero (turned off), so the material represented by $\boxed{\text{graph (a)}}$ would make the best permanent magnet.

(b) The core for an electromagnet should not be a permanent magnet. B is nearly zero when B_0 has returned to zero for case (c). So, the material represented by $\boxed{\text{graph (c)}}$ would make the best core.

77. Strategy Use Eqs. (19-1) and (19-5).

Solution

(a) Compute the magnitude of the force on the particle.

$$F = |q|(v\sin\theta)B = (0.020\times10^{-6}\text{ C})(2.0\text{ m/s})\sin 60°(0.50\text{ T}) = \boxed{1.7\times10^{-8}\text{ N}}$$

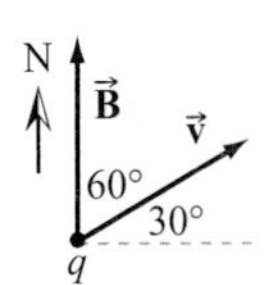

(b) The charge on the particle is positive. According to $\vec{\mathbf{F}} = q\vec{\mathbf{v}}\times\vec{\mathbf{B}}$ and the RHR, the force on the particle is directed out of the page or $\boxed{\text{up}}$.

81. Strategy Use Eq. (19-6) and Newton's second law.

Solution $\Sigma F_r = qvB = ma_r = mv^2/r$, so $v = qrB/m$. The period is $T = C/v = 2\pi r/v$. Substitute for v in the equation for T.

$$T = \frac{2\pi r}{\frac{qrB}{m}} = \boxed{\frac{2\pi m}{qB}}$$, which is independent of the particle's speed.

85. Strategy Use Eqs. (17-10), (19-5), (19-7), and (19-10), and conservation of energy.

Solution

(a) According to RHR 1 and $\vec{\mathbf{F}} = q\vec{\mathbf{v}} \times \vec{\mathbf{B}},$ the ions must be $\boxed{\text{positive}}$; they are missing an electron.

(b) The east plate must be negatively charged to accelerate the ions to the right, so the $\boxed{\text{west}}$ plate must be positively charged.

(c) According to RHR 1 and $\vec{\mathbf{F}}_B = q\vec{\mathbf{v}} \times \vec{\mathbf{B}},$ the magnetic force on the positively charged ions is upward between the plates, so $\vec{\mathbf{E}}$ must be downward to select the correct velocity $(\vec{\mathbf{F}}_E + \vec{\mathbf{F}}_B = 0).$ Since $\vec{\mathbf{E}}$ is downward, the $\boxed{\text{north}}$ plate is positively charged.

(d) Using energy conservation, $\frac{1}{2}mv^2 = e\Delta V_1.$

For a parallel plate capacitor, $\Delta V_2 = Ed.$ Now, $\frac{v^2}{r} = \frac{evB}{m},$ so $v = \frac{erB}{m}.$

For a velocity selector, $E = vB$, or $v = E/B$. So, $E = \frac{erB^2}{m}$ and

$$\Delta V_2 = \frac{erB^2 d}{m} = \frac{(1.602\times10^{-19}\text{ C})(0.10\text{ m})(0.20\text{ T})^2(0.010\text{ m})}{(12\text{ u})(1.6605\times10^{-27}\text{ kg/u})} = \boxed{320\text{ V}}$$

Find ΔV_1.

$$\Delta V_1 = \frac{mv^2}{2e} = \frac{m}{2e}\left(\frac{erB}{m}\right)^2 = \frac{er^2B^2}{2m} = \frac{(1.602\times10^{-19}\text{ C})(0.10\text{ m})^2(0.20\text{ T})^2}{2(12\text{ u})(1.6605\times10^{-27}\text{ kg/u})} = \boxed{1.6\text{ kV}}$$

(e) Compute the correct values of the potential differences.

$$\Delta V_1 = \frac{(1.602\times10^{-19}\text{ C})(0.10\text{ m})^2(0.20\text{ T})^2}{2(14\text{ u})(1.6605\times10^{-27}\text{ kg/u})} = \boxed{1.4\text{ kV}}$$

$$\Delta V_2 = \frac{(1.602\times10^{-19}\text{ C})(0.10\text{ m})(0.20\text{ T})^2(0.010\text{ m})}{(14\text{ u})(1.6605\times10^{-27}\text{ kg/u})} = \boxed{280\text{ V}}$$

89. (a) Strategy The currents have the same magnitude and are equidistant from the point at which the net magnetic field's direction is evaluated. So, the magnitudes of the four fields are the same.

Solution According to RHR 2 and symmetry, the field directions are the following:

	Current	Direction at the center of the square
1	top left	toward the top right wire
2	top right	toward the bottom right wire
3	bottom left	toward the bottom right wire
4	bottom right	toward the top right wire

1 or 4
2 or 3
y
x

Since the magnitudes of the fields are equal, the vertical component of the net field is zero and the horizontal component is $\boxed{\text{to the right}}$.

(b) Strategy Each field is at a 45° angle to the horizontal (either above or below). The y-components cancel and the x-components add. The field for a long thin wire is $B = \mu_0 I/(2\pi r)$.

Solution Find the magnitude of the field.

$$B_x = \frac{\mu_0 I}{2\pi r}\cos 45° = \frac{\mu_0 I}{2\sqrt{2}\pi r}. \text{ Thus, } B_{\text{net},\,x} = \frac{4\mu_0 I}{2\sqrt{2}\pi r} = \frac{\sqrt{2}\mu_0 I}{\pi r} = B_{\text{net}}, \text{ where}$$

$$r = \sqrt{\left(\frac{s}{2}\right)^2 + \left(\frac{s}{2}\right)^2} = \sqrt{\frac{s^2}{2}} = \frac{s}{\sqrt{2}}. \text{ So, } B_{\text{net}} = \frac{2\mu_0 I}{\pi s} = \frac{2(4\pi \times 10^{-7}\text{ T}\cdot\text{m/A})(10.0\text{ A})}{\pi(0.10\text{ m})} = \boxed{80\ \mu\text{T}}.$$

93. Strategy The velocity of the electrons is to the right. Use Eq. (19-5).

Solution The magnetic field points upward, toward the south pole of the magnet. According to $\vec{\mathbf{F}} = -e\vec{\mathbf{v}} \times \vec{\mathbf{B}}$ and RHR 1, the electrons are deflected (and thus the beam moves) $\boxed{\text{into the page}}$.

$\vec{\mathbf{B}}$ $\vec{\mathbf{F}}$ $\vec{\mathbf{v}}$

97. (a) Strategy From Example 19.5, $v = eBr/m$, and $f = v/C = v/(2\pi r)$.

Solution Find the frequency of oscillation.

$$f = \frac{eBr}{2\pi rm} = \frac{eB}{2\pi m} = \frac{(1.602\times10^{-19}\text{ C})(1.3\text{ T})}{2\pi(1.673\times10^{-27}\text{ kg})} = \boxed{20\text{ MHz}}$$

(b) Strategy Use Eq. (6-6) and $v = eBr/m$.

Solution Find the kinetic energy.

$$K = \frac{1}{2}mv^2 = \frac{1}{2}m\left(\frac{eBr}{m}\right)^2 = \frac{e^2B^2r^2}{2m} = \frac{(1.602\times10^{-19}\text{ C})^2(1.3\text{ T})^2(0.16\text{ m})^2}{2(1.673\times10^{-27}\text{ kg})} = \boxed{3.3\times10^{-13}\text{ J}}$$

(c) Strategy $\Delta U = q\Delta V = e\Delta V$; equate this with the final kinetic energy and solve for ΔV.

Solution Find the equivalent voltage.

$$\Delta V = \frac{\Delta U}{e} = \frac{K}{e} = \frac{eB^2r^2}{2m} = \frac{(1.602\times10^{-19}\text{ C})(1.3\text{ T})^2(0.16\text{ m})^2}{2(1.673\times10^{-27}\text{ kg})} = \boxed{2.1\text{ MV}}$$

(d) Strategy and Solution The energy of a proton increases by $e\Delta V$ each time it crosses the gap, or $2e\Delta V$ for each revolution (two gap crossings), so its total energy is $2Ne\Delta V$, where N is the number of revolutions. Equate this with the final kinetic energy and solve for N.

$$2Ne\Delta V = K, \text{ so } N = \frac{K}{2e\Delta V} = \frac{3.32\times10^{-13}\text{ J}}{2(1.602\times10^{-19}\text{ C})(10.0\times10^3\text{ V/rev})} = \boxed{100\text{ rev}}.$$

Chapter 20

ELECTROMAGNETIC INDUCTION

Problems

1. **Strategy** Use Eqs. (20-2a), (19-5), and (19-12a).

Solution

(a) The motional emf is $\mathscr{E} = vBL,$ so $I = \dfrac{\mathscr{E}}{R} = \boxed{\dfrac{vBL}{R}}.$

(b) By the RHR and $\vec{\mathbf{F}} = -e\vec{\mathbf{v}} \times \vec{\mathbf{B}},$ the direction of the force on the electrons in the rod is down. So, the direction of the current is $\boxed{\text{CCW}}$.

(c) By the RHR and $\vec{\mathbf{F}} = I\vec{\mathbf{L}} \times \vec{\mathbf{B}},$ the direction of the magnetic force on the rod is $\boxed{\text{left}}$.

(d) $F = ILB = \dfrac{vBL}{R} LB = \boxed{\dfrac{vB^2L^2}{R}}$

3. **Strategy** Use the result of Problem 1d, $P = Fv,$ and Eqs. (20-2a) and (18-21b).

Solution

(a) According to Problem 1d, the magnitude of the magnetic force on the rod is vB^2L^2/R. The net force must be zero for constant velocity. So, $F_{\text{ext}} = F_{\text{B}} = \boxed{\dfrac{vB^2L^2}{R}}.$

(b) $\dfrac{\Delta W}{\Delta t} = P = Fv = \boxed{\dfrac{v^2B^2L^2}{R}}$

(c) $\mathscr{E} = vBL,$ so $P = \dfrac{\mathscr{E}^2}{R} = \boxed{\dfrac{v^2B^2L^2}{R}}.$

(d) Energy is conserved since the rate at which the external force does work is equal to the power dissipated in the resistor.

5. (a) Strategy Use Newton's second law, Eq. (19-12b), and Eq. (20-2a).

Solution $\Sigma F_y = F_B - mg = 0$, so $F_B = mg$ when the rod is falling with constant velocity. The magnitude of the magnetic force is $F_B = ILB = \frac{\mathscr{E}}{R} LB = \frac{vBL}{R} LB = \frac{vL^2B^2}{R}$. Set this equal to mg and solve for the terminal speed.

$$\frac{vL^2B^2}{R} = mg, \text{ so } v = \frac{mgR}{L^2B^2} = \frac{(0.0150\text{ kg})(9.80\text{ m/s}^2)(8.00\ \Omega)}{(1.30\text{ m})^2(0.450\text{ T})^2} = \boxed{3.44\text{ m/s}}.$$

(b) Strategy Use the potential energy in a uniform gravitational field and Eqs. (18-21b) and (20-2a).

Solution The change in gravitational energy per second is

$$\frac{\Delta U}{\Delta t} = \frac{mg\Delta y}{\Delta t} = -mgv = -\frac{m^2g^2R}{L^2B^2} = -\frac{(0.0150\text{ kg})^2(9.80\text{ m/s}^2)^2(8.00\ \Omega)}{(1.30\text{ m})^2(0.450\text{ T})^2} = -0.505\text{ W},$$

so the magnitude of the change is 0.505 W. The power dissipated is

$$P = \frac{\mathscr{E}^2}{R} = \frac{v^2B^2L^2}{R} = \left(\frac{mgR}{L^2B^2}\right)^2 \frac{B^2L^2}{R} = \frac{m^2g^2R}{L^2B^2} = 0.505\text{ W}.$$

The magnitude of the change in gravitational potential energy per second and the power dissipated in the resistor are the same, 0.505 W.

9. Strategy Use Eqs. (19-5) and (20-2a).

Solution

(a) By $\vec{\mathbf{F}} = -e\vec{\mathbf{v}} \times \vec{\mathbf{B}}$, the electrons are forced toward the center of the disk, so $\boxed{\text{positive}}$ charge accumulates on the edge of the disk.

(b) The motional emf is $\mathscr{E} = vBL$ where the average speed is $v = \frac{1}{2}\omega R$ and L is the radius R.

$$\Delta V = vBL = \left(\frac{\omega R}{2}\right) B(R) = \boxed{\frac{1}{2}\omega BR^2}$$

13. Strategy According to Lenz's law, the direction of an induced current in a loop always opposes the *change* in magnetic flux that induces the current. According to Faraday's law, the rate of change of the magnetic flux is equal to the induced emf.

Solution

(a) According to the RHR, $\vec{\mathbf{B}}$ is directed into the page at the loop. Since the current is decreasing and $B \propto I$ for a long straight wire, $\vec{\mathbf{B}}$ is decreasing, thus a $\boxed{\text{CW}}$ current will flow in the loop to generate a magnetic field directed into the page.

(b) At the uppermost point of the loop, the current in it is parallel to the current in the wire. Parallel currents attract and opposite currents repel, but since the half of the loop with the parallel components of current is closest to the long straight wire, the force on the circle is toward the wire. Thus, the external force must be $\boxed{\text{away from the long straight wire}}$.

(c) By Faraday's law, $\left|\frac{\Delta \Phi_B}{\Delta t}\right| = |\mathscr{E}| = iR = (84\times10^{-3}\text{ A})(24\ \Omega) = \boxed{2.0\text{ Wb/s}}$.

17. Strategy According to Lenz's law, the direction of an induced current in a loop always opposes the *change* in magnetic flux that induces the current.

Solution

(a) When the switch is closed, the CW flowing current in the large coil increases from zero to its maximum value. This current generates a magnetic field directed to the right that increases as the current increases. A current in the small coil is induced creating a field that opposes the increasing magnetic field and, thus, its direction of flow is $\boxed{\text{to the right}}$.

(b) The situation is opposite to that found in part (a) (I and B decrease). So, the current flows $\boxed{\text{to the left}}$.

(c)

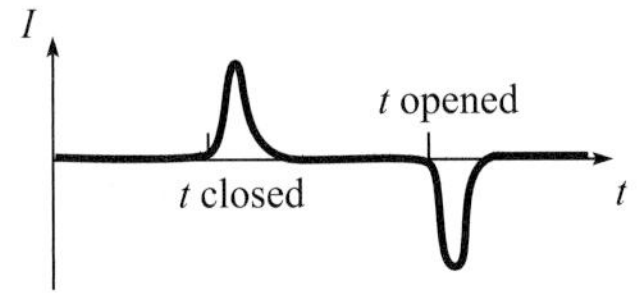

21. (a) Strategy When the motor first starts there is no back emf.

Solution Compute the current when the motor first starts up.

$$I = \frac{\mathcal{E}_{\text{ext}}}{R} = \frac{120.0\text{ V}}{16\ \Omega} = \boxed{7.5\text{ A}}$$

(b) Strategy When the motor is at full speed, the current is smaller due to the back emf.

Solution Compute the current when the motor is at full speed.

$$I = \frac{\mathcal{E}_{\text{ext}} - \mathcal{E}_{\text{back}}}{R} = \frac{120.0\text{ V} - 72\text{ V}}{16\ \Omega} = \boxed{3.0\text{ A}}$$

(c) Strategy When the motor is at less than full speed, the back emf is less than its maximum value.

Solution Compute the back emf.

$$I = \frac{\mathcal{E}_{\text{ext}} - \mathcal{E}_{\text{back}}}{R},\ \text{so}\ \mathcal{E}_{\text{back}} = \mathcal{E}_{\text{ext}} - IR = 120.0\text{ V} - (4.0\text{ A})(16\ \Omega) = \boxed{56\text{ V}}.$$

25. Strategy Use Eq. (20-10).

Solution Find the secondary current amplitude.

$$\frac{I_1}{I_2} = \frac{\mathcal{E}_2}{\mathcal{E}_1} = \frac{N_2}{N_1},\ \text{so}\ I_2 = \frac{N_1}{N_2} I_1 = 100(1.0\times10^{-3}\text{ A}) = \boxed{0.10\text{ A}}.$$

29. Strategy Use Eq. (20-9).

Solution Compute the number of turns required for the secondary coil.

$$N_2 = \frac{\mathcal{E}_2}{\mathcal{E}_1} N_1 = \frac{220\text{ V}}{170\text{ V}}(1000) = \boxed{1300}$$

33. Strategy The induced emf in a solid conductor subjected to a changing magnetic flux causes eddy currents to flow simultaneously along many different paths. These eddy currents dissipate energy. According to Lenz's law, the direction of an induced current in a loop always opposes the *change* in magnetic flux that induces the current.

Solution Assuming that the marble does not contact the sides of the pipe, the reading of the scale doesn't change and only reads the weight of the pipe, 12.0 N.

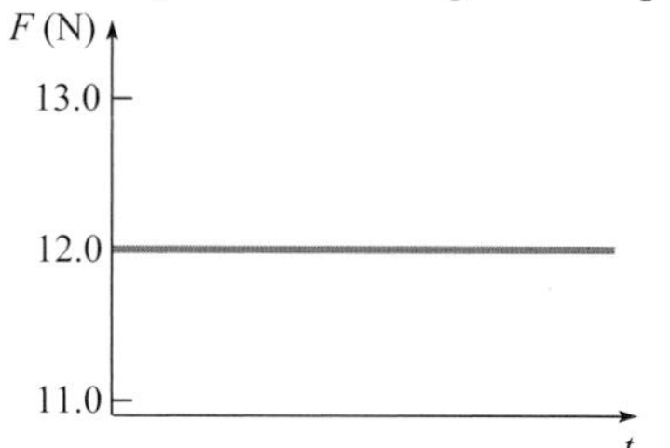

When the magnet reaches terminal velocity, the net force on it is zero. So, the scale supports both the weight of the pipe and the magnet, 12.0 N + 0.3 N = 12.3 N.

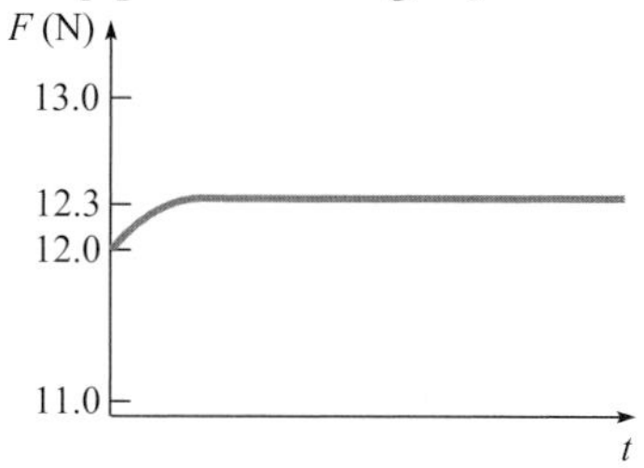

35. Strategy The total flux linkage is $N\Phi = LI$, so the flux through one winding is Φ. Use Eq. (20-15b).

Solution Find the magnetic flux through one winding.

$$\Phi = \frac{LI}{N} = \frac{\mu_0 N^2 \pi \left(\frac{d}{2}\right)^2 I}{\ell N} = \frac{1}{4}\mu_0 \pi n d^2 I = \frac{1}{4}(4\pi\times10^{-7}\text{ H/m})\pi(160\text{ cm}^{-1})(10^2\text{ cm/m})(0.0075\text{ m})^2(0.20\text{ A})$$

$$= \boxed{1.8\times10^{-7}\text{ Wb}}$$

37. Strategy The inductance of a solenoid is inversely proportional to its length.

Solution

$$L \propto \frac{1}{\ell},\text{ so } \frac{L_2}{L_1} = \frac{\ell_1}{\ell_2} = \frac{\ell}{0.50\ell} = 2.0$$

The inductance is increased to 2.0 times its initial value.

41. Strategy $\mathcal{E}_{av}$ is given by Faraday's law and $N\Phi = LI$.

Solution Find the average emf in the solenoid.

$$\mathcal{E}_{av} = N\frac{\Delta\Phi}{\Delta t} = L\frac{\Delta I}{\Delta t} = (0.080\text{ H})\frac{(160.0-20.0)\times10^{-3}\text{ A}}{7.0\text{ s}} = \boxed{1.6\text{ mV}}$$

43. Strategy Since the inductors are in parallel, the emfs induced in each must be equal. An equivalent inductor replacing L_1 and L_2 would have the same induced emf. The current through the equivalent emf must be the sum of the currents through the individual inductors. The relation between induced emf and a changing current is given by Faraday's law.

Solution Calculate the equivalent inductance of two ideal inductors in parallel.

$\frac{\Delta I_{eq}}{\Delta t} = \frac{\Delta I_1}{\Delta t} + \frac{\Delta I_2}{\Delta t}$, $\mathscr{E}_{eq} = \mathscr{E}_1 = \mathscr{E}_2$, and $\mathscr{E} = -L\frac{\Delta I}{\Delta t}$. Find L_{eq} in terms of L_1 and L_2.

$L_{eq}\frac{\Delta I_{eq}}{\Delta t} = L_1\frac{\Delta I_1}{\Delta t} = L_2\frac{\Delta I_2}{\Delta t}$, so $\frac{\Delta I_2}{\Delta t} = \frac{L_1}{L_2}\frac{\Delta I_1}{\Delta t}$. $\frac{\Delta I_{eq}}{\Delta t} = \frac{\Delta I_1}{\Delta t} + \frac{\Delta I_2}{\Delta t} = \frac{\Delta I_1}{\Delta t}\left(1+\frac{L_1}{L_2}\right)$, so

$L_{eq}\frac{\Delta I_{eq}}{\Delta t} = L_{eq}\frac{\Delta I_1}{\Delta t}\left(1+\frac{L_1}{L_2}\right) = L_1\frac{\Delta I_1}{\Delta t}$.

Solve for L_{eq}.

$L_{eq}\left(1+\frac{L_1}{L_2}\right) = L_{eq}\left(\frac{L_1+L_2}{L_2}\right) = L_1$, so $\boxed{L_{eq} = \frac{L_1 L_2}{L_1+L_2}}$.

45. (a) Strategy Immediately after the switch is closed, the current through the inductor is zero, so the entire current flows through the 5.0-Ω and 10.0-Ω resistors as if they are in series. Use Kirchhoff's loop rule to find the current.

Solution Find the current.

$6.0\text{ V} - I(5.0\ \Omega) - I(10.0\ \Omega) = 0$, so $I = \frac{6.0\text{ V}}{15.0\ \Omega} = 0.40\text{ A}$.

Compute the voltages across the resistors.

$V_{5.0} = (0.40\text{ A})(5.0\ \Omega) = \boxed{2.0\text{ V}}$ and $V_{10.0} = (0.40\text{ A})(10.0\ \Omega) = \boxed{4.0\text{ V}}$.

(b) Strategy and Solution After the switch has been closed for a long time, the current in the inductor is no longer changing, so it acts like a short circuit. Thus, no current flows through the 10.0-Ω resistor, so the voltage across it is $\boxed{0}$. The entire voltage of the battery is dropped across the 5.0-Ω resistor, so the voltage across it is $\boxed{6.0\text{ V}}$.

(c) Strategy Since the inductor acts like a short and the 10.0-Ω resistor is bypassed, the current through the inductor is the current that would flow through the circuit if it only contained the battery and the 5.0-Ω resistor.

Solution Compute the current.

$I = \frac{6.0\text{ V}}{5.0\ \Omega} = \boxed{1.2\text{ A}}$

47. (a) Strategy Use the definition of resistance and Eq. (20-17).

Solution Find the maximum current that flowed through the inductor.

$I = \frac{\mathscr{E}}{R} = \frac{6.0\text{ V}}{12\ \Omega} = 0.50\text{ A}$

Calculate the stored energy.

$U = \frac{1}{2}LI^2 = \frac{1}{2}(0.30\text{ H})(0.50\text{ A})^2 = \boxed{38\text{ mJ}}$

(b) Strategy $P = I\mathscr{E}$ and $I = 0.50$ A (at $t = 0$).

Solution Compute the instantaneous rate of change of the inductor's energy.
$\mathscr{E}$ = the voltage drop across the resistors = (0.50 A)(30 Ω) = 15 V

Thus, $P = -(0.50\text{ A})(15\text{ V}) = \boxed{-7.5\text{ W}}$, where $P < 0$ since the energy of the inductor is decreasing.

(c) Strategy Assume that $U_f \approx 0$. Use the definition of average power.

Solution Compute the average rate of change of the inductor's energy.

$$P_{av} = \frac{\Delta U}{\Delta t} = \frac{0 - 38\times10^{-3}\text{ J}}{1.0\text{ s}} = \boxed{-38\text{ mW}}$$

(d) Strategy Use Eq. (20-24).

Solution Solve for t when $I = 0.0010I_0$.

$$I = I_0 e^{-t/\tau}$$

$$\ln e^{t/\tau} = \ln\frac{I_0}{I}$$

$$t = \tau\ln\frac{I_0}{I} = \frac{L}{R_{eq}}\ln\frac{1}{0.0010} = -\frac{0.30\text{ H}}{18\ \Omega + 12\ \Omega}\ln 0.0010 = \boxed{69\text{ ms}}$$

Since 69 ms << 1.0 s, the assumption in part (c) is valid.

49. Strategy Use the definition of resistance and Eqs. (18-17) and (18-21b).

Solution

(a) At the instant the switch is closed, the current through the inductor is zero, so $I_2 = \boxed{0}$.

$I_1 = \frac{\mathscr{E}_b}{R_1} = \frac{45\text{ V}}{27{,}000\ \Omega} = \boxed{1.7\text{ mA}}$, $V_{3.0} = I_2 R_{3.0} = (0)(3.0\text{ k}\Omega) = \boxed{0}$, $V_{27} = I_1 R_{27} = \frac{\mathscr{E}_b}{R_{27}}R_{27} = \boxed{45\text{ V}}$, and

$P = \frac{\mathscr{E}_b^{\ 2}}{R_1} = \frac{(45\text{ V})^2}{27{,}000\ \Omega} = \boxed{75\text{ mW}}$.

Initially there is no current through the inductor, so the induced emf is equal to that of the source, $\boxed{45\text{ V}}$.

(b) After the switch has been closed a long time, the current reaches a constant value and the emf in the inductor is $\boxed{\text{zero}}$.

$I_1 = \frac{\mathscr{E}_b}{R_{27}} = \frac{45\text{ V}}{27{,}000\ \Omega} = \boxed{1.7\text{ mA}}$ and $I_2 = \frac{\mathscr{E}_b}{R_{3.0}} = \frac{45\text{ V}}{3.0\times10^3\ \Omega} = \boxed{15\text{ mA}}$.

The resistors are in parallel, so $V_{3.0} = V_{27} = \mathscr{E}_b = \boxed{45\text{ V}}$.

$$P = \frac{\mathscr{E}_b^2}{R_{eq}} = \frac{(45\text{ V})^2}{\left(\frac{1}{27\times10^3\ \Omega} + \frac{1}{3.0\times10^3\ \Omega}\right)^{-1}} = \boxed{0.75\text{ W}}$$

51. (a) Strategy When the current is no longer changing, the emf in the coil is zero.

Solution Compute the current in the coil.

$$I_0 = \frac{\mathscr{E}_b}{R} = \frac{6.0\text{ V}}{33\ \Omega} = \boxed{180\text{ mA}}$$

(b) Strategy Use Eq. (20-17).

Solution Calculate the energy stored in the coil.

$$U=\frac{1}{2}LI_0{}^2=\frac{1}{2}L\left(\frac{\mathscr{E}_\text{b}}{R}\right)^2=\frac{(0.15\text{ H})(6.0\text{ V})^2}{2(33\ \Omega)^2}=\boxed{2.5\text{ mJ}}$$

(c) Strategy Use Eq. (18-21b).

Solution Calculate the rate at which energy is dissipated.

$$P=\frac{\mathscr{E}_\text{b}{}^2}{R}=\frac{(6.0\text{ V})^2}{33\ \Omega}=\boxed{1.1\text{ W}}$$

(d) Strategy and Solution Since the current is no longer changing, the induced emf is $\boxed{\text{zero}}$.

53. (a) Strategy Use Eqs. (20-21) and (20-22).

Solution Find t when $I=0.67I_\text{f}$.

$$I=I_\text{f}(1-e^{-t/\tau})$$
$$0.67I_\text{f}=I_\text{f}(1-e^{-tR/L})$$
$$\ln e^{-tR/L}=\ln 0.33$$
$$t=-\frac{L}{R}\ln 0.33=-\frac{0.00067\text{ H}}{130\ \Omega}\ln 0.33=\boxed{5.7\times10^{-6}\text{ s}}$$

(b) Strategy The energy in the inductor is given by Eq. (20-17). This energy is at its maximum when the current flowing through the inductor is at its maximum; that is, when the current is equal to $\mathscr{E}/R$.

Solution Compute the maximum energy stored in the inductor.

$$U=\frac{1}{2}LI^2=\frac{1}{2}L\frac{\mathscr{E}^2}{R^2}=\frac{1}{2}(0.00067\text{ H})\left(\frac{24\text{ V}}{130\ \Omega}\right)^2=\boxed{1.1\times10^{-5}\text{ J}}$$

(c) Strategy Since $U=\frac{1}{2}LI^2$ and $I=I_\text{f}(1-e^{-t/\tau})$, $U(t)=U_\text{f}(1-e^{-t/\tau})^2$.

Solution Find t when $U=0.67U_\text{f}$.

$$U=U_\text{f}(1-e^{-t/\tau})^2$$
$$0.67U_\text{f}=U_\text{f}(1-e^{-t/\tau})^2$$
$$\pm\sqrt{0.67}=1-e^{-t/\tau}$$
$$\ln e^{-tR/L}=\ln(1\pm\sqrt{0.67})$$
$$t=-\frac{L}{R}\ln(1\pm\sqrt{0.67})$$
$$t=-\frac{0.00067\text{ H}}{130\ \Omega}\ln(1\pm\sqrt{0.67})=\boxed{8.8\times10^{-6}\text{ s}}$$ or -3.1×10^{-6} s, which is extraneous since $t\geq 0$.

This is more than in part (a) because the energy stored in the inductor is proportional to the current squared. It takes longer for the *square* of the current to be 67% of the maximum *square* of the current than for the current itself to be 67% of the maximum current.

57. Strategy According to Lenz's law, the direction of an induced current in a loop always opposes the *change* in magnetic flux that induces the current.

Solution

(a) Since the flux is provided by the current flowing in the right-hand coil, and that current never changes, the flux never changes. So, there is a magnetic flux in the left-hand coil, but no change in flux, so $\boxed{\text{no}}$ current flows.

(b) For the same reasons given in part (a), $\boxed{\text{no}}$ current flows.

61. Strategy According to Faraday's law, the magnitude of the induced emf around a loop is equal to the rate of change of the magnetic flux through the loop. According to Lenz's law, the direction of an induced current in a loop always opposes the *change* in magnetic flux that induces the current.

Solution As the bar magnet travels from 1 to 2, $\vec{\mathbf{B}}$ is increasing and to the left at the coil. As viewed from the left, a CW current is induced in the coil that generates a magnetic field to the right to oppose the increasing magnetic field due to the bar magnet. Thus, the current is negative and increasing in magnitude. As the bar magnet travels from 2 to 3, $\vec{\mathbf{B}}$ is decreasing and to the left at the coil. The induced current flows CCW to oppose the decreasing field. Thus, the current is positive and decreasing in magnitude.

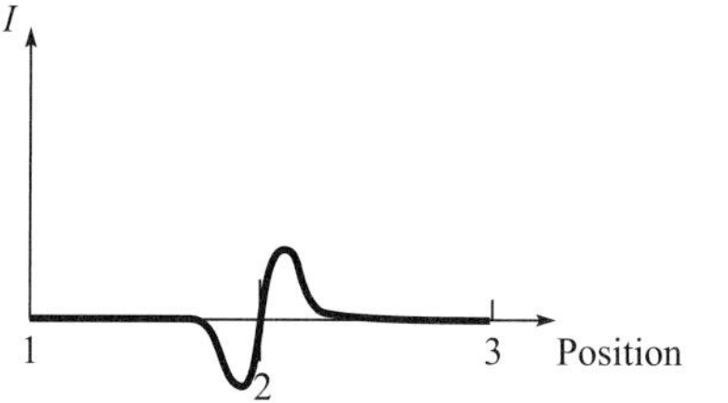

65. Strategy Since $a \ll R$, the field inside the toroid can be considered uniform and perpendicular to the cross-sectional area a^2. So, we can consider the toroid to be a square solenoid with length $\ell = 2\pi R$. Use Eq. (20-15b).

Solution Find the self-inductance of the toroid.

$$L = \frac{\mu_0 N^2 A}{\ell} = \boxed{\frac{\mu_0 N^2 a^2}{2\pi R}}$$

69. Strategy The maximum emf for an ac generator is $\mathscr{E}_{\text{max}} = \omega NBA$.

Solution Solve for ω.

$$\omega = \frac{\mathscr{E}_{\text{max}}}{NBA} = \frac{1.0 \text{ V}}{1000.0(0.50\times10^{-3} \text{ T})\pi(0.050 \text{ m})^2} = \boxed{250 \text{ rad/s}}$$

73. **(a)** **Strategy** We can model the airplane as a metal rod moving through a magnetic field. Then, the motional emf is $\mathscr{E} = vBL$, where B is the component of the field perpendicular to the motion of the plane and to the wing. Therefore, only the upwardly directed component of the magnetic field contributes to the motional emf.

Solution Compute the motional emf.

$\mathscr{E} = vBL = (180\text{ m/s})(0.38\times10^{-3}\text{ T})(46\text{ m}) = \boxed{3.1\text{ V}}$.

(b) **Strategy** Consider the magnetic forces on electrons in the plane's wings.

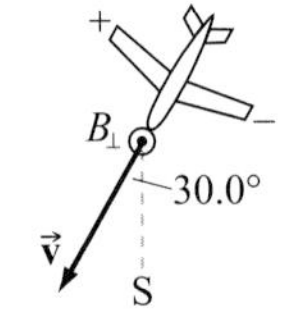

Solution The force on an electron in the wing is $\vec{\mathbf{F}} = -e\vec{\mathbf{v}}\times\vec{\mathbf{B}}$. So, according to the RHR, the electrons are forced along the southernmost wing. Negative charge builds up on the southernmost wingtip and positive on the northernmost. Thus, $\boxed{\text{the northernmost wingtip}}$ is positively charged.

77. **Strategy** When the magnet is shut off, a current flows in the coil such that it generates a magnetic field of equal magnitude and direction as existed between the magnet's poles prior to the power being shut off. $\Delta Q = 9.0\text{ mC}$ of charge flows in a time Δt, so $I_{av} = \Delta Q/\Delta t$. By Faraday's law, the magnitude of the induced emf is $\mathscr{E} = N\,\Delta\Phi/\Delta t$. The average emf is $\mathscr{E}_{av} = N\,\Phi_0/\Delta t = I_{av}R$, where $N\Phi_0 = NBA = NB\pi r^2$ is the initial magnetic flux linkage through the coil.

Solution Substitute and solve for N.

$$I_{av}R = \frac{\Delta Q}{\Delta t}R = N\frac{\Phi_0}{\Delta t} = \frac{NB\pi r^2}{\Delta t},\text{ so } N = \frac{R\Delta Q}{B\pi r^2} = \frac{(25\ \Omega)(9.0\times10^{-3}\text{ C})}{(2.6\text{ T})\pi(0.018\text{ m})^2} = \boxed{85}.$$

81. **Strategy and Solution**

(a) According to Faraday's law, a changing magnetic flux through the coil induces an emf in the coil. The changing flux requires a changing magnetic field, and this in turn requires a changing current in the wire. Therefore, the induction ammeter will not work equally well for ac and dc currents; ac currents are required. The answer is $\boxed{\text{no}}$.

(b) Since equal magnitude currents flowing in opposite directions produce magnetic fields that cancel, there is no changing flux through the coil and, thus, no emf is measured by the induction ammeter. Therefore, the current drawn by the appliance cannot be measured. The answer is $\boxed{\text{no}}$.

Chapter 21

ALTERNATING CURRENT

Problems

1. **Strategy and Solution** The current reverses direction twice per cycle and there are 60 cycles per second, so the current reverses direction $\boxed{\text{120 times per second}}$.

3. **Strategy** 1500 W is the average power dissipated by the heater and $P_{\text{av}} = I_{\text{rms}} V_{\text{rms}}$.

   Solution Calculate I, the peak current.

   $$I = \sqrt{2} I_{\text{rms}} = \sqrt{2}\left(\frac{P_{\text{av}}}{V_{\text{rms}}}\right) = \sqrt{2}\left(\frac{1500\text{ W}}{120\text{ V}}\right) = \boxed{18\text{ A}}$$

5. **Strategy** Find the resistance of the hair dryer. Then, compute the power dissipated in Europe.

   Solution

   $P_{\text{av}} = \dfrac{V_{\text{rms}}^2}{R}$, so $R = \dfrac{V_{\text{US}}^2}{P_{\text{US}}}$. The power dissipated by the dryer in Europe is

   $$P_{\text{R}} = \frac{V_{\text{E}}^2}{R} = \frac{V_{\text{E}}^2}{V_{\text{US}}^2} P_{\text{US}} = \left(\frac{240}{120}\right)^2 (1500\text{ W}) = \boxed{6000\text{ W}}.$$

   $\boxed{\text{The heating element of the hair dryer will burn out because it is not designed for this amount of power.}}$

7. **(a) Strategy** 4200 W is the average power drawn by the heater.

   Solution Compute the rms current.

   $$I_{\text{rms}} = \frac{P_{\text{av}}}{V_{\text{rms}}} = \frac{4200\text{ W}}{120\text{ V}} = \boxed{35\text{ A}}$$

   (b) Strategy Since $P = V^2/R$ and $R_2 = R_1$, $P \propto V^2$. Form a proportion.

   Solution Calculate P_2.

   $$\frac{P_2}{P_1} = \left(\frac{V_2}{V_1}\right)^2, \text{ so } P_2 = \left(\frac{105}{120}\right)^2 (4200\text{ W}) = \boxed{3.2\text{ kW}}.$$

9. **Strategy** Use the definition of rms.

   Solution Compute the amplitude.

   $\mathscr{E}_{\text{m}} = \sqrt{2}\mathscr{E}_{\text{rms}} = \sqrt{2}(4.0\text{ V}) = 5.7\text{ V}$

   The instantaneous sinusoidal emf oscillates between $\boxed{-5.7\text{ V and } 5.7\text{ V}}$.

13. Strategy and Solution The peak current in the circuit is given by $I = \omega Q$ and $Q = CV$, so $I = \omega CV$. Therefore, if the capacitance is increased by a factor of 3.0 and the driving frequency is increased by a factor of 2.0, $\boxed{\text{the current is increased by a factor of 6.0}}$.

15. Strategy The reactance is $X_C = 1/(\omega C)$ where $\omega = 2\pi f$.

Solution

(a) Solve for *f*.

$$X_C = \frac{1}{2\pi fC}, \text{ so } f = \frac{1}{2\pi X_C C} = \frac{1}{2\pi(6.63\times10^3\ \Omega)(0.400\times10^{-6}\ \text{F})} = \boxed{60.0\ \text{Hz}}.$$

(b) Compute the reactance.

$$X_C = \frac{1}{\omega C} = \frac{1}{2\pi(30.0\ \text{Hz})(0.400\times10^{-6}\ \text{F})} = \boxed{13.3\ \text{k}\Omega}$$

17. Strategy $V_{\text{rms}} = I_{\text{rms}} X_C$ where $X_C = 1/(\omega C)$.

Solution Solve for the capacitance, *C*.

$$V_{\text{rms}} = I_{\text{rms}} X_C = \frac{I_{\text{rms}}}{\omega C}, \text{ so } C = \frac{I_{\text{rms}}}{2\pi f V_{\text{rms}}} = \frac{2.3\times10^{-3}\ \text{A}}{2\pi(60.0\ \text{Hz})(115\ \text{V})} = \boxed{53\ \text{nF}}.$$

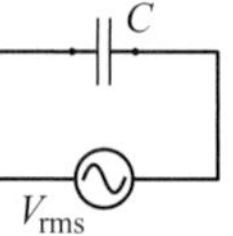

21. Strategy Use Eqs. (18-15), (21-6), and (21-7).

Solution

(a) Find *I*.

$$I = \frac{V}{X_{\text{Ceq}}} = \omega C_{\text{eq}} V = \frac{2\pi f V}{\frac{1}{C_1} + \frac{1}{C_2} + \frac{1}{C_3}} = \frac{2\pi(6300\ \text{Hz})(12.0\ \text{V})}{\frac{1}{2.0\times10^{-6}\ \text{F}} + \frac{1}{3.0\times10^{-6}\ \text{F}} + \frac{1}{6.0\times10^{-6}\ \text{F}}} = 0.475\ \text{A}$$

Find *V* as a function of *C*.

$$V = IX_C = \frac{I}{\omega C} = \frac{0.475\ \text{A}}{2\pi(6300\ \text{Hz})C}$$

The table below gives the results for each capacitor.

C (μF)	V (V)
2.0	6.0
3.0	4.0
6.0	2.0

(b) From part (a), $I = \boxed{0.48\ \text{A}}$.

25. Strategy The inductive reactance is given by $X_L = \omega L$ and the inductance of a solenoid is given by $L = \mu_0 N^2 \pi r^2 / \ell$.

Solution Find the reactance of the solenoid.

$$X_L = \omega L = 2\pi f \left(\frac{\mu_0 N^2 \pi r^2}{\ell} \right) = \frac{2\pi^2 (15.0\times10^3 \text{ Hz})(4\pi\times10^{-7} \text{ T}\cdot\text{m/A})(240)^2 (0.010 \text{ m})^2}{0.080 \text{ m}} = \boxed{27\ \Omega}$$

27. Strategy $V_{rms} = I_{rms} X_L$ where $X_L = \omega L = 2\pi f L$.

Solution Solve for *f*.

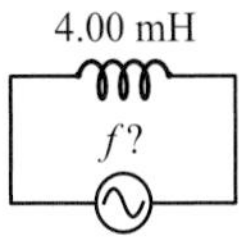

4.00 mH

f?

$V_{rms} = I_{rms} X_L = I_{rms}(2\pi f L)$, so

$$f = \frac{V_{rms}}{2\pi I_{rms} L} = \frac{151.0 \text{ V}}{2\pi(0.820 \text{ A})(4.00\times10^{-3} \text{ H})} = \boxed{7.33 \text{ kHz}}.$$

29. Strategy Since the capacitor and inductor are in series, the same current flows through both.

Solution

(a) $v_C(t)$ lags $i(t)$ by 90° and $v_L(t)$ leads $i(t)$ by 90°, so the phase difference is $\Delta\phi = \phi_L - \phi_C = 90° - (-90°) = \boxed{180°}$.

(b) $v_L(t)$ leads $v_C(t)$ by 180°, which means that they are opposite in sign. The ac voltmeter reads the difference, $\boxed{4.0 \text{ V}}$.

33. Strategy $\mathscr{E}_{rms} = I_{rms} Z$ and $Z = \sqrt{R^2 + X_L^2}$. Only the resistance dissipates power, so use $P_{av} = I_{rms}^2 R$.

Solution Find I_{rms}.

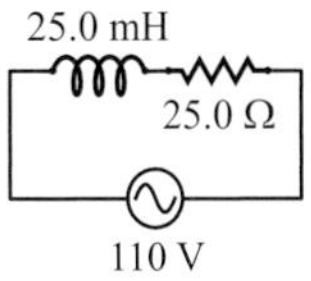

25.0 mH

25.0 Ω

110 V

$P_{av} = I_{rms}^2 R$, so $I_{rms} = \sqrt{\frac{P_{av}}{R}}$.

Find *f*.

$$\mathscr{E}_{rms} = I_{rms} Z = I_{rms}\sqrt{R^2 + X_L^2} = I_{rms}\sqrt{R^2 + \omega^2 L^2}$$

$$\frac{\mathscr{E}_{rms}^2}{I_{rms}^2} = R^2 + \omega^2 L^2$$

$$\omega^2 L^2 = \frac{\mathscr{E}_{rms}^2}{I_{rms}^2} - R^2$$

$$\omega = \frac{1}{L}\sqrt{\left(\frac{\mathscr{E}_{rms}}{I_{rms}}\right)^2 - R^2}$$

$$f = \frac{1}{2\pi L}\sqrt{\left(\frac{\mathscr{E}_{rms}}{\sqrt{P_{av}/R}}\right)^2 - R^2} = \frac{1}{2\pi L}\sqrt{\frac{\mathscr{E}_{rms}^2 R}{P_{av}} - R^2} = \frac{1}{2\pi(0.0250 \text{ H})}\sqrt{\frac{(110 \text{ V})^2(25.0\ \Omega)}{50.0 \text{ W}} - (25.0\ \Omega)^2}$$

$$= \boxed{470 \text{ Hz}}$$

37. (a) Strategy Use Eqs. (21-14) with $X_C = 0$, Eqs. (21-9) and (21-10), and Ohm's law.

Solution Find I.

145.0 Ω 22.0 mH

$\mathscr{E}$

$$I = \frac{\mathscr{E}_m}{Z} = \frac{\mathscr{E}_m}{\sqrt{R^2 + X_L^2}} = \frac{\mathscr{E}_m}{\sqrt{R^2 + \omega^2 L^2}} = \frac{1.20\times10^3\text{ V}}{\sqrt{(145.0\ \Omega)^2 + 4\pi^2(1250\text{ Hz})^2(22.0\times10^{-3}\text{ H})^2}}$$

$= 5.32$ A

Calculate the voltage amplitudes.

$V_L = IX_L = I\omega L = 2\pi(5.32\text{ A})(1250\text{ Hz})(22.0\times10^{-3}\text{ H}) = \boxed{919\text{ V}}$

$V_R = IR = (5.32\text{ A})(145.0\ \Omega) = \boxed{771\text{ V}}$

(b) Strategy and Solution $\boxed{\text{No}}$, the voltage amplitudes do not add to give the amplitude of the source voltage; the voltages across the resistor and the inductor are 90° out of phase. Similar to the impedance, the source voltage is the square root of the sum of squares, $\boxed{\mathscr{E}_m = \sqrt{V_L^2 + V_R^2}}$, as can be seen in a phasor diagram.

(c) Strategy Use the results of parts (a) and (b).

Solution Sketch the phasor diagram.

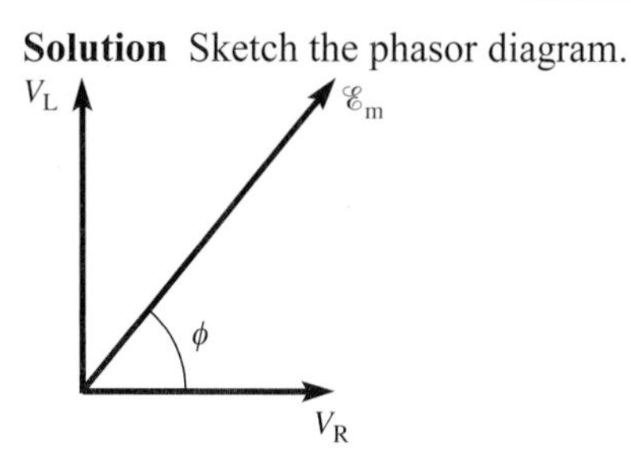

39. Strategy Use Eq. (21-14b) with $X_L = 0$, and Eq. (21-7).

Solution Find the impedance.

300.0 Ω 2.5 μF

$\mathscr{E}$

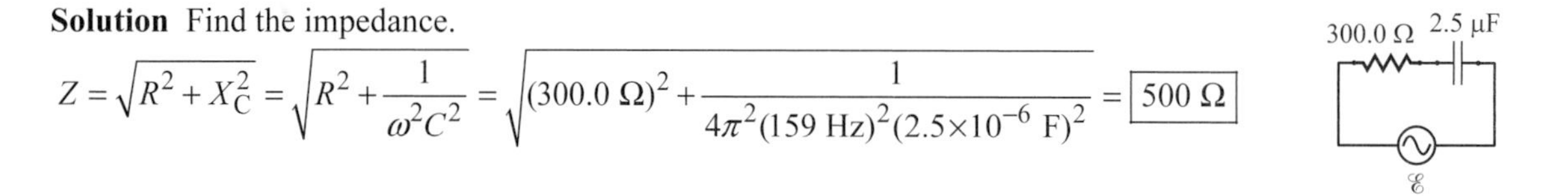

$$Z = \sqrt{R^2 + X_C^2} = \sqrt{R^2 + \frac{1}{\omega^2 C^2}} = \sqrt{(300.0\ \Omega)^2 + \frac{1}{4\pi^2(159\text{ Hz})^2(2.5\times10^{-6}\text{ F})^2}} = \boxed{500\ \Omega}$$

41. (a) Strategy Set $V_R = IR$ equal to $V_C = IX_C = I/(\omega C) = I/(2\pi f C)$ and solve for the frequency.

Solution Find the frequency.

3.3 kΩ 2.0 μF

$\mathscr{E}$

$$V_R = IR = V_C = IX_C = \frac{I}{\omega C} = \frac{I}{2\pi f C},\ \text{so } f = \frac{1}{2\pi RC} = \frac{1}{2\pi(3300\ \Omega)(2.0\times10^{-6}\text{ F})} = \boxed{24\text{ Hz}}.$$

(b) Strategy In a phasor diagram, the source is the hypotenuse of a right triangle and V_R and V_C are the legs, so $\mathscr{E}_m = \sqrt{V_R^2 + V_C^2}$ according to the Pythagorean theorem.

Solution Find $V_C / \mathscr{E}_m = V_R / \mathscr{E}_m$.

V_R $\mathscr{E}_m$ V_C

$$\mathscr{E}_m{}^2 = V_R^2 + V_C^2 = V_R^2 + V_R^2 = 2V_R^2,\ \text{so } \frac{V_R^2}{\mathscr{E}_m{}^2} = \frac{1}{2} \text{ and } \frac{V_R}{\mathscr{E}_m} = \frac{1}{\sqrt{2}} = \frac{V_C}{\mathscr{E}_m}.$$

Therefore, the rms voltages across the components are not half of the rms voltage of the source: $\boxed{\frac{V_R}{\mathscr{E}_m} = \frac{V_C}{\mathscr{E}_m} = \frac{1}{\sqrt{2}}}$.

(c) Strategy The voltage across a capacitor lags the current through it, so the source voltage lags the current. Use the power factor to find ϕ.

Solution Solve for the phase angle.

$\cos\phi = \frac{R}{Z}$, so

$$\phi = \cos^{-1}\frac{R}{Z} = \cos^{-1}\frac{R}{\sqrt{R^2 + X_C^2}} = \cos^{-1}\frac{IR}{\sqrt{(IR)^2 + (IX_C)^2}} = \cos^{-1}\frac{V_R}{\sqrt{V_R^2 + V_C^2}} = \cos^{-1}\frac{V_R}{\mathscr{E}_m} = \cos^{-1}\frac{1}{\sqrt{2}} = \frac{\pi}{4}.$$

Therefore, $\boxed{I \text{ leads } \mathscr{E} \text{ by } \frac{\pi}{4} \text{ rad} = 45°}$.

(d) Strategy Use the power factor to find Z.

Solution Find the impedance.

$$\cos\phi = \frac{R}{Z}, \text{ so } Z = \frac{R}{\cos\phi} = \frac{3.3\text{ k}\Omega}{\cos\frac{\pi}{4}} = \boxed{4.7\text{ k}\Omega}.$$

43. (a) Strategy The power factor is $\cos\phi = R/Z$ where $Z = \sqrt{R^2 + (X_L - X_C)^2}$.

Solution Find the power factor.

$$\cos\phi = \frac{R}{Z} = \frac{R}{\sqrt{R^2 + \left(\omega L - \frac{1}{\omega C}\right)^2}} = \frac{40.0\ \Omega}{\sqrt{(40.0\ \Omega)^2 + \left[(1.00\times10^4\text{ rad/s})(0.0220\text{ H}) - \frac{1}{(1.00\times10^4\text{ rad/s})(0.400\times10^{-6}\text{ F})}\right]^2}}$$

$$= \boxed{0.800}$$

(b) Strategy In Example 21.4, I was found to be 2.0 mA. Use Eq. (21-4).

Solution Power is not dissipated in the inductor or capacitor (assuming they are ideal), so $\boxed{P_{av,\,C} = P_{av,\,L} = 0}$. Compute the average power dissipated in the resistor.

$$P_{av} = I_{rms}{}^2 R = \left(\frac{I}{\sqrt{2}}\right)^2 R = \frac{1}{2}(0.0020\text{ A})^2(40.0\ \Omega) = \boxed{P_{av,\,R} = 8.0\times10^{-5}\text{ W}}$$

45. Strategy The average power is given by $P_{av} = I_{rms}\mathscr{E}_{rms}\cos\phi$ where $I_{rms} = \mathscr{E}_{rms}/Z$ and $\cos\phi = R/Z$.

Solution Find the average power dissipated.

$$P_{av} = I_{rms}\mathscr{E}_{rms}\cos\phi = \frac{\mathscr{E}_{rms}}{Z}\mathscr{E}_{rms}\frac{R}{Z} = \left(\frac{\mathscr{E}_{rms}}{Z}\right)^2 R = \frac{(\mathscr{E}/\sqrt{2})^2 R}{R^2 + \left(\omega L - \frac{1}{\omega C}\right)^2}$$

$$= \frac{(12\text{ V})^2(220\ \Omega)}{2\left\{(220\ \Omega)^2 + \left[2\pi(2500\text{ Hz})(0.15\times10^{-3}\text{ H}) - \frac{1}{2\pi(2500\text{ Hz})(8.0\times10^{-6}\text{ F})}\right]^2\right\}} = \boxed{0.33\text{ W}}$$

220 Ω
$\mathscr{E}$
0.15 mH
8.0 μF

49. (a) Strategy Substitute $t = 0$ into the given equation and interpret the result by referring to the phasor diagram.

Solution

$$V_1 \sin \omega t + V_2 \sin(\omega t + \phi_2) = V \sin(\omega t + \phi)$$
$$V_1 \sin \omega(0) + V_2 \sin[\omega(0) + \phi_2] = V \sin[\omega(0) + \phi]$$
$$\boxed{V_2 \sin \phi_2 = V \sin \phi}$$

The result indicates that the y-component of V_2 is equal to the y-component of V.

(b) Strategy Substitute $t = \pi/(2\omega)$ into the given equation and interpret the result by referring to the phasor diagram. Use the trigonometric identity $\sin(\theta + \pi/2) = \cos\theta$ to simplify.

Solution

$$V_1 \sin \omega\left(\frac{\pi}{2\omega}\right) + V_2 \sin\left[\omega\left(\frac{\pi}{2\omega}\right) + \phi_2\right] = V \sin\left[\omega\left(\frac{\pi}{2\omega}\right) + \phi\right]$$
$$V_1 \sin\frac{\pi}{2} + V_2 \sin\left(\frac{\pi}{2} + \phi_2\right) = V \sin\left(\frac{\pi}{2} + \phi\right)$$
$$\boxed{V_1 + V_2 \cos\phi_2 = V \cos\phi}$$

The result indicates that the sum of the x-components of V_1 and V_2 is equal to the x-component of V.

53. Strategy Find the phase angle and the voltages across each device. Then, draw the phasor diagram and give the resonant frequency for the circuit.

Solution

(a) Find the reactances.

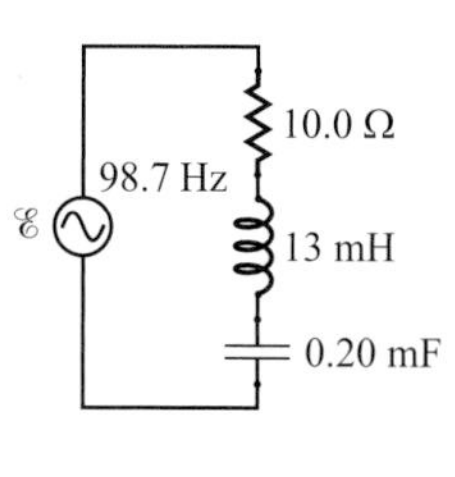

$$X_L = \omega L = 2\pi(98.7\text{ Hz})(13\times10^{-3}\text{ H}) = 8.06\ \Omega \text{ and}$$
$$X_C = \frac{1}{\omega C} = \frac{1}{2\pi(98.7\text{ Hz})(0.20\times10^{-3}\text{ F})} = 8.06\ \Omega.$$

Find the phase angle.

$$\phi = \tan^{-1}\frac{X_L - X_C}{R} = \tan^{-1}\frac{8.06\ \Omega - 8.06\ \Omega}{10.0\ \Omega} = \boxed{0°}$$

Since the phase angle is zero, the impedance is equal to the resistance. Find the current in the circuit.

$$I = \frac{\mathcal{E}_m}{R} = \frac{9.0\text{ V}}{10.0\ \Omega} = 0.90\text{ A}$$

Find the voltages across each device.

$V_L = IX_L = (0.90\text{ A})(8.06\ \Omega) = 7.3\text{ V}$, $V_C = IX_C = (0.90\text{ A})(8.06\ \Omega) = 7.3\text{ V}$, and
$V_R = IR = (0.90\text{ A})(10.0\ \Omega) = 9.0\text{ V} = \mathcal{E}_m$.

(b) Draw a phasor diagram using the results of part (a). Note that $V_L - V_C = 0$ V.

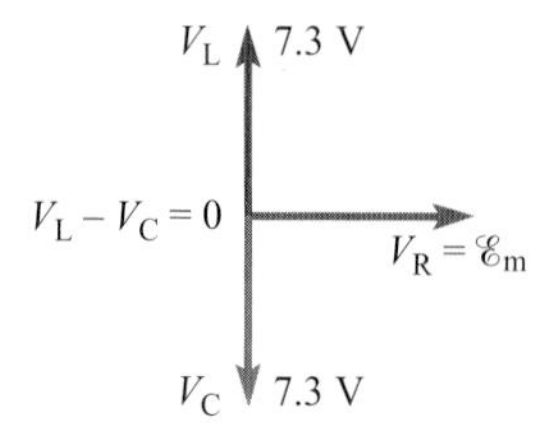

(c) Since the reactances are equal for a frequency of 98.7 Hz, $\boxed{98.7\text{ Hz}}$ is the resonant frequency.

55. (a) Strategy Use Eq. (21-18).

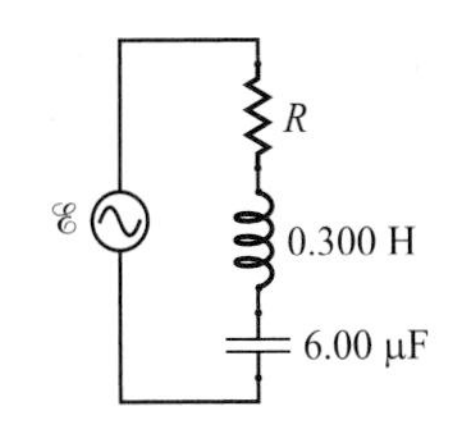

Solution Compute the resonant frequency.

$$\omega_0 = \sqrt{\frac{1}{LC}} = \sqrt{\frac{1}{(0.300\text{ H})(6.00\times10^{-6}\text{ F})}} = \boxed{745\text{ rad/s}}$$

(b) Strategy At resonance, $X_L = X_C$. Use Eq. (21-14a).

Solution Compute the resistance.

$$R = Z = \frac{\mathscr{E}_m}{I} = \frac{440\text{ V}}{0.560\text{ A}} = \boxed{790\ \Omega}$$

(c) Strategy At resonance, the voltages across the capacitor and inductor are 180° out of phase and equal in magnitude, so they cancel.

Solution The peak voltage across the resistor is $V_R = \mathscr{E}_m = \boxed{440\text{ V}}$.

Across the inductor, the peak voltage is

$$V_L = IX_L = I\omega_0 L = \frac{IL}{\sqrt{LC}} = I\sqrt{\frac{L}{C}} = (0.560\text{ A})\sqrt{\frac{0.300\text{ H}}{6.00\times10^{-6}\text{ F}}} = \boxed{125\text{ V}}.$$

Since $X_L = X_C$, $V_C = V_L = \boxed{125\text{ V}}$.

57. Strategy Use Eq. (21-18).

Solution Find the resonant frequency.

$$\omega_0 = \sqrt{\frac{1}{LC}} = \sqrt{\frac{1}{(40.0\times10^{-3}\text{ H})(0.0500\text{ F})}} = \boxed{22.4\text{ rad/s}},\text{ or } f_0 = \frac{\omega_0}{2\pi} = \boxed{3.56\text{ Hz}}.$$

61. Strategy If the currents are equal in amplitude, the impedances of the woofer and tweeter branches are equal. Use Eqs. (21-7), (21-10), and (21-14b).

Solution Solve for the capacitance.

$$Z_{\text{woofer}} = \sqrt{R^2 + (\omega L)^2} = Z_{\text{tweeter}} = \sqrt{R^2 + \left(\frac{1}{\omega C}\right)^2},\text{ so } \omega L = \frac{1}{\omega C}\text{ and}$$

$$C = \frac{1}{\omega^2 L} = \frac{1}{4\pi^2(180\text{ Hz})^2(1.20\times10^{-3}\text{ H})} = \boxed{650\ \mu\text{F}}.$$

65. Strategy $P = [v(t)]^2/R$ and $[v(t)]^2$ reaches its maximum value twice per cycle.

Solution The instantaneous power is at its maximum value two times per cycle, which is twice the frequency, so it is at maximum $2(60\text{ Hz}) = \boxed{120\text{ times per second}}$.

69. Strategy Use Eqs. (20-10) and (21-4).

Solution

(a) Calculate the turns ratio.

$$\frac{N_2}{N_1}=\frac{\mathscr{E}_2}{\mathscr{E}_1}=\frac{240\times10^3\text{ V}}{420\text{ V}}=\boxed{570}$$

(b) Calculate the rms current using the turns ratio.

$$I_1=\frac{N_2}{N_1}I_2=570(60.0\times10^{-3}\text{ A})=\boxed{34\text{ A}}$$

(c) Calculate the average power.

$$P_{\text{av}}=I_{\text{rms}}V_{\text{rms}}=(0.0600\text{ A})(240{,}000\text{ V})=\boxed{14\text{ kW}}$$

73. Strategy Use Eq. (21-7).

Solution Find the capacitance.

$$X_{\text{C}}=\frac{1}{\omega C},\text{ so } C=\frac{1}{\omega X_{\text{C}}}=\frac{1}{2\pi(520\text{ Hz})(6.20\ \Omega)}=\boxed{49\ \mu\text{F}}.$$

77. Strategy Use Eq. (21-17).

Solution Solve for the rms current using the average power.

$$P_{\text{av}}=I_{\text{rms}}\mathscr{E}_{\text{rms}}\cos\phi,\text{ so } I_{\text{rms}}=\frac{P_{\text{av}}}{\mathscr{E}_{\text{rms}}\cos\phi}.$$

(a) $$I_{\text{rms}}=\frac{4.50\times10^3\text{ W}}{(220\text{ V})(1.00)}=\boxed{20\text{ A}}$$

(b) $$I_{\text{rms}}=\frac{4.50\times10^3\text{ W}}{(220\text{ V})(0.80)}=\boxed{26\text{ A}}$$

81. (a) Strategy Use Eq. (21-14b).

12.0 Ω 0.26 μF 15.2 mH $\mathscr{E}$

Solution Find the impedance.

$$Z=\sqrt{R^2+(X_{\text{L}}-X_{\text{C}})^2}=\sqrt{R^2+\left(\omega L-\frac{1}{\omega C}\right)^2}$$

$$=\sqrt{(12.0\ \Omega)^2+\left[2\pi(2.50\times10^3\text{ Hz})(15.2\times10^{-3}\text{ H})-\frac{1}{2\pi(2.50\times10^3\text{ Hz})(0.26\times10^{-6}\text{ F})}\right]^2}=\boxed{13\ \Omega}$$

(b) Strategy Use Eq. (21-14a) with rms values.

Solution Find the rms current.

$$I_{\text{rms}}=\frac{\mathscr{E}_{\text{rms}}}{Z}=\frac{240\text{ V}}{13.5\ \Omega}=\boxed{18\text{ A}}$$

(c) Strategy Use Eq. (21-16).

Solution Find the phase angle.

$$\cos\phi = \frac{R}{Z}, \text{ so } \phi = \cos^{-1}\frac{R}{Z} = \cos^{-1}\frac{12.0\ \Omega}{13.5\ \Omega} = \boxed{27°}.$$

(d) Strategy Use Eqs. (21-7) and (21-10). Form a proportion.

Solution Compare the reactances.

$$\frac{X_C}{X_L} = \frac{1/(\omega C)}{\omega L} = \frac{1}{\omega^2 LC} = \frac{1}{4\pi^2(2.50\times10^3\text{ Hz})^2(0.0152\text{ H})(0.26\times10^{-6}\text{ F})} > 1$$

Since $X_C > X_L$, the capacitor dominates the inductor, so $\boxed{\text{the current leads the voltage}}$ since the current through a capacitor leads the voltage across it.

(e) Strategy $I_{rms} = 17.83$ A to four significant figures. Use Eqs. (21-6) and (21-9) and Ohm's law (using rms values).

Solution Find the rms voltage for each circuit element.

$$V_{R\text{ rms}} = I_{rms}R = (17.83\text{ A})(12.0\ \Omega) = \boxed{210\text{ V}}$$

$$V_{L\text{ rms}} = I_{rms}X_L = I_{rms}\omega L = (17.83\text{ A})2\pi(2.50\times10^3\text{ Hz})(0.0152\text{ H}) = \boxed{4.3\text{ kV}}$$

$$V_{C\text{ rms}} = I_{rms}X_C = \frac{I_{rms}}{\omega C} = \frac{17.83\text{ A}}{2\pi(2.50\times10^3\text{ Hz})(0.26\times10^{-6}\text{ F})} = \boxed{4.4\text{ kV}}$$

85. Strategy Use Eqs. (21-4) and (21-17).

Solution

(a) The average power is $P_{av} = I_{rms}\mathcal{E}_{rms}$. Find I_{rms}.

$$I_{rms} = \frac{P_{av}}{\mathcal{E}_{rms}} = \frac{12\times10^6\text{ W}}{250\times10^3\text{ V}} = \boxed{48\text{ A}}$$

(b) The average power is $P_{av} = I_{rms}\mathcal{E}_{rms}\cos\phi$. Find I_{rms}.

$$I_{rms} = \frac{P_{av}}{\mathcal{E}_{rms}\cos\phi} = \frac{12\times10^6\text{ W}}{(250\times10^3\text{ V})(0.86)} = \boxed{56\text{ A}}$$

(c) Since the current is greater, $\boxed{\text{the power lost in transmission is greater}}$ due to I^2R losses in the transmission line. The power company would want to charge more to make up for this loss.

REVIEW AND SYNTHESIS: CHAPTERS 19–21

Review Exercises

1. **Strategy** The maximum torque occurs when the plane of the loop is parallel to the magnetic field of the solenoid. Use Eqs. (19-13a) and (19-17).

   Solution Find the maximum possible magnetic torque on the loop.

   $\tau = N_1 I_1 A_1 B_s = N_1 I_1 A_1 \mu_0 n_s I_s = (100)(2.20\text{ A})\pi(0.0800\text{ m})^2(4\pi\times10^{-7}\text{ T}\cdot\text{m/A})(8500\text{ m}^{-1})(25.0\text{ A})$
   $= \boxed{1.2\text{ N}\cdot\text{m}}$

   Since the magnetic field inside the solenoid is along the axis of the solenoid, the magnetic torque on the loop is at its maximum value $\boxed{\text{when the plane of the loop is parallel to the axis of the solenoid}}$.

5. **Strategy** Use Eq. (19-14) and $\vec{\mathbf{F}}_\text{B} = q\vec{\mathbf{v}}\times\vec{\mathbf{B}}$. Let the positive y-direction be up.

   Solution According to the RHR, the magnetic field generated by the power line at point P is in the positive y-direction. Looking at the side view, we see that the angle between the velocity of the muon and the magnetic field is $180° - 25° = 155°$. The charge of the muon is negative, so according to the RHR and $\vec{\mathbf{F}}_\text{B} = q\vec{\mathbf{v}}\times\vec{\mathbf{B}}$, the direction of the force on the muon is out of the plane of the paper in the side view (or to the right in the end-on view). Compute the magnitude of the force.

   $$F_\text{B} = evB\sin\theta = ev\frac{\mu_0 I}{2\pi r}\sin\theta = \frac{(1.602\times10^{-19}\text{ C})(7.0\times10^7\text{ m/s})(4\pi\times10^{-7}\text{ T}\cdot\text{m/A})(16.0\text{ A})\sin155°}{2\pi(0.850\text{ m})}$$
   $$= 1.8\times10^{-17}\text{ N}$$

   Thus, $\vec{\mathbf{F}}_\text{B} = \boxed{1.8\times10^{-17}\text{ N out of the plane of the paper in the side view (or to the right in the end on view)}}$.

9. **Strategy** Use Lenz's law and $\vec{\mathbf{F}} = I\vec{\mathbf{L}}\times\vec{\mathbf{B}}$.

   Solution According to the RHR, the magnetic field points into the page at the loop.

   (a) A counterclockwise current is induced because the flux through the loop is increasing as it nears the wire. This counterclockwise current generates a field directed out of the page. According to the RHR, the magnetic force on the part of the loop closest to the wire is away from the wire and that on the part of the loop farthest from the wire is toward the wire. Since the magnetic field is greater nearer the wire, the net force on the loop is away from the wire.

   (b) No current is induced because there is no change in flux through the loop. Since there is no current in the loop, there is no magnetic force acting on the loop.

   (c) A clockwise current is induced because the flux through the loop is decreasing as it moves away from the wire. This clockwise current generates a field directed into the page. According to the RHR, the magnetic force on the part of the loop closest to the wire is toward the wire and that on the part of the loop farthest from the wire is away from the wire. Since the magnetic field is greater nearer the wire, the net force on the loop is toward the wire.

13. (a) Strategy At resonance, $X_C = X_L$, so $I \propto R^{-1}$. Thus, when the resistance is doubled, the current is cut in half. Use Eq. (21-4).

Solution If the initial power dissipated is $P_i = I_i^2 R_i$, then the final power dissipated is

$P_f = I_f^2 R_f = \left(\frac{I_i}{2}\right)^2 (2R_i) = \frac{1}{2} I_i^2 R_i = \frac{1}{2} P_i$. Therefore, the power is cut in half when the resistance of an *RLC* circuit is doubled.

(b) Strategy When the circuit is not at resonance, the situation is more complicated because the current is inversely proportional to the impedance *Z*, and the impedance is not simply equal to the resistance. Use Eq. (21-14b).

Solution Find the initial and final impedances in terms of the initial resistance.

$$Z_i = \sqrt{R^2 + (X_L - X_C)^2} = \sqrt{R^2 + (2R - R)^2} = \sqrt{2}R$$

$$Z_f = \sqrt{(2R)^2 + (X_L - X_C)^2} = \sqrt{4R^2 + (2R - R)^2} = \sqrt{5}R$$

Form a proportion to find the final current in terms of the initial current.

$$\frac{I_f}{I_i} = \frac{Z_i}{Z_f} = \frac{\sqrt{2}R}{\sqrt{5}R} = \sqrt{\frac{2}{5}}, \text{ so } I_f = \sqrt{\frac{2}{5}} I_i.$$

Form a proportion to find the final power dissipated in terms of the initial power dissipated.

$$\frac{P_f}{P_i} = \frac{I_f^2 R_f}{I_i^2 R_i} = \frac{\left(\sqrt{\frac{2}{5}} I_i\right)^2 (2R)}{I_i^2 R} = \frac{4}{5}, \text{ so } P_f = \frac{4}{5} P_i.$$

Thus, the power is 4/5 of its original value.

17. Strategy Since the measurement is made relatively close to the electron beam, the beam can be approximated as a long, straight wire. Use Eq. (19-14) and the definition of current.

Solution Let *N* be the number of electrons, then the amount of charge passing the point in 1.30 microseconds is $\Delta q = Ne$. Find the magnetic field strength Kieran measures.

$$B = \frac{\mu_0 I}{2\pi r} = \frac{\mu_0}{2\pi r}\left(\frac{\Delta q}{\Delta t}\right) = \frac{\mu_0}{2\pi r}\left(\frac{Ne}{\Delta t}\right) = \frac{(4\pi \times 10^{-7} \text{ T}\cdot\text{m/A})(1.40 \times 10^{11})(1.602 \times 10^{-19} \text{ C})}{2\pi(0.0200 \text{ m})(1.30 \times 10^{-6} \text{ s})} = \boxed{1.73 \times 10^{-7} \text{ T}}$$

MCAT Review

1. Strategy and Solution The current flows counterclockwise through the apparatus. By the RHR, the magnetic field generated by the current is directed upward. The correct answer is A.

2. Strategy Refer to the data in the table.

Solution Two rows of data are given for a mass of 0.01 kg. The current is increased by 150% from 10.0 A to 15.0 A. Due to the increase in the current, the exit speed of the projectile increases by 150% from 2.0 km/s to 3.0 km/s. Thus, the exit speed is directly proportional to the current. So, if the current were decreased by a factor of two, the exit speed would be decreased by a factor of two, as well. The correct answer is D.

3. **Strategy and Solution** Since, for a given current, the force is constant along the entire length of the railgun, lengthening the rails would increase the exit speed because of the longer distance over which the force is present. The correct answer is $\boxed{\text{D}}$.

4. **Strategy and Solution** Since the resistance of the rails is directly proportional to their resistivity, lowering the resistivity of the rails would decrease the power required to maintain the current that flows through them. The correct answer is $\boxed{\text{B}}$.

5. **Strategy** The average power is equal to the change in kinetic energy divided by the time interval.

   Solution Find the average power supplied by the railgun.

$$P_{\text{av}} = \frac{\Delta K}{\Delta t} = \frac{\frac{1}{2}mv^2 - 0}{\Delta t} = \frac{mv^2}{2\Delta t} = \frac{(0.10\text{ kg})(10.0\text{ m/s})^2}{2(2.0\text{ s})} = 2.5\text{ W}$$

   The correct answer is $\boxed{\text{B}}$.

6. **Strategy and Solution** $F = ma \propto I^2$, so $a \propto I^2/m$. The speed of the projectile is directly proportional to its acceleration, so $v \propto I^2/m$. Form a proportion using data from the table to find the approximate speed of the 0.08-kg projectile.

   $\frac{v_2}{v_1} = \frac{I_2^2/m_2}{I_1^2/m_1} = \left(\frac{I_2}{I_1}\right)^2 \frac{m_1}{m_2}$, so $v_2 = \left(\frac{I_2}{I_1}\right)^2 \frac{m_1}{m_2} v_1$. For (1), we use the data given in the third row of the table.

$$v_2 = \left(\frac{I_2}{I_1}\right)^2 \frac{m_1}{m_2} v_1 = \left(\frac{20.0}{10.0}\right)^2 \frac{0.02}{0.08}(1.4\text{ km/s}) = 1.4\text{ km/s}$$

   The correct answer is $\boxed{\text{C}}$.

7. **Strategy and Solution** The power from the power plant is given by $P = IV$, so, for a given amount of power, increasing the voltage decreases the current required. Since the power lost as heat is given by $P = I^2R$, reducing the current reduces the power lost as heat. The correct answer is $\boxed{\text{A}}$.

8. **Strategy and Solution** The magnetic field lines due to the section of current-carrying wire are circles (which is the only possibility, given the symmetry of the situation). The direction of the field is determined by the RHR. Pointing the thumb of the right hand in the direction of the current in the wire, then curling the fingers inward toward the palm, the direction of the field is indicated by the direction of the curl of the fingers—in this case counterclockwise, as seen from above. The correct answer is $\boxed{\text{D}}$.

Chapter 22

ELECTROMAGNETIC WAVES

Problems

1. **Strategy** Apply the Ampère-Maxwell law.

   Solution The Ampère-Maxwell law is

   $$\Sigma B_{\parallel}\Delta l = \mu_0\left(I + \epsilon_0 \frac{\Delta\Phi_E}{\Delta t}\right) = \mu_0 I \text{ since } \frac{\Delta\Phi_E}{\Delta t} = 0.$$

   By the RHR, the magnetic field lines are circles concentric with the central axis of the wire. At a distance $r \geq R$ from the central axis, the circumference of a circle is $2\pi r$. Thus, $\Sigma B_{\parallel}\Delta l = B(2\pi r) = \mu_0 I$, and $B = \boxed{\frac{\mu_0 I}{2\pi r}}$.

3. **Strategy** Apply the Ampère-Maxwell law. For an uniform electric field between two plates of area A and charge q, the magnitude of the field is given by $E = \sigma/\epsilon_0 = q/(\epsilon_0 A)$.

   Solution The Ampère-Maxwell law is

   $$\Sigma B_{\parallel}\Delta l = \mu_0\left(I + \epsilon_0 \frac{\Delta\Phi_E}{\Delta t}\right) = \mu_0\epsilon_0 \frac{\Delta\Phi_E}{\Delta t} \text{ since } I = 0.$$

   By the RHR, the magnetic field lines are circles concentric with the central axis of the wire. At a distance $r \leq R$ from the central axis, the circumference of a circle is $2\pi r$. Thus, $\Sigma B_{\parallel}\Delta l = B(2\pi r)$. Find $\frac{\Delta\Phi_E}{\Delta t}$ in terms of I.

   The electric field in the gap is $E = \frac{\sigma}{\epsilon_0} = \frac{q}{\epsilon_0 A_{\text{wire}}} = \frac{q}{\epsilon_0 \pi R^2}$. So, $\frac{\Delta\Phi_E}{\Delta t} = \frac{\Delta(EA_{\text{circle}})}{\Delta t} = \frac{\Delta q \pi r^2}{\Delta t \epsilon_0 \pi R^2} = \frac{Ir^2}{\epsilon_0 R^2}$, and the Ampère-Maxwell law gives $2\pi r B = \mu_0\epsilon_0 \times \frac{Ir^2}{\epsilon_0 R^2}$, or $B = \boxed{\frac{\mu_0 I r}{2\pi R^2}}$.

5. **Strategy and Solution** A rod-shaped dipole antenna must be an electric dipole antenna. At a point due south of the transmitter, the EM waves are traveling due south. The electric field at this point is oriented vertically, like the antenna. The magnetic field is perpendicular to both the electric field and the direction the EM waves travel, so the magnetic field must be oriented $\boxed{\text{east-west}}$.

7. **Strategy** Assume that the magnetic field is uniform at the antenna. Use Faraday's law and Eq. (20-5) for magnetic flux.

   Solution From Faraday's Law, we have

   $$\mathscr{E} = -N\frac{\Delta\Phi_B}{\Delta t}.$$

   The maximum emf will occur when $\Delta\Phi_B$, the change in flux through the coil, is a maximum. The flux through the coil is $\Phi_B = BA\cos\theta = \Phi_{\text{max}}\cos\theta$, where θ is the angle between the magnetic field and the normal to the plane of the coil and $\Phi_{\text{max}} = BA$. So,

   $\Delta\Phi_B = \Delta(BA\cos\theta)$ and the emf is $\mathscr{E} = -N\frac{\Delta(BA\cos\theta)}{\Delta t} = -N\frac{\Delta\Phi_{\text{max}}}{\Delta t}\cos\theta = \mathscr{E}_{\text{max}}\cos\theta.$

   The emf is a maximum when $\theta = 0$. Otherwise, the emf is reduced by a factor of $\cos\theta$.

9. Strategy The frequency, wavelength, and speed of EM radiation are related by $\lambda f = c$.

Solution Compute the wavelength of the radio waves.

$$\lambda = \frac{c}{f} = \frac{3.00\times10^8\ \text{m/s}}{90.9\times10^6\ \text{Hz}} = \boxed{3.3\ \text{m}}$$

11. Strategy The frequency, wavelength, and speed of EM radiation are related by $\lambda f = c$. Use Figure 22.7 to determine the part of the EM spectrum to which the waves belong.

Solution

(a) Compute the wavelength.

$$\lambda = \frac{c}{f} = \frac{3.00\times10^8\ \text{m/s}}{60.0\ \text{Hz}} = \boxed{5.00\times10^6\ \text{m}}$$

(b) $\boxed{\text{The radius of the Earth is } 6.4\times10^6\ \text{m, which is close in value to the wavelength.}}$

(c) According to Figure 22.7, the waves are $\boxed{\text{radio waves}}$.

13. Strategy Divide the maximum frequency of visible light by the minimum frequency to determine the number of octaves. The frequency, wavelength, and speed of EM radiation are related by $\lambda f = c$. Use Figure 22.7.

Solution

(a) $\dfrac{750\ \text{THz}}{430\ \text{THz}} = 1.7 \approx 2^1$

The human eye can perceive $\boxed{\text{about one octave}}$ of visible light.

(b) Microwaves range from about 1 mm to 30 cm. The corresponding frequencies are

$$f = \frac{c}{\lambda} = \frac{3.00\times10^8\ \text{m/s}}{1\times10^{-3}\ \text{m}} = 3\times10^{11}\ \text{Hz} \text{ and } f = \frac{3.00\times10^8\ \text{m/s}}{30\times10^{-2}\ \text{m}} = 1\times10^9\ \text{Hz}.$$

Then, $\dfrac{3\times10^{11}\ \text{Hz}}{1\times10^9\ \text{Hz}} = 300 \approx 2^{8.2}$. So, the microwave region is $\boxed{\text{approximately 8 octaves}}$ wide.

17. Strategy The speed of light in matter is given by $v = c/n$, where n is the index of refraction.

Solution Compute the index of refraction of topaz.

$$n = \frac{c}{v} = \frac{3.00\times10^8\ \text{m/s}}{1.85\times10^8\ \text{m/s}} = \boxed{1.62}$$

19. Strategy The time of travel is equal to the distance traveled divided by the rate of travel.

Solution Compute the time it takes for light to travel 50.0 cm.

$$\Delta t = \frac{d}{c} = \frac{50.0\times10^{-2}\ \text{m}}{3.00\times10^8\ \text{m/s}} = 1.67\times10^{-9}\ \text{s} = \boxed{1.67\ \text{ns}}$$

21. (a) Strategy The wavelength is shorter in matter than it is in vacuum. Use Eq. (22-5).

Solution Compute the wavelength of light inside the glass.

$$\lambda_g = \frac{\lambda_v}{n} = \frac{692\text{ nm}}{1.52} = \boxed{455\text{ nm}}$$

(b) Strategy The frequency in glass is the same as the frequency in air. Use $c = f\lambda$.

Solution Compute the frequency of light inside the glass.

$$f_g = f_a = \frac{c}{\lambda_a} = \frac{3.00\times10^8\text{ m/s}}{692\times10^{-9}\text{ m}} = \boxed{4.34\times10^{14}\text{ Hz}}$$

25. Strategy The electric and magnetic fields oscillate with the same frequency, and their amplitudes are proportional to each other. Use Eqs. (22-6) and (22-7). The direction of propagation of EM waves is determined by $\vec{\mathbf{E}}\times\vec{\mathbf{B}}$.

Solution

(a) Compute the amplitude of the electric field.

$$E_m = cB_m = (3.00\times10^8\text{ m/s})(2.5\times10^{-11}\text{ T}) = \boxed{7.5\text{ mV/m}}$$

The frequency of the electric field is $\boxed{3.0\text{ MHz}}$, the same as that of the magnetic field.

(b) Compute the magnitude of the electric field.

$$E = cB = (3.00\times10^8\text{ m/s})(1.5\times10^{-11}\text{ T}) = \boxed{4.5\text{ mV/m}}$$

Since the magnetic field is in the $+z$-direction and the wave is traveling in the $-y$-direction, by $\vec{\mathbf{E}}\times\vec{\mathbf{B}}$ and the RHR, the electric field at $y = 0$ and $t = 0$ must point $\boxed{\text{in the }+x\text{-direction}}$.

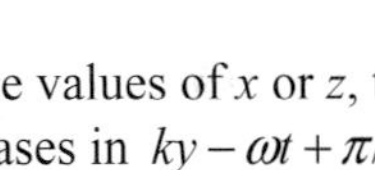

27. (a) Strategy and Solution Since the electric field depends on the value of y but not on the values of x or z, the wave moves parallel to the y-axis. The direction can be found by noting that as t increases in $ky-\omega t+\pi/6$, y must increase to maintain the relative phase. So, the wave is moving in the $\boxed{+y\text{-direction.}}$

(b) Strategy The amplitudes of the electric and magnetic fields are proportional to each other. Use Eq. (22-7). The direction of propagation of EM waves is determined by $\vec{\mathbf{E}}\times\vec{\mathbf{B}}$.

Solution Since the electric field is in the $+z$-direction when $t = 0$ and $y = 0$ and the wave is traveling in the $+y$-direction, by $\vec{\mathbf{E}}\times\vec{\mathbf{B}}$ and the RHR, the magnetic field at $t = 0$ and $y = 0$ must point in the $+x$-direction. The components are

$$\boxed{B_x = \frac{E_m}{c}\sin(ky-\omega t+\pi/6),\ B_y = B_z = 0}.$$

29. (a) Strategy Intensity is related to average energy density by $I = \langle u\rangle c$.

Solution Compute the average energy density.

$$\langle u\rangle = \frac{I}{c} = \frac{1400\text{ W/m}^2}{3.00\times10^8\text{ m/s}} = \boxed{4.7\times10^{-6}\text{ J/m}^3}$$

(b) Strategy The rms values for the electric and magnetic fields are related to the average energy density by $\langle u\rangle = \epsilon_0 {E_{\text{rms}}}^2 = {B_{\text{rms}}}^2/\mu_0$.

Solution Find the rms values of the electric and magnetic fields.

$$E_{\text{rms}} = \sqrt{\frac{\langle u\rangle}{\epsilon_0}} = \sqrt{\frac{4.7\times10^{-6}\ \text{J/m}^3}{8.854\times10^{-12}\ \text{C}^2/(\text{N}\cdot\text{m}^2)}} = \boxed{730\ \text{V/m}} \text{ and}$$

$$B_{\text{rms}} = \sqrt{\mu_0\langle u\rangle} = \sqrt{(4\pi\times10^{-7}\ \text{T}\cdot\text{m/A})(4.7\times10^{-6}\ \text{J/m}^3)} = \boxed{2.4\times10^{-6}\ \text{T}}.$$

33. Strategy Use Eqs. (22-10), (22-12), and (22-13).

Solution Find the average power of the source.

$$\begin{aligned}\langle P\rangle &= IA = I(4\pi r^2) = \langle u\rangle c(4\pi r^2) = \epsilon_0 {E_{\text{rms}}}^2 c(4\pi r^2)\\ &= [8.854\times10^{-12}\ \text{C}^2/(\text{N}\cdot\text{m}^2)](0.055\ \text{V/m})^2(3.00\times10^8\ \text{m/s})4\pi(22{,}000\ \text{m})^2 = \boxed{49\ \text{kW}}\end{aligned}$$

35. Strategy Assume the Sun is an isotropic source. Intensity is related to the average power radiated by $I = \langle P\rangle/A$ where $A = 4\pi r^2$.

Solution

(a) Find the total average power output of the Sun.

$$\langle P\rangle = AI = 4\pi r^2 I = 4\pi(1.50\times10^{11}\ \text{m})^2(1400\ \text{W/m}^2) = \boxed{4.0\times10^{26}\ \text{W}}$$

(b) Find the intensity of the sunlight incident on Mercury.

$$I = \frac{\langle P\rangle}{A} = \frac{3.96\times10^{26}\ \text{W}}{4\pi(5.8\times10^{10}\ \text{m})^2} = \boxed{9400\ \text{W/m}^2}$$

37. Strategy Use Eqs. (22-10) and (22-13). Replace each quantity with its SI units.

Solution Using Eq. (22-10), the units of $\langle u\rangle$ are

$$[\langle u\rangle] = \left[\epsilon_0 E_{\text{rms}}^2\right] = \frac{\text{C}^2}{\text{N}\cdot\text{m}^2}\cdot\frac{\text{V}^2}{\text{m}^2} = \frac{\text{C}^2}{\text{J}\cdot\text{m}}\cdot\frac{\text{J}^2}{\text{C}^2\cdot\text{m}^2} = \text{J/m}^3.$$

So, the units of $I = \langle u\rangle c$ are $[\langle u\rangle c] = \frac{\text{J}}{\text{m}^3}\cdot\frac{\text{m}}{\text{s}} = \frac{\text{J/s}}{\text{m}^2} = \text{W/m}^2$, which are the correct units for intensity.

41. Strategy Use Eq. (22-16b).

Solution

(a) The intensity of the light passing through the first polarizing sheet is

$I_1 = I_0\cos^2\theta_1 = I_0\cos^2\theta.$

The intensity of the light passing through the second sheet is

$$I_2 = I_1\cos^2\theta_2 = I_1\cos^2(90° - \theta) = I_0\cos^2\theta\cos^2(90° - \theta) = I_0\cos^2\theta\sin^2\theta = \boxed{\frac{1}{4}I_0\sin^2 2\theta},$$

since $\cos(90° - \theta) = \sin\theta$ and $2\cos\theta\sin\theta = \sin 2\theta$.

(b) The sine function reaches a maximum when its argument is 90°.

$2\theta = 90°$, so $\theta = \boxed{45°}$.

45. Strategy and Solution

(a) Microwaves that are transmitted are linearly polarized perpendicular to the strips. Since the microwaves are vertically polarized, a plate with horizontal strips will transmit the microwaves best. Plate $\boxed{\text{(a)}}$ will have the best transmission, since its strips are horizontal.

(b) The worst transmitter will be the best reflector. The strips in plate (c) are vertical, so plate $\boxed{\text{(c)}}$ will be the worst transmitter and the best reflector.

(c) The best transmitter is (a), so (a) transmits intensity I_1; the worst, is (c). Plate (b) is rotated 30.0° with respect to (a), so $I_{\text{second best}} = I_b = I_1 \cos^2 30.0° = \boxed{0.750 I_1}$.

47. Strategy The calculation of the relative velocity is the same as in Example 22.9, except that v_{rel} is negative (source and observer receding).

Solution The speeder is going faster than the police car. Since the police car is going 38.0 m/s, the speeder is going $38.0 \text{ m/s} + 7.0 \text{ m/s} = \boxed{45.0 \text{ m/s}}$.

49. Strategy A Doppler shift of this magnitude almost certainly requires a relativistic relative velocity. Use Eq. (22-17).

Solution A receiver measuring a longer wavelength would also observe a smaller frequency. This implies a longer period between wave crests and, thus, the source and observer are moving $\boxed{\text{farther apart}}$. So, v_{rel} is negative. Let $v_{rel} = -v$ where $v > 0$, for simplicity. Find v.

$$f_o = f_s\sqrt{\frac{1-\frac{v}{c}}{1+\frac{v}{c}}}$$

$$\left(\frac{f_o}{f_s}\right)^2 = \frac{1-\frac{v}{c}}{1+\frac{v}{c}}$$

$$\left(\frac{f_o}{f_s}\right)^2 + \left(\frac{f_o}{f_s}\right)^2 \frac{v}{c} = 1 - \frac{v}{c}$$

$$\frac{v}{c}\left[\left(\frac{f_o}{f_s}\right)^2 + 1\right] = 1 - \left(\frac{f_o}{f_s}\right)^2$$

$$v = c\frac{1-(f_o/f_s)^2}{1+(f_o/f_s)^2} = c\frac{1-(\lambda_s/\lambda_o)^2}{1+(\lambda_s/\lambda_o)^2} = (3.00\times10^8 \text{ m/s})\frac{1-(1/2)^2}{1+(1/2)^2} = \boxed{1.80\times10^8 \text{ m/s}}$$

53. Strategy The time delay in the conversation was equal to the travel time for an EM wave to go from the Earth to the Moon and back.

Solution Compute the time delay.

$$\Delta t = \frac{2d}{c} = \frac{2(3.845\times10^8 \text{ m})}{3.00\times10^8 \text{ m/s}} = \boxed{2.56 \text{ s}}$$

57. Strategy (a) Refer to Figure 22.7. (b) The length of the pulse is equal to the distance EM radiation travels in 20.0 ps. (c) The speed of the EM wave is reduced to $v = c/n$ in water. (d) Divide the length of a pulse by the wavelength to find the number that fit in one pulse. (e) The total EM energy in one pulse is equal to the power times the time of the pulse.

Solution

(a) According to Figure 22.7, the pulse is in the $\boxed{\text{infrared}}$ part of the EM spectrum.

(b) $\text{length} = c\Delta t = (3.00\times10^8\ \text{m/s})(20.0\times10^{-12}\ \text{s}) = \boxed{6.00\ \text{mm}}$

(c) $\text{length} = \dfrac{c}{n}\Delta t = \dfrac{3.00\times10^8\ \text{m/s}}{1.33}(20.0\times10^{-12}\ \text{s}) = \boxed{4.51\ \text{mm}}$

(d) $\dfrac{\text{length of pulse}}{\text{wavelength}} = \dfrac{6.00\times10^{-3}\ \text{m}}{1060\times10^{-9}\ \text{m}} = \boxed{5660}$

(e) $U = P\Delta t = (60.0\times10^{-3}\ \text{W})(20.0\times10^{-12}\ \text{s}) = \boxed{1.20\ \text{pJ}}$

61. Strategy The angular speed is $\omega = \Delta\theta/\Delta t$. For the returning light to pass through the next notch, the angular separation will be $\Delta\theta = 2\pi/5$. $\Delta t = 2d/c$ is the round-trip time for the light beam. The three lowest angular speeds are integer multiples of ω.

Solution Find the lowest angular speed.

$$\omega = \frac{\Delta\theta}{\Delta t} = \frac{2\pi/5}{2d/c} = \frac{\pi c}{5d} = \frac{\pi(3.00\times10^8\ \text{m/s})}{5(8.6\times10^3\ \text{m})} = 2.2\times10^4\ \text{rad/s}$$

The three lowest angular speeds are $\boxed{2.2\times10^4\ \text{rad/s},\ 4.4\times10^4\ \text{rad/s, and}\ 6.6\times10^4\ \text{rad/s}}$.

65. Strategy The cross-sectional area of the beam is $A = \pi r^2$. The intensity of the beam is equal to the power per unit area.

Solution

(a) Compute the intensity just outside the laser.

$$I = \frac{P}{A} = \frac{10.0\ \text{W}}{\pi(0.0020\ \text{m})^2} = \boxed{8.0\times10^5\ \text{W/m}^2}$$

(b) Compute the intensity at the surface of the Moon.

$$I = \frac{10.0\ \text{W}}{\pi\left(\frac{85{,}000\ \text{m}}{2}\right)^2} = \boxed{1.8\times10^{-9}\ \text{W/m}^2}$$

Chapter 23

REFLECTION AND REFRACTION OF LIGHT

Problems

1. **Strategy** Every point on a wavefront is considered a source of spherical wavelets. A surface tangent to the wavelets at a later time is the wavefront at that time.

   Solution The wavefronts due to an isotropic point source are spherical.

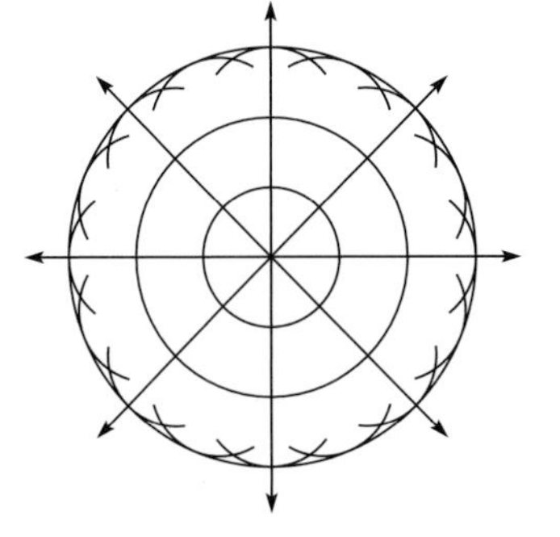

3. **Strategy** Every point on a wavefront is considered a source of spherical wavelets. A surface tangent to the wavelets at a later time is the wavefront at that time.

   Solution The planar wavefront incident on the reflecting wall at normal incidence is transmitted through the opening and reflected from the reflecting wall beyond the opening.

On the incident side are two planar waves. On the transmitted side is one hemispherical wave.

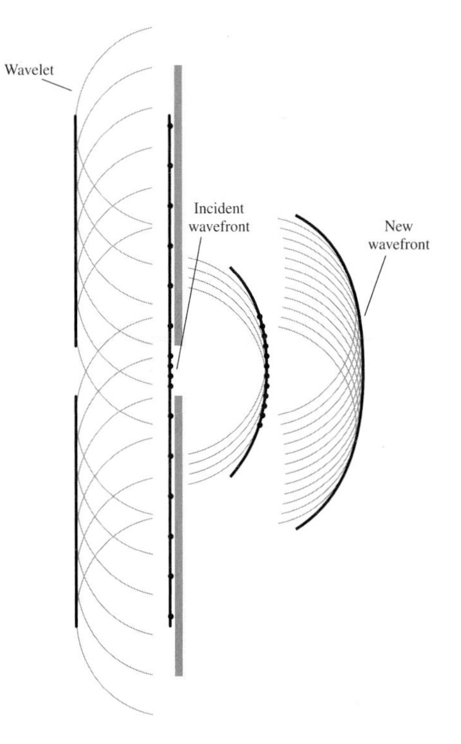

5. **Strategy** With respect to the normal at the point of incidence, the angle of incidence equals the angle of reflection and the reflected ray lies in the same plane, the plane of incidence, as the incident ray and the normal. Every point on a wavefront is considered a source of spherical wavelets. A surface tangent to the wavelets at a later time is the wavefront at that time.

   Solution The ray diagram and wavefronts for a spherical wave (from a point source) reflecting from a planar surface.

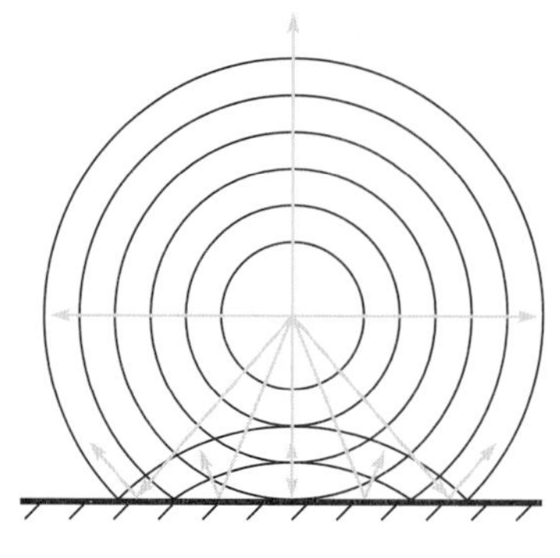

7. **Strategy** Redraw the figure, including the angle of incidence and the angle of reflection. Use the laws of reflection.

   Solution From the figure, we see that the angle of incidence is $\theta_i = 90° - 50° = 40°$. Using the laws of reflection, we find that the angle of reflection is $\theta_r = \theta_i = 40°$. The angles θ_i, θ_r, and δ must add to 180°. Solve for δ.

   $\theta_i + \theta_r + \delta = 40° + 40° + \delta = 180°$, so $\delta = \boxed{100°}$.

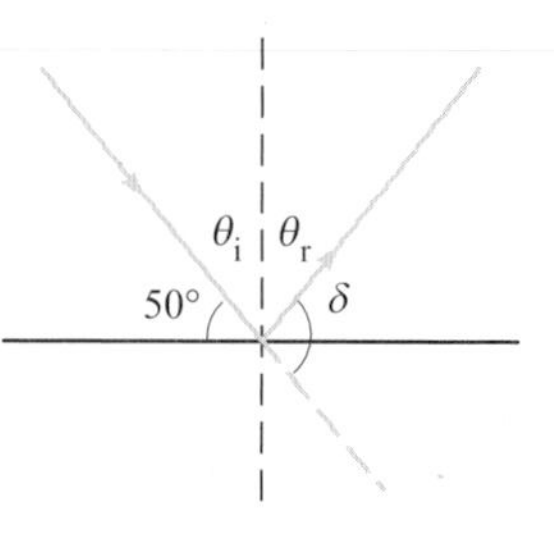

9. **Strategy** Draw a figure after Figure 23.7, labeling angles and lengths as necessary.

   Solution

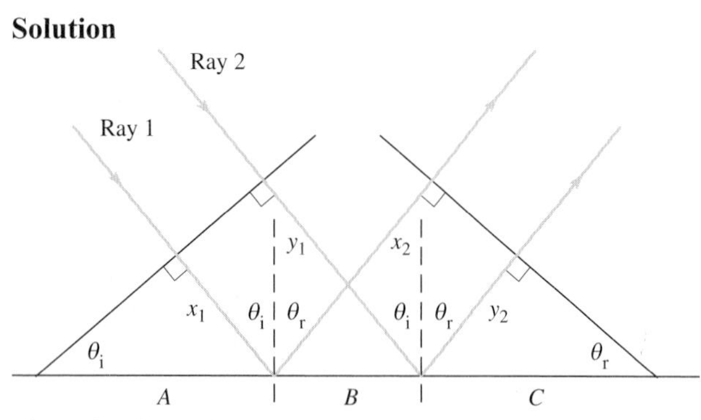

Since the time of travel from one wavefront to the other is the same for both rays, the distance traveled must also be the same.

Distance traveled by ray 1 = Distance traveled by ray 2

$$x_1 + x_2 = y_1 + y_2 \quad (1)$$

These distances can be expressed in terms of θ_i, θ_r, A, B, and C.

$x_1 = A\sin\theta_i$, $x_2 = (B+C)\sin\theta_r$, $y_1 = (A+B)\sin\theta_i$, and $y_2 = C\sin\theta_r$.

Substitute into (1) and simplify.

$$A\sin\theta_i + (B+C)\sin\theta_r = (A+B)\sin\theta_i + C\sin\theta_r$$
$$A\sin\theta_i + B\sin\theta_r + C\sin\theta_r = A\sin\theta_i + B\sin\theta_i + C\sin\theta_r$$
$$B\sin\theta_r = B\sin\theta_i$$
$$\sin\theta_r = \sin\theta_i$$
$$\theta_r = \theta_i$$

11. Strategy Use Snell's law, Eq. (23-4).

Solution Find the angle the Sun's rays in air make with the vertical.

$$n_\text{i} \sin\theta_\text{i} = n_\text{t} \sin\theta_\text{t}, \text{ so } \theta_\text{i} = \sin^{-1}\left(\frac{n_\text{t}}{n_\text{i}}\sin\theta_\text{t}\right) = \sin^{-1}\left(\frac{1.333}{1.000}\sin 42.0°\right) = \boxed{63.1°}.$$

13. Strategy Due to refraction of the light coming from the dolphin, the speed of the dolphin appears to be less than the actual speed by a factor of $1/n$.

Solution The dolphin appears to be moving at a speed of

$$\frac{15 \text{ m/s}}{1.33} = \boxed{11 \text{ m/s}}.$$

15. Strategy Draw a diagram. Use Snell's law, Eq. (23-4).

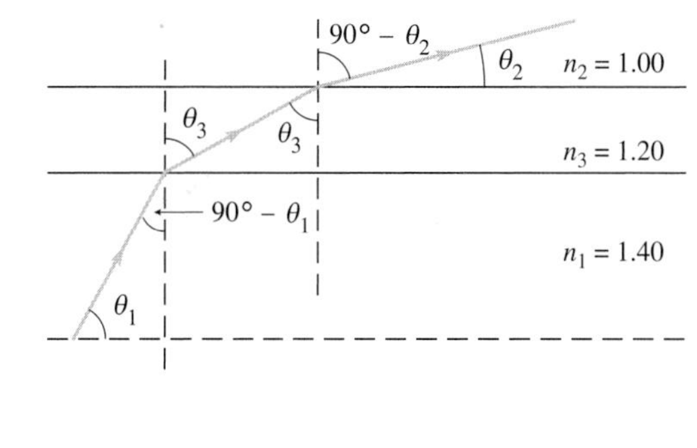

Solution Find θ_1.

Relate θ_1 and θ_3.

$n_1 \sin(90° - \theta_1) = n_3 \sin\theta_3$

Relate θ_2 and θ_3.

$n_2 \sin(90° - \theta_2) = n_3 \sin\theta_3$

Eliminate $n_3 \sin\theta_3$ and solve for θ_1.

$n_1 \sin(90° - \theta_1) = n_2 \sin(90° - \theta_2)$

$$90° - \theta_1 = \sin^{-1}\left[\frac{n_2}{n_1}\sin(90° - \theta_2)\right]$$

$$\theta_1 = 90° - \sin^{-1}\left[\frac{1.00}{1.40}\sin(90° - 5.00°)\right] = \boxed{44.6°}$$

17. Strategy Draw a diagram. Use Snell's law, Eq. (23-4).

Solution Initially, the coin is just hidden from view, so the angle of the observer's eye with respect to the vertical is $\theta_1 = \tan^{-1}\frac{6.5}{8.9}$. After the water is poured into the mug, the observer can just see the near end of the coin. Find θ_2.

$$n_1 \sin\theta_1 = n_1 \sin\left(\tan^{-1}\frac{6.5}{8.9}\right) = n_2 \sin\theta_2, \text{ so } \theta_2 = \sin^{-1}\left[\frac{n_1}{n_2}\sin\left(\tan^{-1}\frac{6.5}{8.9}\right)\right].$$

From the diagram, we see that $\tan\theta_2 = \frac{6.5 \text{ cm} - d}{8.9 \text{ cm}}$, where d is the diameter of the coin.

Find d.

$$\frac{6.5 \text{ cm} - d}{8.9 \text{ cm}} = \tan\left\{\sin^{-1}\left[\frac{n_1}{n_2}\sin\left(\tan^{-1}\frac{6.5}{8.9}\right)\right]\right\}, \text{ so}$$

$$d = 6.5 \text{ cm} - (8.9 \text{ cm})\tan\left\{\sin^{-1}\left[\frac{1.000}{1.333}\sin\left(\tan^{-1}\frac{6.5}{8.9}\right)\right]\right\} = \boxed{2.1 \text{ cm}}.$$

19. Strategy Draw a diagram. Use geometry and Snell's law, Eq. (23-4).

Solution Find θ_1 in terms of β.

$$\beta + 2\alpha = 180° \rightarrow \alpha = \frac{180° - \beta}{2} = 90° - \frac{\beta}{2}$$

$$\theta_1 = 90° - \alpha = 90° - \left(90° - \frac{\beta}{2}\right) = \frac{\beta}{2}$$

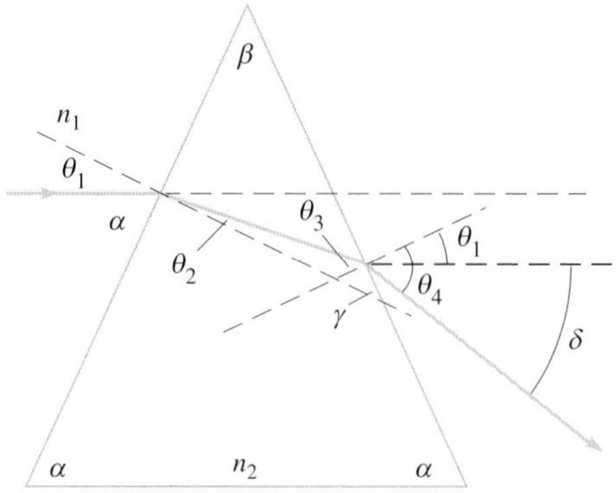

Find θ_2.

$$n_1 \sin\theta_1 = n_2 \sin\theta_2, \text{ so } \sin\theta_2 = \frac{n_1}{n_2}\sin\theta_1 = \frac{n_1}{n_2}\sin\frac{\beta}{2}.$$

Since β is small, use the approximation $\sin\theta \approx \theta$.

$$\sin\theta_2 \approx \frac{n_1}{n_2}\frac{\beta}{2}$$

Since $n_1 \sim n_2$ and $\frac{\beta}{2}$ is small, $\sin\theta_2$ is small, so apply the approximation for θ_2.

$$\theta_2 \approx \frac{n_1}{n_2}\frac{\beta}{2}$$

Using $n_1 = 1$ and $n_2 = n$ gives $\theta_2 = \frac{\beta}{2n}$.

Find θ_3 in terms of n and β.

From the figure, $90° + \beta + \gamma = 180°$, which implies $\gamma = 90° - \beta$.

$$\gamma + (90° + \theta_3) + \theta_2 = 90° - \beta + 90° + \theta_3 + \theta_2 = 180°, \text{ so } \theta_3 = \beta - \theta_2 = \beta - \frac{\beta}{2n} = \beta\left(1 - \frac{1}{2n}\right).$$

Use Snell's law to find θ_4, with $n_1 = 1$ and $n_2 = n$.

$$\sin\theta_4 = n\sin\theta_3 = n\sin\left[\beta\left(1 - \frac{1}{2n}\right)\right]$$

Now, $\beta\left(1 - \frac{1}{2n}\right) < \beta$, so it is small, and θ_4 is small as well. Thus, $\theta_4 \approx n\beta\left(1 - \frac{1}{2n}\right) = \beta\left(n - \frac{1}{2}\right)$.

Now find δ.

$$\delta + \theta_1 = \theta_4, \text{ so } \delta = \theta_4 - \theta_1 = \beta\left(n - \frac{1}{2}\right) - \frac{\beta}{2} = \beta n - \frac{\beta}{2} - \frac{\beta}{2} = \boxed{\beta(n-1)}.$$

21. Strategy Draw a diagram. Use geometry and Snell's law, Eq. (23-4).

Solution For the longest visible wavelengths, $n_2 = 1.517$.

$n_1 \sin\theta_1 = n_2 \sin\theta_2$, so

$$\theta_2 = \sin^{-1}\left(\frac{n_1}{n_2}\sin\theta_1\right) = \sin^{-1}\left(\frac{1.000}{1.517}\sin 55.0°\right) = 32.7°.$$

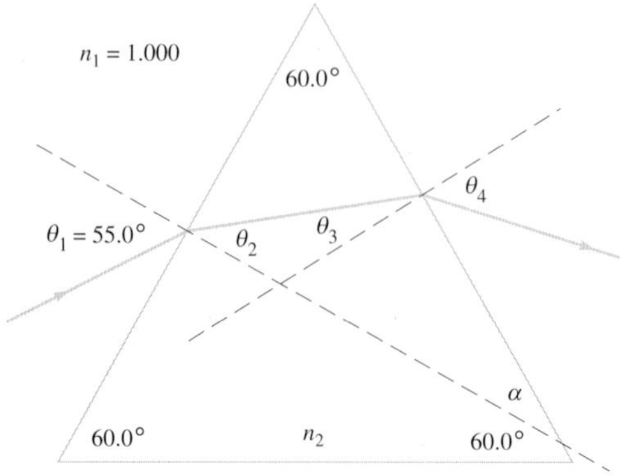

Find θ_3.

$60.0° + 90.0° + \alpha = 180.0°$, so $\alpha = 30.0°$.

$$\theta_2 + (90.0° + \theta_3) + \alpha = 180.0°$$

$$\theta_2 + \theta_3 + 30.0° = 90.0°$$

$$\theta_3 = 60.0° - \theta_2 = 60.0° - 32.7° = 27.3°$$

Find θ_4.

$n_2 \sin\theta_3 = n_1 \sin\theta_4$, so

$$\theta_4 = \sin^{-1}\left(\frac{n_2}{n_1}\sin\theta_3\right) = \sin^{-1}\left(\frac{1.517}{1.000}\sin 27.3°\right) = 44.1°.$$

For the shortest visible wavelengths, set $n_2 = 1.538$ and follow the same process. Find θ_2.

$$\theta_2 = \sin^{-1}\left(\frac{1.000}{1.538}\sin 55.0°\right) = 32.18°$$

Find θ_3.

$$\theta_3 = 60.0° - 32.18° = 27.82°$$

Find θ_4.

$$\theta_4 = \sin^{-1}\left(\frac{1.538}{1.000}\sin 27.82°\right) = 45.9°$$

The range of refraction angles is $\boxed{44.1° \le \theta \le 45.9°}$.

23. Strategy From Table 23.1, the index of refraction for diamond is 2.419, for air it is 1.000, and for water it is 1.333. Total internal reflection occurs when the angle of incidence is greater than or equal to the critical angle. Use Eq. (23-5a).

Solution

(a) Calculate the critical angle for diamond surrounded by air.

$$\theta_c = \sin^{-1}\frac{n_t}{n_i} = \sin^{-1}\frac{1.000}{2.419} = \boxed{24.42°}$$

(b) Calculate the critical angle for diamond under water.

$$\theta_c = \sin^{-1}\frac{n_t}{n_i} = \sin^{-1}\frac{1.333}{2.419} = \boxed{33.44°}$$

(c) Under water, the larger critical angle means that fewer light rays are totally reflected at the bottom surfaces of the diamond. Thus, less light is reflected back toward the viewer.

25. Strategy Total internal reflection occurs when the angle of incidence is greater than or equal to the critical angle. Use Eqs. (23-5).

Solution

(a) Find the index of refraction of the glass.

$$\theta_c = \sin^{-1}\frac{n_t}{n_i}, \text{ so } \sin\theta_c = \frac{n_t}{n_i} \text{ and } n_i = \frac{1.000}{\sin 40.00°} = \boxed{1.556}.$$

(b) $\boxed{\text{No}}$; rays from the defect could reach all points above the glass since $\boxed{\text{for } 0 \le \theta_i \le \theta_c,\ 0 \le \theta_t \le 90°}$.

29. Strategy Total internal reflection occurs when the angle of incidence is greater than or equal to the critical angle. Use Eqs. (23-5) and Snell's law, Eq. (23-4).

Solution When the light is incident on the Plexiglas tank, some is transmitted at angle θ_1, so $n\sin\theta_i = n_1\sin\theta_1$ where $n = 1.00$ for air and $n_1 = 1.51$ for Plexiglas. At the Plexiglas carbon tetrachloride interface, θ_1 is the incident angle and θ_2 is the transmitted angle, so $n_2\sin\theta_2 = n_1\sin\theta_1$ where $n_2 = 1.461$ for carbon tetrachloride. The ray passes through the carbon tetrachloride and is incident on the bottom tank-liquid interface at angle θ_2. Here the light must experience total internal reflection, so $\theta_2 = \theta_c = \sin^{-1}\frac{n_1}{n_2}$, or $\frac{n_1}{n_2} = \sin\theta_2$. Find θ_i.

$$n\sin\theta_i = n_1\sin\theta_1 = n_2\sin\theta_2 = n_2\frac{n_1}{n_2} = n_1,\text{ so } \theta_i = \sin^{-1}\frac{n_1}{n} = \sin^{-1}\frac{1.51}{1.00} = \sin^{-1}1.51,\text{ or } \sin\theta_i = 1.51,$$

which is impossible since $\sin\theta \le 1$ for all θ. Thus, there is [no] angle θ for which light is transmitted into the carbon tetrachloride but not into the Plexiglas at the bottom of the tank.

33. (a) Strategy The reflected light is totally polarized when the angle of incidence equals Brewster's angle. Use Eq. (23-6).

Solution Compute Brewster's angle.

$$\theta_B = \tan^{-1}\frac{n_t}{n_i} = \tan^{-1}\frac{1.309}{1.000} = 52.62°$$

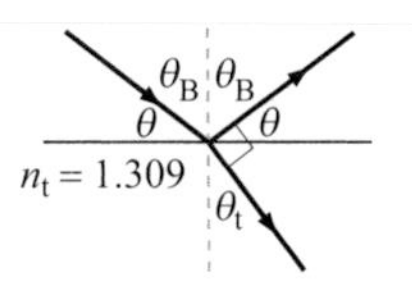

The angle with respect to the horizontal is the complement of this angle.

$\theta = 90° - 52.62° = \boxed{37.38°}$

(b) Strategy and Solution For Brewster's angle, the reflected light is polarized [perpendicular to the plane of incidence.]

(c) Strategy When the angle of incidence is Brewster's angle, the incident and transmitted rays are complementary.

Solution Find the angle of transmission.

$\theta_t = 90° - \theta_i = 90° - 52.62° = 37.38°$

The angle below the horizontal is the complement of this angle.

$90° - 37.38° = \boxed{52.62°}$

35. Strategy The equation derived in Example 23.4 can be used for this problem (with $n_w = n_d$) since $n_d > n_a$.

Solution Find the depth of the defect.

$$\frac{\text{apparent depth}}{\text{actual depth}} = \frac{n_a}{n_d} = \frac{1.000}{2.419},\text{ so actual depth} = \text{apparent depth}\cdot 2.419 = 2.0\text{ mm}\cdot 2.419 = \boxed{4.8\text{ mm}}.$$

37. Strategy Use Snell's law with $n_a = 1.000$, $n_w = 1.333$, and $\theta_a = 90°$.

Solution Find θ_w.

$n_w \sin\theta_w = n_a \sin\theta_a$, so

$$\theta_w = \sin^{-1}\left(\frac{n_a}{n_w}\sin 90°\right) = \sin^{-1}\left[\frac{n_a}{n_w}(1)\right] = \sin^{-1}\left(\frac{1.000}{1.333}\right) = 48.6°.$$

A right triangle is formed by the actual location of the penny, the apparent location of the penny, and the bottom of the bowl directly below the apparent location. Let y be the distance from the latter location to the actual location of the penny, then $y = (8.0\text{ cm})\tan 48.6° = 9.1\text{ cm}$. Since the penny appears to be 3.0 cm from the edge of the bowl, the horizontal distance between the penny and the edge of the bowl is 9.1 cm + 3.0 cm = $\boxed{12.1\text{ cm}}$.

39. Strategy For a plane mirror, a point source and its image are at the same distance from the mirror (on opposite sides) and both lie on the same normal line. Draw a ray diagram. Use geometry and the laws of reflection.

Solution Suppose the mirror is hung at the proper height (see the figure). The top of Daniel's head is at point A, his eyes are at point B, and his shoes are at point D. Lines AD and CE are perpendicular, and $\theta_i = \theta_r$, so triangles BCE and DCE are congruent and $BC = CD = \frac{1}{2}BD$.

Similarly, triangles CDE and FED are congruent, so

$$EF = CD = \frac{1}{2}BD = \frac{1}{2}\cdot 1.82\text{ m} = \boxed{0.91\text{ m}}.$$

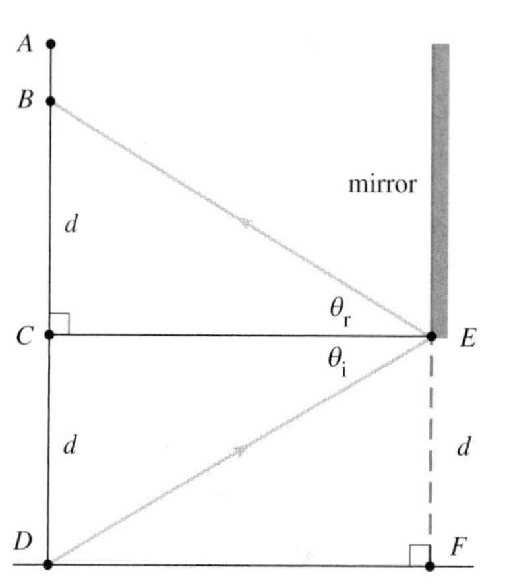

41. Strategy For a plane mirror, a point source and its image are at the same distance from the mirror (on opposite sides) and both lie on the same normal line.

Solution Since Gustav is holding the match, the distance to the image is twice the distance from Gustav to the mirror, so the distance from Gustav to the mirror is half this distance.

$$\frac{1}{2}\cdot 4\text{ m} = \boxed{2\text{ m}}$$

43. Strategy For a plane mirror, a point source and its image are at the same distance from the mirror (on opposite sides) and both lie on the same normal line.

Solution Maurizio sees three images by looking straight into each mirror. He sees three other images by looking where each pair of mirrors meet (left wall and right wall, left wall and ceiling, right wall and ceiling). He sees one more image by looking at the corner where all three mirrors meet. $\boxed{\text{He sees 7 images total.}}$

45. Strategy (a) Use Figure 23.27a and similar triangles. (b) Use any pair of rays in Figure 23.27a. Draw a diagram.

Solution

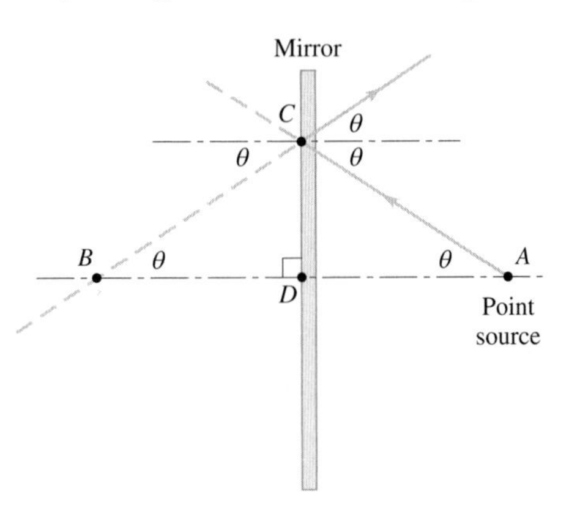

(a) In the figure, a ray from the point source strikes the mirror at point C and reflects. The line $\overleftrightarrow{AD}$ is normal to the surface of the mirror. When the reflected ray is traced back behind the mirror, it crosses $\overleftrightarrow{AD}$ at point B. By geometry and the laws of reflection, $\angle CAB$ and $\angle ABC$ have the same measure. Thus, the two right triangles CAD and CBD are similar and $AD = BD$. Since AD is the same for all rays, we conclude that all rays, when extended behind the mirror, will meet at point B.

(b) Point B is the image point, since all rays appear to come from it. Thus, the image point lies on a line through the object and perpendicular to the mirror, $\overleftrightarrow{AD}$. Since $BD = AD$ as shown in part (a), the object and image distances are equal.

47. Strategy The mirror is concave, so $p = 20.0$ cm and $f = 5.00$ cm. Find the image distance q using the mirror equation.

Solution Find the position of the image.

$$\frac{1}{p}+\frac{1}{q}=\frac{1}{f}, \text{ so } q=\frac{pf}{p-f}=\frac{(20.0\text{ cm})(5.00\text{ cm})}{20.0\text{ cm}-5.00\text{ cm}}=6.67\text{ cm}.$$

The image is formed $\boxed{\text{6.67 cm in front of the mirror}}$.

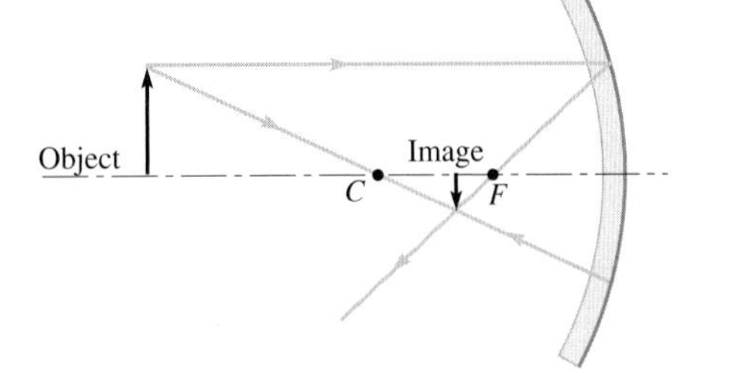

49. Strategy Since the image is real, the image distance is positive and the image is inverted (negative). The image is twice the size of the object, so $h' = -2h$. Use the magnification and mirror equations.

Solution Relate the object and image locations.

$$\frac{h'}{h}=\frac{-2h}{h}=-\frac{q}{p}, \text{ so } q=2p.$$

Find the distance of the object from the mirror.

$$\frac{1}{p}+\frac{1}{q}=\frac{1}{p}+\frac{1}{2p}=\frac{3}{2p}=\frac{1}{f}, \text{ so } p=\frac{3}{2}f=\frac{3}{2}\left(\frac{R}{2}\right)=\frac{3}{4}(25.0\text{ cm})=18.8\text{ cm}.$$

The object is $\boxed{\text{18.8 cm in front of the mirror}}$.

Object
Image
C
F
18.8 cm

53. Strategy Refer to the derivation of the magnification for a concave mirror. Draw a ray diagram using a ray that is not one of the three principal rays.

Solution The ray diagram.

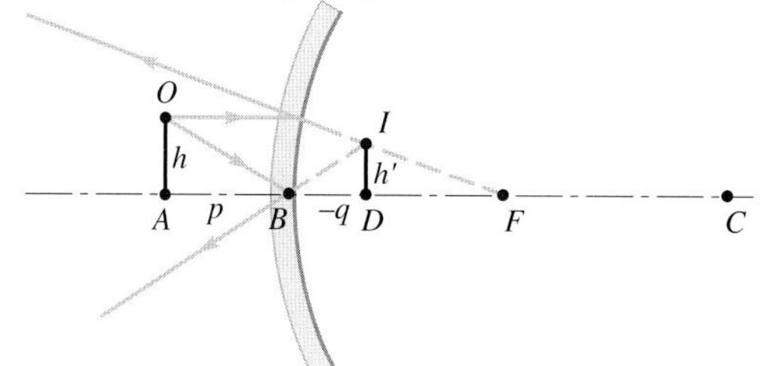

In the diagram, the two right triangles OAB and IDB are similar, so

$$\frac{h}{p}=\frac{h'}{-q} \text{ or } \frac{h'}{h}=\frac{-q}{p}.$$

The negative sign is included since $h/p>0$; $q<0$ since the image is behind the mirror, so $h'/q<0$, and $-h'/q>0$.

Combining this with the magnification definition $h'=mh$ gives

$$m=\frac{h'}{h}=-\frac{q}{p}.$$

57. (a) Strategy Use the thin lens equation to find the object distance. Then, draw the ray diagram.

Solution Find the object distance.

$$\frac{1}{p}+\frac{1}{q}=\frac{1}{f}, \text{ so } p=\frac{qf}{q-f}=\frac{(5.00\text{ cm})(3.50\text{ cm})}{5.00\text{ cm}-3.50\text{ cm}}=\boxed{11.7\text{ cm}}.$$

The diagram is shown.

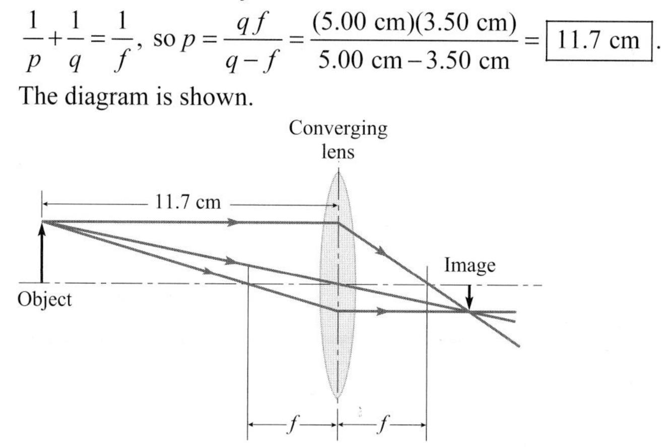

(b) Strategy and Solution Light rays actually pass through the image location, so the image is $\boxed{\text{real}}$.

(c) Strategy Use the magnification equation.

Solution Compute the magnification of the image.

$$m=-\frac{q}{p}=-\frac{5.00\text{ cm}}{11.667\text{ cm}}=\boxed{-0.429}$$

59. Strategy Sketch a ray diagram for a converging lens using the three principal rays. Place the object twice the focal length from the lens.

Solution As can be seen in the figure, when an object is placed twice the focal length away from a converging lens, an inverted, real, and same-sized image is formed.

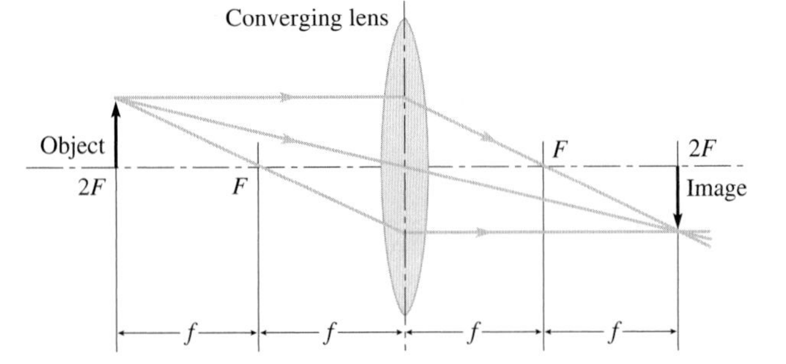

61. Strategy Sketch a ray diagram for a converging lens using the three principal rays. Place the object a distance equal to the focal length from the lens.

Solution As can be seen in the figure, when an object is placed at the focal point of a converging lens, the rays emerging from the lens are parallel to each other, so the image is at infinity.

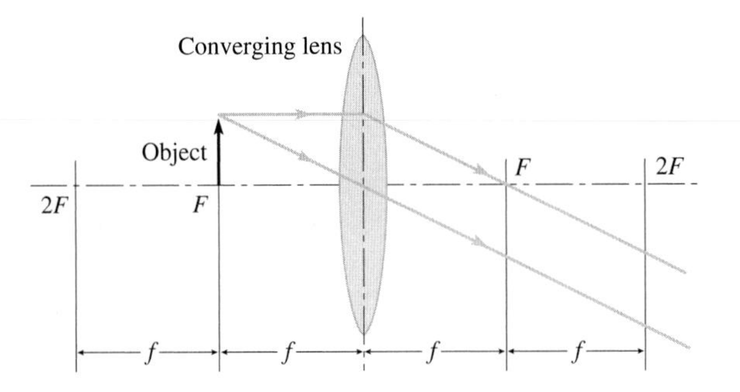

63. Strategy Draw a ray diagram using the three principal rays.

Solution Find the height and position of the image.

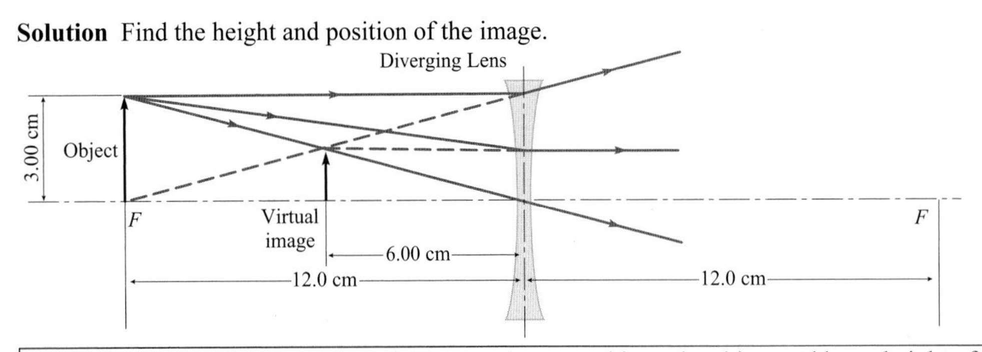

The image is located 6.00 cm from the lens on the same side as the object and has a height of 1.50 cm.

65. Strategy Use the thin lens and magnification equations.

Solution

(a) Solve the thin lens equation for q.

$$\frac{1}{p}+\frac{1}{q}=\frac{1}{f}, \text{ so } q=\frac{pf}{p-f}.$$

For $f = 8.00$ cm and $p = 5.00$ cm, we have

$$q=\frac{(5.00 \text{ cm})(8.00 \text{ cm})}{5.00 \text{ cm}-8.00 \text{ cm}}=-13.3 \text{ cm}.$$

Since q is negative, the image is virtual and on the same side of the lens as the object. The magnification is $m = -\frac{q}{p} = -\frac{-13.33\text{ cm}}{5.00\text{ cm}} = 2.67.$

Since m is positive and greater than 1, the image is upright and enlarged. The same process is used for the other object distances, and the results are summarized in the table.

p (cm)	q (cm)	m	Real or virtual	Orientation	Relative size
5.00	−13.3	2.67	virtual	upright	enlarged
14.0	18.7	−1.33	real	inverted	enlarged
16.0	16.0	−1.00	real	inverted	same
20.0	13.3	−0.667	real	inverted	diminished

(b) The image height can be found from the magnification equation.

$h' = mh$

For $p = 5.00\text{ cm}$, $h = 4.00\text{ cm}$, and $m = 2.67$ we have $h' = 2.67 \cdot 4.00\text{ cm} = \boxed{10.7\text{ cm}}$.

For $p = 20.0\text{ cm}$, $h = 4.00\text{ cm}$, and $m = -0.667$ we have $h' = -0.667 \cdot 4.00\text{ cm} = \boxed{-2.67\text{ cm}}$.

69. Strategy Use the thin lens and magnification equations. First find p and q. In this arrangement, p and q are positive so $m = -q/p$ is negative and the image is inverted.

Solution

(a) We calculate m using the given heights.

$m = \frac{h'}{h} = \frac{-60.0\text{ cm}}{2.40\text{ cm}} = -25.0 = \frac{-q}{p}$, so $q = 25.0p$.

Substitute this expression for q and $f = 12.0\text{ cm}$ into the thin lens equation.

$\frac{1}{p} + \frac{1}{q} = \frac{1}{p} + \frac{1}{25.0p} = \frac{1.04}{p} = \frac{1}{f} = \frac{1}{12.0\text{ cm}}$, so $p = 1.04(12.0\text{ cm}) = 12.48\text{ cm}$ and $q = 25.0p = 312\text{ cm}$.

The distance from the slide to the screen is $p + q = 12.48\text{ cm} + 312\text{ cm} = \boxed{3.24\text{ m}}$

(b) Since $\frac{1}{f} = \frac{1}{p} + \frac{1}{q}$, moving the screen away makes q larger, so p must be smaller. To maintain focus, the lens is moved $\boxed{\text{closer}}$ to the slide.

73. (a) Strategy The image is virtual, so the image distance is negative. Use the magnification equation.

Solution Find the object distance.

$m = \frac{h'}{h} = -\frac{q}{p}$, so $p = -\frac{h}{h'}q = -\frac{8.0\text{ cm}}{3.5\text{ cm}}(-4.0\text{ cm}) = \boxed{9.1\text{ cm}}$.

(b) Strategy and Solution The image is upright, virtual, smaller than the object, and closer to the mirror than the object. The mirror is $\boxed{\text{convex}}$.

(c) Strategy The radius of curvature is twice the absolute value of the focal length. Use the mirror equation.

Solution Find the focal length of the mirror.

$$\frac{1}{p}+\frac{1}{q}=\frac{1}{f}, \text{ so } f=\left(\frac{1}{p}+\frac{1}{q}\right)^{-1}=\left(-\frac{h'}{hq}+\frac{1}{q}\right)^{-1}=q\left(1-\frac{h'}{h}\right)^{-1}=(-4.0\text{ cm})\left(1-\frac{3.5}{8.0}\right)^{-1}=\boxed{-7.1\text{ cm}}.$$

Compute the radius of curvature.

$$R=2|f|=2(7.1\text{ cm})=\boxed{14\text{ cm}}$$

77. Strategy Use Snell's law, Eq. (23-4). Draw a diagram.

Solution In the figure, the displacement is labeled d. The distance the ray travels in the glass is H. To find d, first find θ_2, θ_3, and H. Find θ_2.

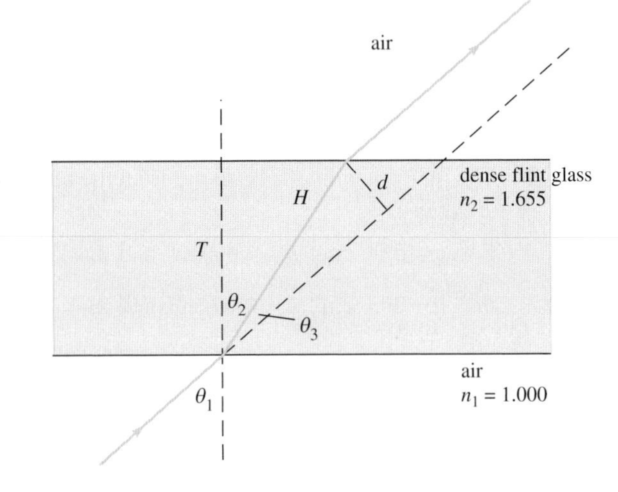

$n_1 \sin\theta_1 = n_2 \sin\theta_2$, so

$$\theta_2=\sin^{-1}\left(\frac{n_1}{n_2}\sin\theta_1\right)=\sin^{-1}\left(\frac{1.000}{1.655}\sin 60.0°\right)=31.6°.$$

Find H using the fact that $H\cos\theta_2 = T$.

$$H=\frac{T}{\cos\theta_2}=\frac{5.00\text{ mm}}{\cos 31.6°}=5.87\text{ mm}$$

Find θ_3 using the fact that $\theta_1=\theta_2+\theta_3$.

$$\theta_3=\theta_1-\theta_2=60.0°-31.6°=28.4°$$

Find d using the fact that $H\sin\theta_3 = d$.

$$d=H\sin\theta_3=(5.87\text{ mm})\sin 28.4°=\boxed{2.79\text{ mm}}$$

81. Strategy Total internal reflection occurs when the angle of incidence is greater than or equal to the critical angle. Use Eq. (23-5a), the laws of reflection, and Snell's law, Eq. (23-4).

Solution

(a) The critical angle is

$$\theta_c=\sin^{-1}\frac{n_t}{n_i}=\sin^{-1}\frac{1.000}{1.333}=48.61°$$

$\theta_i=\theta_r$, and since $75° > 48.61° = \theta_c$, $\boxed{\text{the light is totally reflected back into the water at a 75° angle with respect to the normal.}}$

(b) Since $25° < 48.61°$, part of the light is reflected back at a $\boxed{25°}$ angle and part is transmitted at an angle given by Snell's law.

$$n_1\sin\theta_1=n_2\sin\theta_2, \text{ so } \theta_2=\sin^{-1}\left(\frac{n_1}{n_2}\sin\theta_1\right)=\sin^{-1}\left(\frac{1.333}{1.000}\sin 25°\right)=\boxed{34°}.$$

83. Strategy and Solution Since the mirror is fixed to the car, the speed relative to the car is the same as the speed relative to the mirror. Since the object and image are equidistant for a plane mirror, the image speed will equal the object speed, which is $\boxed{8.0\text{ km/h}}$ relative to the car.

85. Strategy For a concave mirror, only a virtual image is upright. For a virtual image, the image distance is negative. Use the mirror and magnification equations.

Solution Find the pin-mirror distance p.

$$\frac{h'}{h}=\frac{-q}{p},\ \text{so } p=\frac{-qh}{h'}=\frac{-(-30.0\text{ cm})(3.00\text{ cm})}{9.00\text{ cm}}=\boxed{10.0\text{ cm}}.$$

The image is behind the mirror, so the pin-image distance is $p+|q|=10.0\text{ cm}+30.0\text{ cm}=\boxed{40.0\text{ cm}}$.

Find the focal length.

$$\frac{1}{p}+\frac{1}{q}=\frac{1}{f},\ \text{so } f=\frac{pq}{p+q}=\frac{(10.0\text{ cm})(-30.0\text{ cm})}{10.0\text{ cm}-30.0\text{ cm}}=15.0\text{ cm}.$$

So, $R=2f=30.0$ cm.

89. Strategy Since the image is virtual, q must be negative. Use the thin lens equation.

Solution Find the focal length.

$$\frac{1}{p}+\frac{1}{q}=\frac{1}{f},\ \text{so } f=\frac{pq}{p+q}=\frac{(10.0\text{ cm})(-30.0\text{ cm})}{10.0\text{ cm}-30.0\text{ cm}}=\boxed{15.0\text{ cm}}.$$

The focal length is positive, so the lens is $\boxed{\text{converging}}$.

93. Strategy Use geometry and Snell's law, Eq. (23-4).

Solution From Snell's law,
$n_{air}\sin i=n\sin r$ and $n\sin i'=n_{air}\sin r'$.
Assuming $i=r'$, then $n_{air}\sin i=n_{air}\sin r'$, and we have $n\sin r=n\sin i'$, which implies $r=i'$.
In the small isosceles triangle in the figure with angles r and i', there is a third unlabeled angle; call it θ. We know that $\theta+r+i'=180°$. Also, $A+180°+\theta=360°$, or $\theta+A=180°$. So, $A=r+i'$. In addition, $r=i'$, thus $A=2r$, or $r=\frac{1}{2}A$. The deviation angle δ is given by $\delta=\delta_1+\delta_2$, where $\delta_1=i-r$ and $\delta_2=r'-i'$. So, $\delta=\delta_1+\delta_2=i-r+r'-i'$.
For the minimum deviation, substitute $\delta_{min}=D$, $i=r'$, $r=i'$, and $r=\frac{1}{2}A$, and solve for i.

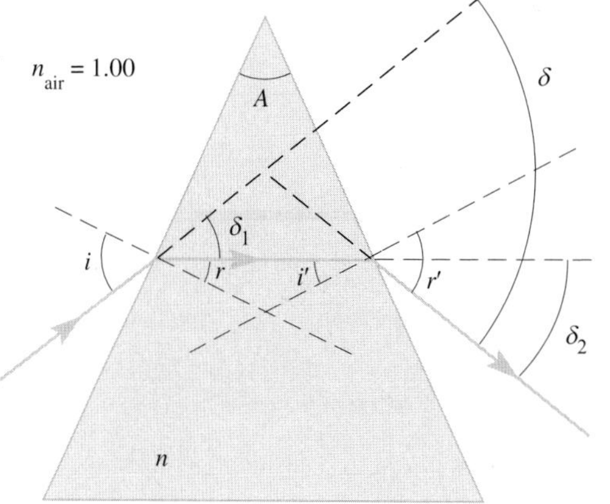

$$D=i-\frac{1}{2}A+i-\frac{1}{2}A=2i-A,\ \text{so } i=\frac{1}{2}(A+D).$$

Substitute for i and $r=\frac{1}{2}A$ into Snell's law and solve for n.

$$n_{air}\sin i=1.00\sin\frac{1}{2}(A+D)=n\sin r=n\sin\frac{1}{2}A,\ \text{so } n=\frac{\sin\frac{1}{2}(A+D)}{\sin\frac{1}{2}A}.$$

97. Strategy The focal length of a concave mirror is positive, so $f = R/2 = 7.0$ cm. Use the mirror equation.

Solution Find the image location.

$$\frac{1}{p}+\frac{1}{q}=\frac{1}{f}, \text{ so } q=\frac{pf}{p-f}=\frac{(9.0\text{ cm})(7.0\text{ cm})}{9.0\text{ cm}-7.0\text{ cm}}=32\text{ cm}.$$

$q > 0$, so the image is $\boxed{\text{32 cm in front of the mirror}}$.

♦

101. Strategy Redraw the diagram, labeling the vertices of similar triangles.

Solution In the figure, triangle ABF and triangle ACG are similar, so

$$\frac{h}{f}=\frac{-d}{D}, \text{ so } f=-\frac{h}{d}D.$$

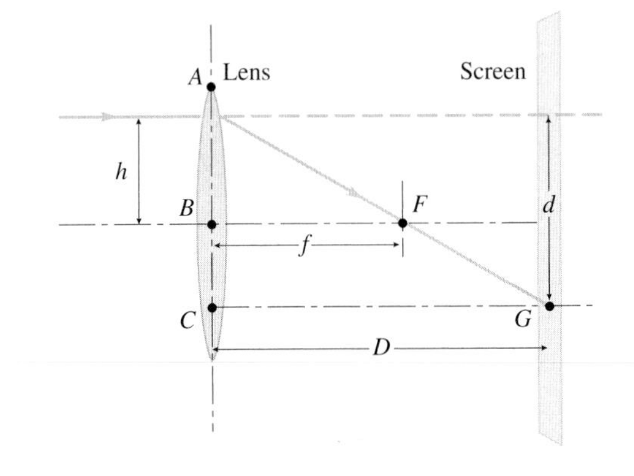

For paraxial rays, the slope of the d vs. h graph is constant. The middle three data points reflect this case. Find the slope.

$$m=\frac{\Delta d}{\Delta h}=\frac{-1.0-1.0}{0.5-(-0.5)}=\frac{-2.0}{1.0}=-2.0$$

For a constant slope, $\frac{\Delta d}{\Delta h}=\frac{d}{h}$, so

$$f=-\frac{1}{-2.0}D=0.50(1.0\text{ m})=\boxed{50\text{ cm}}.$$

Chapter 24

OPTICAL INSTRUMENTS

Problems

1. **Strategy** Use the lens and total transverse magnification equations.

Solution

(a) Find the image due to the first lens.

$$\frac{1}{p_1}+\frac{1}{q_1}=\frac{1}{f_1}, \text{ so } q_1=\frac{1}{\frac{1}{f_1}-\frac{1}{p_1}}=\frac{1}{\frac{1}{5.0\text{ cm}}-\frac{1}{12.0\text{ cm}}}=8.6\text{ cm}.$$

Find the object distance for the second lens.

$$p_2=s-q_1=2.0\text{ cm}-8.6\text{ cm}=-6.6\text{ cm}$$

Find the location of the final image.

$$\frac{1}{p_2}+\frac{1}{q_2}=\frac{1}{f_2}, \text{ so } q_2=\frac{1}{\frac{1}{f_2}-\frac{1}{p_2}}=\frac{1}{\frac{1}{4.0\text{ cm}}-\frac{1}{-6.6\text{ cm}}}=2.5\text{ cm}.$$

The final image is 2.5 cm past the 4.0-cm lens. The image is real since q_2 is positive.

(b) Compute the overall magnification.

$$m=m_1m_2=-\frac{q_1}{p_1}\left(-\frac{q_2}{p_2}\right)=-\frac{8.6\text{ cm}}{12.0\text{ cm}}\times-\frac{2.5\text{ cm}}{-6.6\text{ cm}}=\boxed{-0.27}$$

3. **(a) Strategy** Use the lens equations.

Solution Find the image location of the first lens.

$$\frac{1}{p_1}+\frac{1}{q_1}=\frac{1}{f_1}, \text{ so } q_1=\left(\frac{1}{f_1}-\frac{1}{p_1}\right)^{-1}.$$ So, the object distance for the second lens is

$$p_2=s-q_1=s-\left(\frac{1}{f_1}-\frac{1}{p_1}\right)^{-1}=0.880\text{ m}-\left(\frac{1}{0.250\text{ m}}-\frac{1}{1.100\text{ m}}\right)^{-1}=0.556\text{ m}.$$

Find the focal length of the second lens.

$$\frac{1}{p_2}+\frac{1}{q_2}=\frac{1}{f_2}, \text{ so } f_2=\left(\frac{1}{p_2}+\frac{1}{q_2}\right)^{-1}=\left(\frac{1}{0.556\text{ m}}+\frac{1}{0.150\text{ m}}\right)^{-1}=0.118\text{ m}=\boxed{11.8\text{ cm}}.$$

(b) Strategy Use the magnification and total transverse magnification equations.

Solution Find the total magnification of this lens combination.

$$m=m_1\times m_2=-\frac{q_1}{p_1}\times-\frac{q_2}{p_2}=\frac{(0.3235)(0.150)}{(1.100)(0.5565)}=\boxed{0.0793}$$

5. Strategy Draw a ray diagram for the system of lenses. Use the lens equations.

Solution The ray diagram:

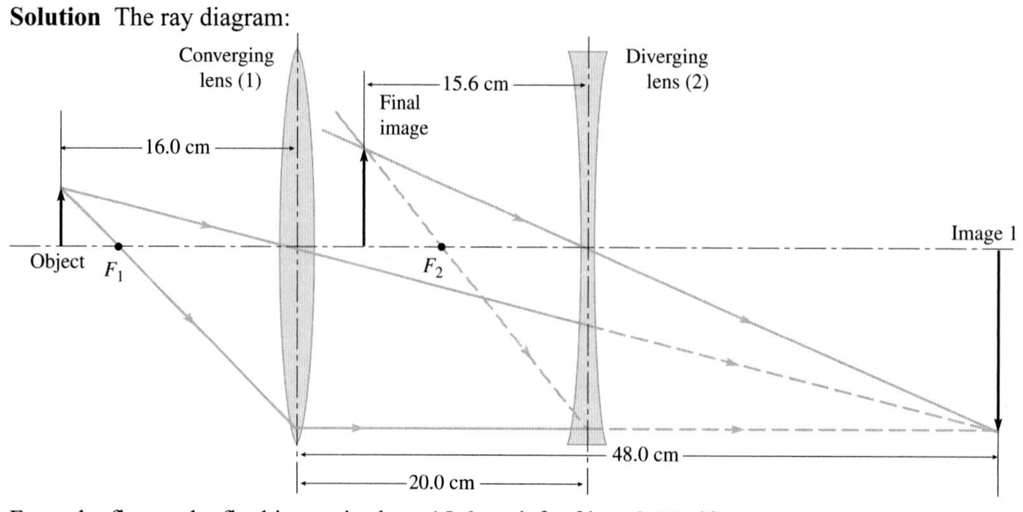

From the figure, the final image is about 15.6 cm left of lens 2. Verify using the lens equations. Find the image due to the first lens (1).

$$\frac{1}{p_1}+\frac{1}{q_1}=\frac{1}{f_1},\ \text{so } q_1=\frac{1}{\frac{1}{f_1}-\frac{1}{p_1}}=\frac{1}{\frac{1}{12.0\text{ cm}}-\frac{1}{16.0\text{ cm}}}=48.0\text{ cm}.$$

For the diverging lens (2), the object distance is $p_2 = s - q_1 = 20.0\text{ cm} - 48.0\text{ cm} = -28.0\text{ cm}$. Find q_2.

$$q_2=\frac{1}{\frac{1}{f_2}-\frac{1}{p_2}}=\frac{1}{\frac{1}{-10.0\text{ cm}}-\frac{1}{-28.0\text{ cm}}}=-15.6\text{ cm}$$

The final image is located $\boxed{\text{15.6 cm to the left of the diverging lens}}$.

9. Strategy Use the lens equations and transverse and total transverse magnification equations.

Solution

(a) Find the image location of the first lens.

$$\frac{1}{p_1}+\frac{1}{q_1}=\frac{1}{f_1},\ \text{so } q_1=\left(\frac{1}{f_1}-\frac{1}{p_1}\right)^{-1}=\left(\frac{1}{3.70\text{ cm}}-\frac{1}{6.00\text{ cm}}\right)^{-1}=9.65\text{ cm}.$$

So, the object distance for the second lens is $p_2 = s - q_1 = (24.65\text{ cm} - 6.00\text{ cm}) - 9.65\text{ cm} = 9.00\text{ cm}$.

Find the focal length of the second lens.

$$\frac{1}{p_2}+\frac{1}{q_2}=\frac{1}{f_2},\ \text{so } f_2=\left(\frac{1}{p_2}+\frac{1}{q_2}\right)^{-1}=\left(\frac{1}{9.00\text{ cm}}+\frac{1}{32.0\text{ cm}-24.65\text{ cm}}\right)^{-1}=\boxed{4.05\text{ cm}}.$$

(b) Since its focal length is positive, the lens is $\boxed{\text{converging}}$.

(c) Find the total magnification of this system.

$$m=m_1\times m_2=-\frac{q_1}{p_1}\times-\frac{q_2}{p_2}=\frac{(9.65)(32.0-24.65)}{(6.00)(9.00)}=\boxed{1.31}$$

(d) Find the image height.

$$m=\frac{h'}{h},\ \text{so } h'=mh=1.314(12.0\text{ cm})=\boxed{15.8\text{ cm}}.$$

11. Strategy Use the lens equations.

Solution

(a) Find the image location of the first lens.

$$\frac{1}{p_1}+\frac{1}{q_1}=\frac{1}{f_1}, \text{ so } q_1=\left(\frac{1}{f_1}-\frac{1}{p_1}\right)^{-1}=\frac{f_1 p_1}{p_1-f_1}=\frac{(3.00\text{ cm})(4.00\text{ cm})}{4.00\text{ cm}-3.00\text{ cm}}=12.0\text{ cm}.$$

This is a real image located between the two lenses; so, to display an image, a screen can be placed $\boxed{\text{12.0 cm to the right of the converging lens}}$.

(b) The object distance for the second lens is $p_2=s-q_1=10.0\text{ cm}-12.0\text{ cm}=-2.0\text{ cm}$.

Find the final image location.

$$\frac{1}{p_2}+\frac{1}{q_2}=\frac{1}{f_2}, \text{ so } q_2=\left(\frac{1}{f_2}-\frac{1}{p_2}\right)^{-1}=\frac{f_2 p_2}{p_2-f_2}=\frac{(-5.00\text{ cm})(-2.0\text{ cm})}{-2.0\text{ cm}-(-5.00\text{ cm})}=3.3\text{ cm}.$$

This is a real image located 3.3 cm to the right of the diverging lens; so, to display an image, a screen can be placed $\boxed{\text{3.3 cm to the right of the diverging lens}}$.

13. Strategy The image distance is the distance from the lens to the film. Use the thin lens equation.

Solution For an object at infinity, $1/p=1/\infty=0$, so the image distance is equal to the focal length, 200.0 mm = 20.00 cm. Find the image distance for an object 2.0 m from the lens.

$$\frac{1}{p}+\frac{1}{q}=\frac{1}{f}, \text{ so } q=\left(\frac{1}{f}-\frac{1}{p}\right)^{-1}=\left(\frac{1}{0.2000\text{ m}}-\frac{1}{2.0\text{ m}}\right)^{-1}=0.222\text{ m}=22.2\text{ cm}.$$

The minimum and maximum distances from the lens to the film are $\boxed{20.00\text{ cm}}$ and $\boxed{22.2\text{ cm}}$, respectively.

17. Strategy Use the thin lens and transverse magnification equations.

Solution

(a) Find the image location. The image is inverted, so $h'=-1.20$ cm.

$$-\frac{q}{p}=\frac{h'}{h}, \text{ so } q=-\frac{h'p}{h}=-\frac{(-1.20\text{ cm})(75.0\text{ m})}{4.00\text{ m}}=22.5\text{ cm}.$$

Find the focal length.

$$\frac{1}{f}=\frac{1}{p}+\frac{1}{q}, \text{ so } f=\frac{1}{\frac{1}{p}+\frac{1}{q}}=\frac{1}{\frac{1}{75.0\text{ m}}+\frac{1}{0.225\text{ m}}}=0.224\text{ m}=\boxed{224\text{ mm}}.$$

(b) Find the image location.

$$\frac{1}{p}+\frac{1}{q}=\frac{1}{f}, \text{ so } q=\frac{1}{\frac{1}{f}-\frac{1}{p}}=\frac{1}{\frac{1}{50.0\text{ mm}}-\frac{1}{75.0\times10^3\text{ mm}}}=50.0\text{ mm}.$$

Find the image size.

$$\frac{h'}{h}=-\frac{q}{p}, \text{ so } h'=-\frac{qh}{p}=-\frac{(50.0\text{ mm})(4.00\text{ m})}{75.0\text{ m}}=-2.67\text{ mm}.$$

The image would be $\boxed{2.67\text{ mm}}$ long.

(c) Find q in terms of p.

$$-\frac{q}{p}=\frac{h'}{h}, \text{ so } q=-\frac{h'p}{h}.$$

Substitute this value into the thin lens equation to find p.

$$\frac{1}{p}+\frac{1}{q}=\frac{1}{p}-\frac{h}{h'p}=\frac{1}{p}\left(1-\frac{h}{h'}\right)=\frac{1}{f}, \text{ so } p=f\left(1-\frac{h}{h'}\right).$$

Using $f = 50.0\text{ mm} = 0.0500\text{ m}$, $h = 4.00\text{ m}$, and $h' = -1.20\text{ cm} = -0.0120\text{ m}$ yields

$$p=(0.0500\text{ m})\left(1-\frac{4.00\text{ m}}{-0.0120\text{ m}}\right)=16.7\text{ m}.$$

The person would have to be $\boxed{16.7\text{ m}}$ away from the hippo.

19. Strategy Use the thin lens and transverse magnification equations. $h' = -20\text{ mm}$ since the image is inverted.

Solution Find the image location.

$$-\frac{q}{p}=\frac{h'}{h}, \text{ so } q=-\frac{h'p}{h}=-\frac{(-20\text{ mm})(300\text{ m})}{300\text{ m}}=20\text{ mm}.$$

Find the focal length.

$$\frac{1}{f}=\frac{1}{p}+\frac{1}{q}, \text{ so } f=\frac{1}{\frac{1}{p}+\frac{1}{q}}=\frac{1}{\frac{1}{300\text{ m}}+\frac{1}{0.020\text{ m}}}=0.020\text{ m}=\boxed{20\text{ mm}}.$$

Since $p >> q$, $f \approx q$, so the result is reasonable.

21. Strategy The slide is inverted with respect to the image, so h is negative. Use the thin lens and transverse magnification equations.

Solution First find the object distance.

$$\frac{1}{p}+\frac{1}{q}=\frac{1}{f}, \text{ so } p=\frac{1}{\frac{1}{f}-\frac{1}{q}}=\frac{1}{\frac{1}{12\text{ cm}}-\frac{1}{5.0\times10^2\text{ cm}}}=12.3\text{ cm}.$$

Find the image height.

$$\frac{h'}{h}=-\frac{q}{p}, \text{ so } h'=-\frac{qh}{p}=-\frac{(5.0\text{ m})(-2.4\text{ cm})}{12.3\text{ cm}}=98\text{ cm}.$$

Find the image width.

$$w'=-\frac{(5.0\text{ m})(-3.6\text{ cm})}{12.3\text{ cm}}=150\text{ cm}$$

The screen must be at least $\boxed{98\text{ cm by }150\text{ cm}}$.

23. (a) Strategy and Solution A focal length range of 1.85 cm to 2.00 cm corresponds to a 2.00 cm lens-retina distance in a normal eye. For a distant object, the focus of the eye can be adjusted to 1.90 cm so that the object is seen clearly, but for close objects, there is not much room for adjustment by the eye muscle to accommodate. Thus, this eye is $\boxed{\text{farsighted}}$.

(b) Strategy Use the thin lens equation.

Solution Solve for the object location.

$$\frac{1}{p}+\frac{1}{q}=\frac{1}{f}, \text{ so } p=\frac{1}{\frac{1}{f}-\frac{1}{q}}.$$

Find p for $q = 1.90\text{ cm}$ and f between 1.85 cm and 1.90 cm.

$$p=\frac{1}{\frac{1}{1.85\text{ cm}}-\frac{1}{1.90\text{ cm}}}=70\text{ cm and } p=\frac{1}{\frac{1}{1.90\text{ cm}}-\frac{1}{1.90\text{ cm}}}=\infty.$$

The eye can focus from $\boxed{70\text{ cm to infinity}}$.

25. Strategy The refractive power of a lens is the reciprocal of the focal length. Use the thin lens equation. Let $p = \infty$ for distant objects and $q = -2.0$ m for a virtual image at Colin's far point.

Solution Find the required refractive power.

$$P = \frac{1}{f} = \frac{1}{q} + \frac{1}{p} = \frac{1}{-2.0\text{ m}} + \frac{1}{\infty} = \boxed{-0.50\text{ D}}$$

27. Strategy The refractive power of a lens is the reciprocal of the focal length. Use the thin lens equation.

Solution

(a) $p = \infty$ and $q = -2.0\text{ m} + 0.020\text{ m}$. Find the necessary refractive power of the eyeglass lenses.

$$P = \frac{1}{f} = \frac{1}{q} + \frac{1}{p} = \frac{1}{-2.0\text{ m} + 0.020\text{ m}} + \frac{1}{\infty} = \boxed{-0.51\text{ D}}$$

(b) The refractive power of the eye by itself can be calculated using $p = 2.0$ m and $q = 2.0\text{ cm} = 0.020$ m.

$$P_{\text{eye}} = \frac{1}{f} = \frac{1}{p} + \frac{1}{q} = \frac{1}{2.0\text{ m}} + \frac{1}{0.020\text{ m}} = 50.5\text{ D}$$

This is the relaxed state. Using the accommodation gives $P_{\text{eye}} = 50.5\text{ D} + 4.0\text{ D} = 54.5\text{ D}$.

This gives an object distance of $p = \dfrac{1}{\frac{1}{f} - \frac{1}{q}} = \dfrac{1}{54.5\text{ m}^{-1} - \frac{1}{0.020\text{ m}}} = 22$ cm.

$\boxed{\text{Without his glasses, his near point is 22 cm.}}$

The refractive power of the eye with the glasses is $P_{\text{e-g}} = P_{\text{eye}} + P_{\text{glasses}} = 50.5 - 0.505 = 50$ D.

The accommodation gives $P_{\text{e-g}} = 50 + 4.0 = 54$ D.

The object distance is $p = \dfrac{1}{\frac{1}{f} - \frac{1}{q}} = \dfrac{1}{54\text{ m}^{-1} - \frac{1}{0.020\text{ m}}} = 25$ cm.

$\boxed{\text{With his glasses, his near point is 25 cm.}}$

29. Strategy The refractive power of a lens is the reciprocal of the focal length. Use Eq. (24-6), where $N = 0.25$ m.

Solution Compute the angular magnification.

$$M = \frac{N}{f} = \frac{0.25\text{ m}}{\frac{1}{5.5\text{ m}^{-1}}} = \boxed{1.4}$$

31. Strategy Use the thin lens equation and the equation for the angular magnification found in Example 24.6.

Solution

(a) $q = -25$ cm and $f = 5.0$ cm. Find the object distance; that is, the distance between the magnifying glass and the beetle.

$$\frac{1}{f} = \frac{1}{p} + \frac{1}{q}, \text{ so } p = \frac{fq}{q - f} = \frac{(5.0\text{ cm})(-25\text{ cm})}{-25\text{ cm} - 5.0\text{ cm}} = \boxed{4.2\text{ cm}}.$$

(b) The angular magnification is $M = \dfrac{N}{p} = \dfrac{25\text{ cm}}{4.2\text{ cm}} = \boxed{6.0}$.

33. (a) Strategy When the magnifying glass focuses the image to its smallest size, the image distance is equal to the focal length. The mean distance of the Sun from the Earth is 1.50×10^{11} m. The mean radius of the Sun is 6.96×10^{8} m. Use the thin lens and transverse magnification equations.

Solution Find the image size.

$$m=-\frac{q}{p}=-\frac{f}{p}=\frac{h'}{h},\ \text{so } h'=-\frac{f}{p}h=-\frac{0.060\text{ m}}{1.50\times10^{11}\text{ m}}(2\times6.96\times10^{8}\text{ m})=-0.56\text{ mm}.$$

The size of the image of the Sun is about $\boxed{0.56\text{ mm in diameter}}$.

(b) Strategy The intensity is inversely proportional to the area.

Solution Find the intensity of the image by forming a proportion.

$$\frac{I_{\text{image}}}{I_{\text{glass}}}=\frac{I_2}{I_1}=\frac{A_1}{A_2}=\left(\frac{d_1}{d_2}\right)^2,\ \text{so } I_2=\left(\frac{d_1}{d_2}\right)^2 I_1=\left(\frac{40\text{ mm}}{0.557\text{ mm}}\right)^2(0.85\text{ kW/m}^2)=\boxed{4.4\times10^3\text{ kW/m}^2}.$$

37. Strategy Use Eq. (24-8).

Solution

(a) Find the angular magnification of the microscope.

$$M_{\text{total}}=-\frac{L}{f_o}\times\frac{N}{f_e}=-\frac{18.0\text{ cm}}{1.44\text{ cm}}\times\frac{25\text{ cm}}{1.25\text{ cm}}=\boxed{-250}$$

(b) Since the angular magnification is inversely proportional to the objective focal length, the magnification is doubled if the focal length is halved. Therefore, the required focal length is $(1.44\text{ cm})/2=\boxed{0.720\text{ cm}}$.

41. Strategy Since the final image is not at infinity, the image from the objective lens is not at the focal point of the eyepiece. The angular magnification due to the eyepiece is equal to the near point divided by the object distance for the eyepiece. Use the thin lens equation and Figure 24.16.

Solution Find the object distance for the eyepiece.

$$M_e=\frac{N}{p_e},\ \text{so } p_e=\frac{N}{M_e}=\frac{25.0\text{ cm}}{5.00}=5.00\text{ cm}.$$

Find the focal length of the eyepiece.

$$\frac{1}{p_e}+\frac{1}{q_e}=\frac{1}{f_e},\ \text{so } f_e=\frac{1}{\frac{1}{p_e}+\frac{1}{q_e}}=\frac{1}{\frac{1}{5.00\text{ cm}}+\frac{1}{-25.0\text{ cm}}}=6.25\text{ cm}.$$

Find the image distance for the objective lens.

$$q_o+p_e=f_o+L+f_e,\ \text{so } q_o=f_o+L+f_e-p_e=1.50\text{ cm}+16.0\text{ cm}+6.25\text{ cm}-5.00\text{ cm}=18.8\text{ cm}.$$

Find the object distance for the objective lens.

$$\frac{1}{p_o}+\frac{1}{q_o}=\frac{1}{f_o},\ \text{so } p_o=\frac{1}{\frac{1}{f_o}-\frac{1}{q_o}}=\frac{1}{\frac{1}{1.50\text{ cm}}-\frac{1}{18.8\text{ cm}}}=\boxed{1.63\text{ cm}}.$$

43. (a) Strategy Use the thin lens equation.

Solution Find the object distance for the eyepiece.

$$\frac{1}{p_e}+\frac{1}{q_e}=\frac{1}{f_e},\ \text{so } p_e=\frac{f_e q_e}{q_e-f_e}=\frac{(2.80\text{ cm})(-25.0\text{ cm})}{-25.0\text{ cm}-2.80\text{ cm}}=2.52\text{ cm}.$$

The distance between the lenses is $q_o+p_e=16.5\text{ cm}+2.52\text{ cm}=\boxed{19.0\text{ cm}}$.

(b) Strategy The angular magnification of the eyepiece is $M_e = N/p_e$. Use the thin lens and the transverse magnification equations and the fact that the total magnification is equal to the product of magnifications due to the eyepiece and the objective.

Solution The transverse magnification for the objective is

$$m_o = -\frac{q_o}{p_o} = -q_o\left(\frac{1}{p_o}\right) = -q_o\left(\frac{1}{f_o} - \frac{1}{q_o}\right).$$

The total magnification is

$$M_{total} = m_o M_e = -q_o\left(\frac{1}{f_o} - \frac{1}{q_o}\right) \times \frac{N}{p_e} = -(16.5 \text{ cm})\left(\frac{1}{0.500 \text{ cm}} - \frac{1}{16.5 \text{ cm}}\right) \times \frac{25.0 \text{ cm}}{2.518 \text{ cm}} = \boxed{-318}.$$

(c) Strategy Use the thin lens equation.

Solution Find the object distance for the objective lens.

$$\frac{1}{p_o} + \frac{1}{q_o} = \frac{1}{f_o}, \text{ so } p_o = \frac{1}{\frac{1}{f_o} - \frac{1}{q_o}} = \frac{1}{\frac{1}{5.00 \text{ mm}} - \frac{1}{165 \text{ mm}}} = \boxed{5.16 \text{ mm}}.$$

45. (a) Strategy and Solution Since the angular magnification due to a telescope is equal to the ratio of the objective and eyepiece focal lengths, the magnification would be 1; that is, there would be no magnification, so you can't really make a telescope.

(b) Strategy Since the lenses have different focal lengths, there will be a magnification. The lens with the smaller strength should be the objective lens, since it has the longer focal length. Use Eqs. (24-9) and (24-10).

Solution Compute the magnification.

$$M = -\frac{f_o}{f_e} = -\frac{\frac{1}{1.3 \text{ m}^{-1}}}{\frac{1}{3.5 \text{ m}^{-1}}} = -2.7$$

Compute the barrel length.

$$f_o + f_e = \frac{1}{1.3 \text{ m}^{-1}} + \frac{1}{3.5 \text{ m}^{-1}} = 1.05 \text{ m}$$

Using a lens from each pair of glasses, the telescope would be 1.05 m long and have an angular magnification of -2.7.

49. (a) Strategy The moon is far enough away that we can approximate its distance as infinite. Then we can use the equation for the barrel length of a telescope, Eq. (24-9), for the distance between the lenses.

Solution Compute the barrel length of the telescope.

$$\text{barrel length} = f_o + f_e = 2.40 \text{ m} + 0.160 \text{ m} = \boxed{2.56 \text{ m}}$$

(b) Strategy Use the transverse magnification equation. $h = 3474$ km, $p = 384,500$ km, and $q \approx f_o = 2.40$ m.

Solution Find the diameter of the image produced.

$$\frac{h'}{h} = -\frac{q}{p}, \text{ so } h' = -\frac{qh}{p} = -\frac{(2.40 \text{ m})(3474 \text{ km})}{384,500 \text{ km}} = -0.0217 \text{ m} = -2.17 \text{ cm}.$$

The diameter of the image is $\boxed{2.17 \text{ cm}}$.

(c) Strategy Use Eq. (24-10).

Solution Compute the angular magnification.

$$M = -\frac{f_o}{f_e} = -\frac{2.40\text{ m}}{0.160\text{ m}} = \boxed{-15}$$

53. Strategy Use the thin lens equation. Let the lenses be labeled from left to right as 1, 2, and 3. Let their respective focal lengths and object and image distances, as well as the three image x-axis locations due to each lens be labeled with the number of each lens.

Solution

(a) Find the image distance for lens 1.

$$\frac{1}{p_1}+\frac{1}{q_1}=\frac{1}{f_1},\text{ so } q_1=\left(\frac{1}{f_1}-\frac{1}{p_1}\right)^{-1}=\left(\frac{1}{2.00\text{ cm}}-\frac{1}{2.50\text{ cm}}\right)^{-1}=10.0\text{ cm.}$$

The image location for lens 1 is 10.0 cm to the right of lens 1, so $x_1 = 2.50\text{ cm} + 10.0\text{ cm} = 12.5\text{ cm}$.

The object distance for the second lens is $p_2 = s - q_1 = (16.5\text{ cm} - 2.50\text{ cm}) - 10.0\text{ cm} = 4.0\text{ cm}$.

Since lens 3 is located at $x = 19.8$ cm, the final image distance is $q_3 = 39.8\text{ cm} - 19.8\text{ cm} = 20.0\text{ cm}$.

Find the object distance for lens 3.

$$\frac{1}{p_3}+\frac{1}{q_3}=\frac{1}{f_3},\text{ so } p_3=\left(\frac{1}{f_3}-\frac{1}{q_3}\right)^{-1}=\left(\frac{1}{4.00\text{ cm}}-\frac{1}{20.0\text{ cm}}\right)^{-1}=5.00\text{ cm.}$$

The object distance for lens 3 is 5.00 cm to the left of lens 3, so $x_3 = 19.8\text{ cm} - 5.00\text{ cm} = 14.8\text{ cm}$. This object is the image due to the unknown lens, so the image distance for the unknown lens is $q_2 = 14.8\text{ cm} - 16.5\text{ cm} = -1.7\text{ cm}$. Find the focal length of the unknown lens.

$$\frac{1}{p_2}+\frac{1}{q_2}=\frac{1}{f_2},\text{ so } f_2=\left(\frac{1}{p_2}+\frac{1}{q_2}\right)^{-1}=\left(\frac{1}{4.0\text{ cm}}+\frac{1}{-1.7\text{ cm}}\right)^{-1}=-3.0\text{ cm.}$$

The focal length is negative, so the unknown lens is a $\boxed{\text{diverging}}$ lens.

(b) According to part (a), the focal length of the unknown lens is $f_2 = \boxed{-3.0\text{ cm}}$.

(c) The ray diagram is shown.

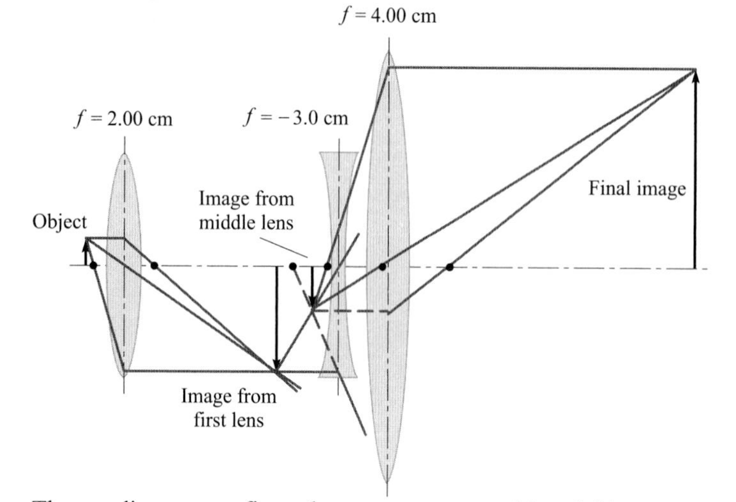

The ray diagram confirms the answers to parts (a) and (b).

57. Strategy Use the thin lens and transverse magnification equations. The image is real, so both the object and image distances are positive.

Solution Find the object distance.

$m=\frac{h'}{h}=-\frac{q}{p}$, so $p=-\frac{h}{h'}q$, but p, q, and h are greater than zero, so $h'<0$.

Find the focal length of the projector lens.

$$\frac{1}{q}+\frac{1}{p}=\frac{1}{q}+\frac{1}{-hq/h'}=\frac{1}{q}\left(1-\frac{h'}{h}\right)=\frac{1}{f},\ \text{so}\ f=q\left(1-\frac{h'}{h}\right)^{-1}=(3.50\text{ m})\left(1-\frac{-2.00\text{ m}}{0.0508\text{ m}}\right)^{-1}=\boxed{8.67\text{ cm}}.$$

61. Strategy Use the thin lens/mirror and total transverse magnification equations.

Solution

(a) Find the image location for the lens.

$$\frac{1}{p_1}+\frac{1}{q_1}=\frac{1}{f_1},\ \text{so}\ q_1=\left(\frac{1}{f_1}-\frac{1}{p_1}\right)^{-1}=\left(\frac{1}{15.0\text{ cm}}-\frac{1}{20.0\text{ cm}}\right)^{-1}=60.0\text{ cm}.$$

The first image is 60.0 cm to the right of the lens, which puts it $75.0\text{ cm}-60.0\text{ cm}=15.0\text{ cm}$ in front of the mirror. This image becomes the object for the mirror, so the object distance is $p_2=15.0$ cm, which is 5.0 cm to the left of the focal point of the mirror. From our study of concave mirrors, we know an object in this position has an image that is $\boxed{\text{real}}$ and inverted. Since the first image was also inverted, the final image will be $\boxed{\text{upright}}$ with respect to the original object.

(b) Use the mirror equation with $p_2=15.0$ cm and $f_2=10.0$ cm to find the final image.

$$q_2=\frac{f_2p_2}{p_2-f_2}=\frac{(10.0\text{ cm})(15.0\text{ cm})}{15.0\text{ cm}-10.0\text{ cm}}=30\text{ cm}$$

The final image is $\boxed{\text{30 cm to the left of the mirror}}$.

(c) Find the total transverse magnification.

$$m=m_1\times m_2=\frac{-q_1}{p_1}\times\frac{-q_2}{p_2}=\frac{60.0\text{ cm}}{20.0\text{ cm}}\times\frac{30\text{ cm}}{15.0\text{ cm}}=\boxed{6.0}$$

65. Strategy Use the lens equations, as well as the transverse and total transverse magnification equations.

Solution

(a) Find the intermediate image distance.

$$\frac{1}{p_1}+\frac{1}{q_1}=\frac{1}{f_1},\ \text{so}\ q_1=\frac{f_1p_1}{p_1-f_1}=\frac{(30.0\text{ cm})(1.8\text{ cm})}{1.8\text{ cm}-30.0\text{ cm}}=-1.9\text{ cm}.$$

The object distance for lens 2 is $p_2=s-q_1=21.0\text{ cm}-(-1.9\text{ cm})=22.9$ cm. Find the final image distance.

$$q_2=\frac{f_2p_2}{p_2-f_2}=\frac{(-15.0\text{ cm})(22.915\text{ cm})}{22.915\text{ cm}-(-15.0\text{ cm})}=-9.07\text{ cm}$$

The intermediate image is $\boxed{\text{1.9 cm to the left of the first lens}}$ and the final image is $\boxed{\text{9.07 cm to the left of the second lens}}$.

(b) The total magnification is $m=m_1\times m_2=\frac{-q_1}{p_1}\times\frac{-q_2}{p_2}=\frac{1.9\text{ cm}}{1.8\text{ cm}}\times\frac{9.07\text{ cm}}{22.9\text{ cm}}=\boxed{0.42}$.

(c) Calculate the height of the final image.

$h'=mh=0.42\times2.00\text{ mm}=\boxed{0.84\text{ mm}}$

67. Strategy For a distant mountain range, the object distance can be approximated as infinite, so the lens-film distance must equal the focal length of the lens, 50.0 mm. For the flower bed at 1.5 m, the lens-film distance is found by solving the thin lens equation for q, using $p = 1.5\text{ m} = 1500\text{ mm}$ and $f = 50.0\text{ mm}$.

Solution Find the distance the lens moves with respect to the film.

$$\frac{1}{p}+\frac{1}{q}=\frac{1}{f},\text{ so } q=\left(\frac{1}{f}-\frac{1}{p}\right)^{-1}=\left(\frac{1}{50.0\text{ mm}}-\frac{1}{1500\text{ mm}}\right)^{-1}=51.7\text{ mm.}$$

The lens must move a distance of $51.7\text{ mm} - 50.0\text{ mm} = \boxed{1.7\text{ mm}}$.

69. (a) Strategy Assume the image formed by the objective is at the focal point of the eyepiece. Use the thin lens equation.

Solution Find the image distance for the objective.

$$q_\text{o} + f_\text{e} = \text{distance between the lenses}$$
$$q_\text{o} + 4.0\text{ cm} = 32.0\text{ cm}$$
$$q_\text{o} = 28.0\text{ cm}$$

Find the object distance for the objective.

$$p_\text{o} = \frac{f_\text{o}q_\text{o}}{q_\text{o}-f_\text{o}} = \frac{(5.0\text{ cm})(28.0\text{ cm})}{28.0\text{ cm}-5.0\text{ cm}} = \boxed{6.1\text{ cm}}$$

(b) Strategy Use Eqs. (24-7) and (24-8).

Solution Calculate the angular magnification for the microscope.

$$M_\text{total} = -\frac{L}{f_\text{o}}\times\frac{N}{f_\text{e}} = -\frac{q_\text{o}-f_\text{o}}{f_\text{o}}\times\frac{N}{f_\text{e}} = -\frac{28.0\text{ cm}-5.0\text{ cm}}{5.0\text{ cm}}\times\frac{25\text{ cm}}{4.0\text{ cm}} = \boxed{-29}$$

73. Strategy Since light rays are reversible, the "reduction" is equal to the inverse of the magnification. (The focal lengths have changed roles.) Use Eq. (24-10).

Solution Compute the factor by which the angular size has been reduced.

$$\frac{1}{M} = -\frac{f_\text{e}}{f_\text{o}} = -\frac{0.015\text{ m}}{2.20\text{ m}} = \boxed{-0.0068}$$

Chapter 25

INTERFERENCE AND DIFFRACTION

Problems

1. (a) Strategy The wavelength and frequency are related by $\lambda f = c$.

Solution Compute the wavelength of a 60-kHz EM wave.

$$\lambda = \frac{c}{f} = \frac{3.00\times10^8 \text{ m/s}}{6.0\times10^4 \text{ Hz}} = \boxed{5.0 \text{ km}}$$

(b) Strategy Compute the path difference to determine the phase difference.

Solution The path difference is $\Delta l = (19 \text{ km} + 12 \text{ km}) - 21 \text{ km} = 10 \text{ km}$, which is equal to two wavelengths. So, the path difference results in constructive interference, but the reflection of the signal from the helicopter results in a half-wavelength phase difference, which cause destructive interference. To summarize:

$\boxed{\text{Destructive interference occurs, since the path difference is 10 km and there is a } \lambda/2 \text{ phase shift}}$.

3. Strategy Since the radio waves interfere destructively and the reflections at the planes introduce $180°$ phase shifts, the path differences must be equal to integral numbers of wavelengths.

Plane
d
h
Roger 102 km Coast Guard

Solution The path differences are $d + h - 102$ km, where $d = \sqrt{(102 \text{ km})^2 + h^2}$. The wavelength is $\lambda = c/f$. Find the wavelength in terms of the path differences.

$\Delta l = \sqrt{(102 \text{ km})^2 + h^2} + h - 102 \text{ km} = m\lambda$, so

$\lambda = \dfrac{\sqrt{(102 \text{ km})^2 + h^2} + h - 102 \text{ km}}{m}$. Now, the heights of the planes are $h = 780$ m, 975 m, and 1170 m, so we have $\lambda = \dfrac{783 \text{ m}}{m_1} = \dfrac{980 \text{ m}}{m_2} = \dfrac{1177}{m_3}$ for the respective heights. Form ratios of the integers m.

$\dfrac{m_2}{m_1} = \dfrac{980}{783} = 1.25 = \dfrac{5}{4}$ and $\dfrac{m_3}{m_1} = \dfrac{1177}{783} = 1.50 = \dfrac{6}{4}$. So, $m_1 = 4$ and $\lambda = \dfrac{783 \text{ m}}{4} = 196$ m. Therefore, the frequency of the signal is $f = \dfrac{c}{\lambda} = \dfrac{3.00\times10^8 \text{ m/s}}{196 \text{ m}} = \boxed{1530 \text{ kHz}}$.

5. (a) Strategy Sketch four sinusoidal waves, each with wavelength 6 cm. From top to bottom, the amplitudes are 2 cm, 2 cm, 3 cm, and 1 cm. The third wave is 180° out of phase with the other three.

Solution The waves are shown.

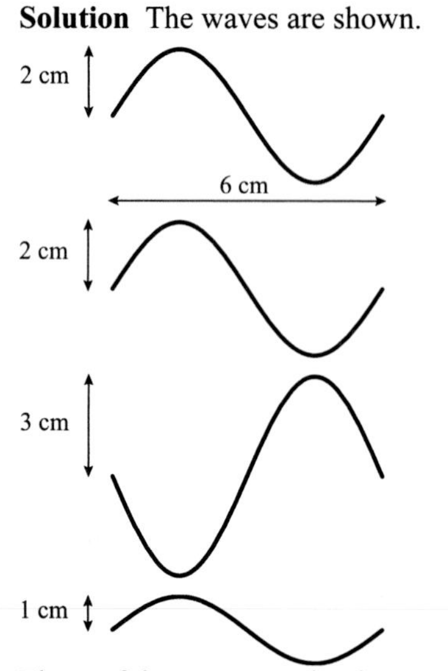

Three of the waves are in phase, so their amplitudes add. The amplitude of the third wave is 180° out of phase, so its amplitude is subtracted. The amplitude of the sum is $2\text{ cm}+2\text{ cm}-3\text{ cm}+1\text{ cm}=\boxed{2\text{ cm}}$.

(b) Strategy The first two waves interfere constructively, so use Eq. (25-3) to find the intensity of the new wave. The second two waves interfere destructively, so use Eq. (25-6) to find the intensity of the new wave; this wave will also be 180° out of phase with the sum of the first two, since the third wave's amplitude is greater than the fourth's, so use Eq. (25-6) again to find the final intensity. The first two waves each have intensity I_0. Use the fact that intensity is directly proportional to amplitude to find the intensity of the third and fourth waves in terms of I_0.

Solution Find the intensity of the combined first two waves.

$$I=I_1+I_2+2\sqrt{I_1I_2}=I_0+I_0+2\sqrt{I_0I_0}=4I_0$$

Find the intensities of the third and fourth waves in terms of I_0. Form proportions.

$$\frac{I_3}{I_0}=\frac{3^2}{2^2}=\frac{9}{4},\text{ so }I_3=\frac{9}{4}I_0.\ \frac{I_4}{I_0}=\frac{1^2}{2^2}=\frac{1}{4},\text{ so }I_4=\frac{1}{4}I_0.$$

Find the intensity of the combined third and fourth waves.

$$I=I_3+I_4-2\sqrt{I_3I_4}=\frac{9}{4}I_0+\frac{1}{4}I_0-2\sqrt{\frac{9}{4}I_0\frac{1}{4}I_0}=\frac{5}{2}I_0-2\sqrt{\frac{9}{16}I_0^2}=I_0$$

Find the intensity of the combination of all four waves.

$$I=4I_0+I_0-2\sqrt{4I_0I_0}=5I_0-4I_0=\boxed{I_0}$$

(c) Strategy Three of the waves are in phase, so their amplitudes add. The amplitude of the fourth wave is 180° out of phase, so its amplitude is subtracted.

Solution The amplitude of the sum is $2\text{ cm}+2\text{ cm}+3\text{ cm}-1\text{ cm}=\boxed{6\text{ cm}}$.

(d) Strategy Repeat the process of part (b), but this time, the combination of the second two waves will be in phase with the combination of the first two.

Solution Find the intensity of the combined first two waves.

$I = I_1 + I_2 + 2\sqrt{I_1 I_2} = I_0 + I_0 + 2\sqrt{I_0 I_0} = 4I_0$

Find the intensities of the third and fourth waves in terms of I_0. Form proportions.

$\frac{I_3}{I_0} = \frac{3^2}{2^2} = \frac{9}{4}$, so $I_3 = \frac{9}{4}I_0$. $\frac{I_4}{I_0} = \frac{1^2}{2^2} = \frac{1}{4}$, so $I_4 = \frac{1}{4}I_0$.

Find the intensity of the combined third and fourth waves.

$I = I_3 + I_4 - 2\sqrt{I_3 I_4} = \frac{9}{4}I_0 + \frac{1}{4}I_0 - 2\sqrt{\frac{9}{4}I_0 \frac{1}{4}I_0} = \frac{5}{2}I_0 - 2\sqrt{\frac{9}{16}I_0^2} = I_0$

Find the intensity of the combination of all four waves.

$I = 4I_0 + I_0 + 2\sqrt{4I_0 I_0} = 5I_0 + 4I_0 = \boxed{9I_0}$

7. Strategy Since the light from each lamp is incoherent, the intensity of the lamps together is just the sum of the intensities.

Solution Find the combined intensity of the lamps.

$I = I_0 + 4I_0 = \boxed{5I_0}$

9. Strategy A path length difference equal to an integral number of wavelengths results in constructive interference.

Solution Ray 1 is incident at the first step and reflected (2) and at the second step and again reflected (3). When rays 2 and 3 reach an observer, ray 3 has traveled a distance $2d - d'$ farther than ray 2. For constructive interference to happen, this distance must be an integral number of wavelengths, so $m\lambda = 2d - d'$.
d and h are the hypotenuse and adjacent leg of a right triangle, respectively, so $h = d\cos\theta$. To find d', we observe that the hypotenuse of the right triangle of which it is a leg is $2d\sin\theta$, so

$d' = (2d\sin\theta)\sin\theta = 2d\sin^2\theta$.

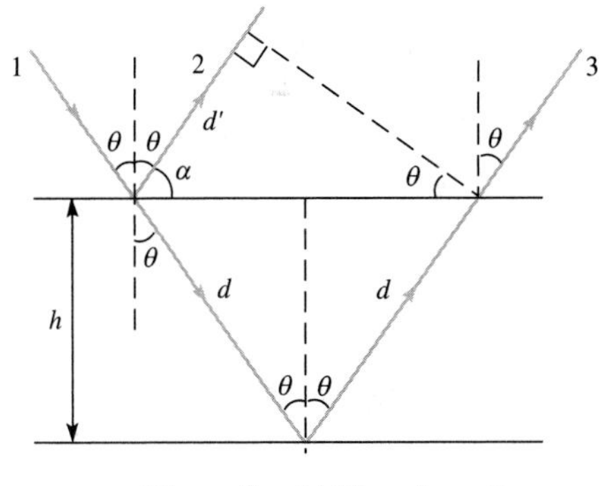

Find λ.

$$m\lambda = 2d - d' = 2d - 2d\sin^2\theta = 2d(1-\sin^2\theta) = \frac{2h}{\cos\theta}\cos^2\theta = 2h\cos\theta, \text{ so } \lambda = \frac{2h\cos\theta}{m} = \frac{(446\text{ nm})\cos\theta}{m}.$$

The table shows the value of the wavelength for the various angles and two values of m.

θ	λ (nm) ($m = 1$)	λ (nm) ($m = 2$)
0°	446	223
10.0°	439	220
20.0°	419	210

The $m = 2$ values are below the visible wavelength range, so the wavelengths are those given in the table above for $m = 1$.

13. Strategy The mirror should be moved in (shorter path length). Since the number of wavelengths traveled in the arm with the vessel decreases, we must decrease the number of wavelengths traveled in the other arm.

Solution As the mirror is moved a distance d, the path length through the arm changed by $2d$, which is equal to 274 wavelengths. Find the distance.

$274\lambda = 2d$, so $d = 137\lambda = 137 \cdot 633\text{ nm} = \boxed{86.7\ \mu\text{m}}$.

15. (a) Strategy When light reflects from a boundary with a medium with a higher index of refraction, it is inverted ($180°$ phase change). Inversion alone will result in destructive interference.

Solution If there is no gap between the plates of glass—they are $\boxed{\text{touching}}$—the light is inverted at the boundary and destructive interference occurs. Therefore, the minimum distance between the two glass plates for one of the dark regions is $\boxed{\text{zero}}$.

(b) Strategy There must be some nonzero thickness of air between the plates for constructive interference to occur. Let this distance between the plates be t.

Solution The light is inverted after is reflects of the bottom layer of glass after it has passed through the thin film of air between the plates, so the path difference due to the light traveling through the thin film of air must be equal to one half wavelength to compensate. The light passes through the thickness of the air twice, before and after reflection. Find t.

$$2t = \frac{\lambda}{2},\text{ so } t = \frac{\lambda}{4} = \frac{550\text{ nm}}{4} = \boxed{140\text{ nm}}.$$

(c) Strategy Refer to parts (a) and (b). For destructive interference, the path difference due to the light traveling through the thin film of air must be equal to one full wavelength.

Solution Find the next largest distance between the plates resulting in a dark region.

$$2t = \lambda,\text{ so } t = \frac{\lambda}{2} = \frac{550\text{ nm}}{2} = \boxed{280\text{ nm}}.$$

17. Strategy Since the film appears red, red light must be experiencing constructive interference. At the air-oil boundary, reflected light is $180°$ out of phase, since $n_{\text{oil}} > n_{\text{air}}$. For transmitted light reflected at the oil-water boundary, the reflective ray is not phase shifted since $n_{\text{water}} < n_{\text{oil}}$. The two reflected rays are $180°$ out of phase, so constructive interference occurs when the relative path length difference, $2t$, is an odd multiple of one half the wavelength in the oil, $\lambda = \lambda_0 / n_{\text{oil}}$.

Solution Find the minimum possible thickness of the film.

$$2t = \left(m + \frac{1}{2}\right)\lambda = \left(m + \frac{1}{2}\right)\frac{\lambda_0}{n_{\text{oil}}}$$

The minimum thickness, t, occurs when $m = 0$.

$$t = \frac{1}{4}\frac{\lambda_0}{n_{\text{oil}}} = \frac{1}{4}\frac{630\text{ nm}}{1.50} = \boxed{105\text{ nm}}$$

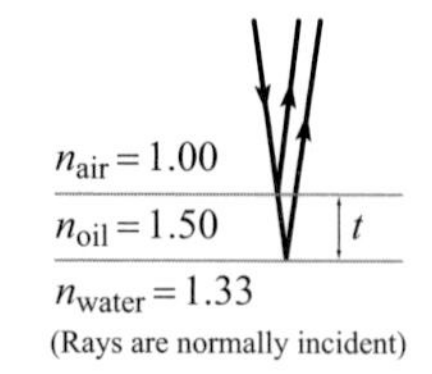

19. Strategy The reflected rays at both boundaries are inverted, since $n_{\text{air}} < n_{\text{film}}$ and $n_{\text{film}} < n_{\text{glass}}$. To minimize reflection, the rays should interfere destructively. So, the path length difference, $2t$, must be equal to an odd multiple of one half the wavelength in the film, $\lambda = \lambda_0 / n_{\text{film}}$.

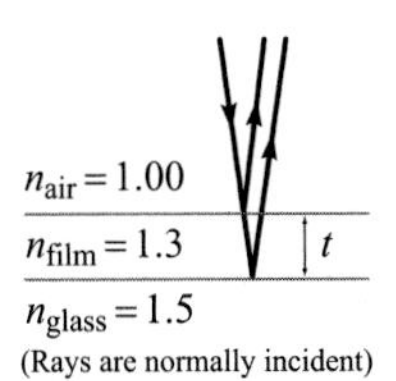

Solution Find the minimum thickness.

$$2t = \left(m + \frac{1}{2}\right)\lambda = \left(m + \frac{1}{2}\right)\frac{\lambda_0}{n_{\text{film}}}$$

Solving for t and substituting $n_{\text{film}} = 1.3$, $\lambda_0 = 500.0\text{ nm}$, and $m = 0$ yields the minimum thickness.

$$t = \left(m + \frac{1}{2}\right)\frac{\lambda_0}{2n_{\text{film}}} = \left(0 + \frac{1}{2}\right)\frac{500.0\text{ nm}}{2 \cdot 1.3} = \boxed{96\text{ nm}}$$

21. (a) Strategy Rays reflected off the front of the film will be inverted since $n_{\text{air}} < n_{\text{film}}$, but rays reflected off the back of the film are not inverted, so the reflected rays will be 180° out of phase plus the phase shift caused by the path length difference. Destructive interference occurs when the path length difference, $2t$, is equal to an integral number of wavelengths in the film, $\lambda = \lambda_0 / n_{\text{film}}$.

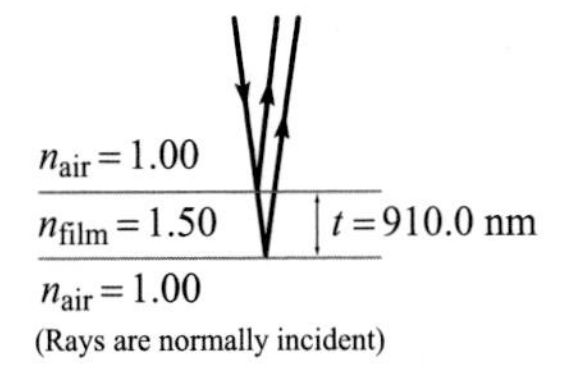

Solution Find the missing wavelengths.

$$2t = m\lambda = m\frac{\lambda_0}{n_{\text{film}}}$$

Solving for λ_0 with $t = 910.0\text{ nm}$ and $n_{\text{film}} = 1.50$ gives

$$\lambda_0 = \frac{2tn_{\text{film}}}{m} = \frac{2730\text{ nm}}{m}.$$

Substituting $m = 4$, 5, and 6 gives wavelengths in the visible spectrum: $\lambda_0 = \boxed{683\text{ nm, }546\text{ nm, and }455\text{ nm}}$.

(b) Strategy The condition for constructive interference is used to find the strongest wavelengths in reflected light. This occurs when the path length difference, $2t$, is an odd multiple of half the wavelength in the film.

Solution Find the strongest wavelengths.

$$2t = \left(m + \frac{1}{2}\right)\lambda = \left(m + \frac{1}{2}\right)\frac{\lambda_0}{n_{\text{film}}}$$

Solving for λ_0 with $t = 910.0\text{ nm}$ and $n_{\text{film}} = 1.50$ gives

$$\lambda_0 = \frac{2tn_{\text{film}}}{m + \frac{1}{2}} = \frac{2730\text{ nm}}{m + \frac{1}{2}}.$$

Substituting $m = 4$, 5, and 6 gives wavelengths in the visible spectrum: $\lambda_0 = \boxed{607\text{ nm, }496\text{ nm, and }420\text{ nm.}}$

25. Strategy Constructive interference occurs when the rays are in phase. Destructive interference occurs when they are 180° out of phase. Consider phase shifts due to reflections and path-length differences.

Solution Let n_i be the index of refraction for the incident side and n_t be the index for the transmitted side.

First case: $n > n_i$ and $n > n_t$

Ray 1 is inverted upon reflection and ray 2 is not. Constructive interference occurs for rays 1 and 2 when the path length difference, $2t$, is equal to an odd multiple of the wavelength in the film, $\lambda = \lambda_0 / n$.

$$2t = \left(m + \frac{1}{2}\right)\lambda = \left(m + \frac{1}{2}\right)\frac{\lambda_0}{n} \qquad (1)$$

The path length difference between rays 3 and 4 is $2t$, and ray 4 is not inverted by either of its two reflections within the film. So, destructive interference occurs if $2t$ is equal to an odd multiple of the wavelength in the film.

Thus, $2t = \left(m + \frac{1}{2}\right)\frac{\lambda_0}{n}$, which is the same as the condition for constructive interference for rays 1 and 2.

For rays 1 and 2 to interfere destructively, $2t$ must equal an integral number of wavelengths in the film, so

$$2t = \frac{m\lambda_0}{n} \qquad (2).$$

For rays 3 and 4 to interfere constructively, $2t$ must equal an integral number of wavelengths in the film, so

$2t = \frac{m\lambda_0}{n}$, which is the same as the condition for destructive interference for rays 1 and 2.

Second case: $n < n_i$ and $n < n_t$

Now, ray 2 is inverted but ray 1 is not. The path length difference must still equal an odd multiple of the wavelength in the film for constructive interference. The condition is given by (1). The condition for destructive interference is still given by (2). Ray 4 is now inverted by both reflections, but two 180° phase changes result in no net change. Therefore, the conditions for constructive and destructive interference are the same as before. So, when rays 1 and 2 interfere constructively/destructively, rays 3 and 4 interfere destructively/constructively.

Third case: $n_t < n < n_i$

Neither ray 1 nor ray 2 is inverted. For constructive interference, the path length difference must be equal to an integral number of wavelengths in the film, so $2t = \frac{m\lambda_0}{n}$.

For destructive interference, the path length difference must be an odd multiple of the wavelength in the film, so

$$2t = \left(m + \frac{1}{2}\right)\frac{\lambda_0}{n}.$$

Ray 4 is inverted by its second reflection, so the path length difference between rays 3 and 4 must be equal to an odd multiple of the wavelength in the film for constructive interference, $2t = \left(m + \frac{1}{2}\right)\frac{\lambda_0}{n}$, and for destructive interference, $2t = \frac{m\lambda_0}{n}$. So, when rays 1 and 2 interfere constructively/destructively, rays 3 and 4 interfere destructively/constructively.

Fourth case: $n_i < n < n_t$

Both rays 1 and 2 are inverted, so the conditions for constructive and destructive interference are the same as found in the third case; 180° phase change for both is the same as no phase change at all. Ray 4 is inverted by its first reflection, so the conditions for constructive and destructive interference for rays 3 and 4 are the same as found in the third case. So, as in the previous three cases, when rays 1 and 2 interfere constructively/destructively, rays 3 and 4 interfere destructively/constructively.

29. Strategy Sketch the diagram according to the instructions in the Problem statement. Compare the results to Eq. (25-10).

Solution

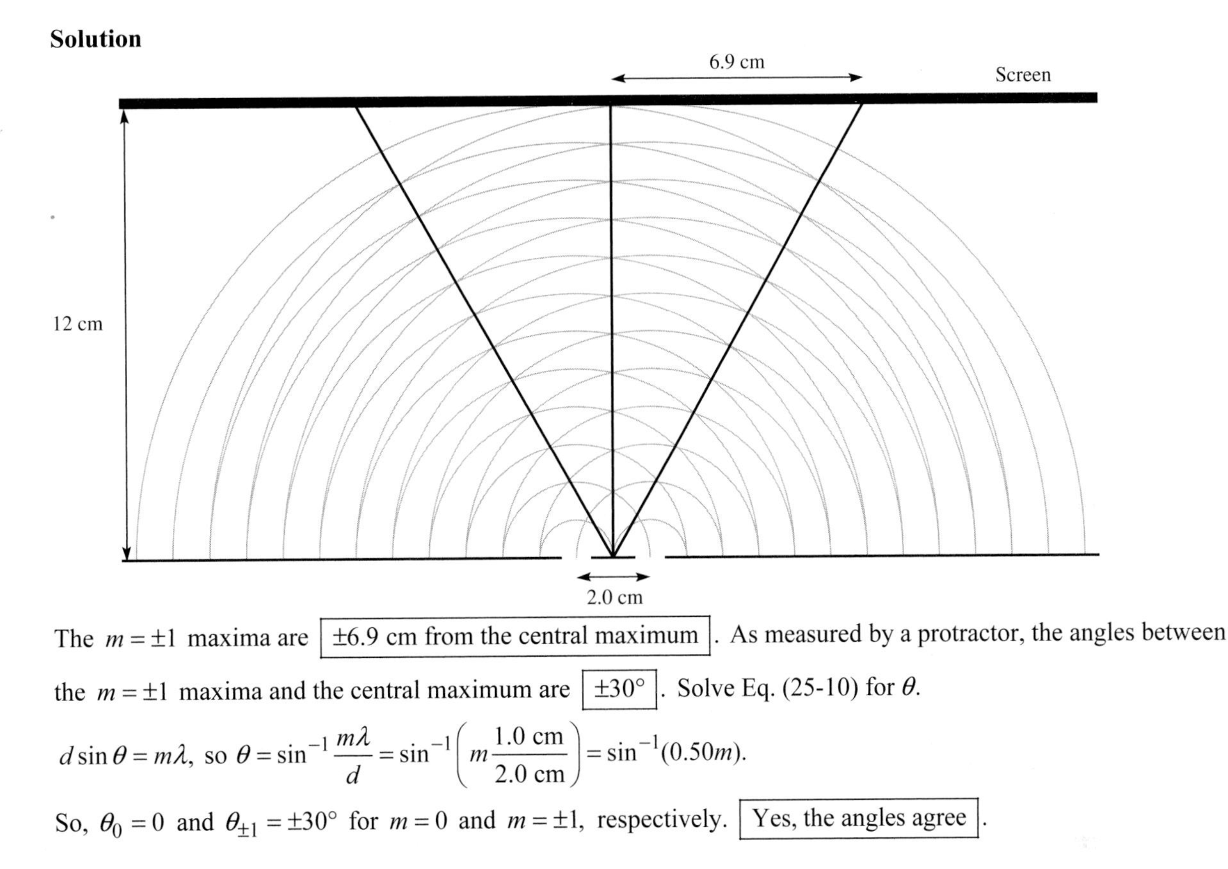

The $m = \pm 1$ maxima are $\boxed{\pm 6.9 \text{ cm from the central maximum}}$. As measured by a protractor, the angles between the $m = \pm 1$ maxima and the central maximum are $\boxed{\pm 30°}$. Solve Eq. (25-10) for θ.

$d \sin\theta = m\lambda$, so $\theta = \sin^{-1}\dfrac{m\lambda}{d} = \sin^{-1}\left(m\dfrac{1.0 \text{ cm}}{2.0 \text{ cm}}\right) = \sin^{-1}(0.50m)$.

So, $\theta_0 = 0$ and $\theta_{\pm 1} = \pm 30°$ for $m = 0$ and $m = \pm 1$, respectively. $\boxed{\text{Yes, the angles agree}}$.

31. Strategy Show that the small-angle approximation is justified. Then, use the result for the slit separation obtained in Example 25.4.

Solution Compare x and D.

$\dfrac{D}{x} = \dfrac{2.50 \text{ m}}{0.760\times10^{-2} \text{ m}} = 329$, so $x \ll D$ and the small angle approximation is justified.

Find the wavelength of the light.

$d = \dfrac{\lambda D}{x}$, so $\lambda = \dfrac{dx}{D} = \dfrac{(0.0150\times10^{-2} \text{ m})(0.760\times10^{-2} \text{ m})}{2.50 \text{ m}} = \boxed{456 \text{ nm}}$.

33. Strategy Check to see if the small angle approximation is justified. If so, use the result of Example 25.4 for the distance d between the slits.

Solution Compare x and D.

$\dfrac{D}{x} = \dfrac{1.5 \text{ m}}{0.0135 \text{ m}} = 111$, so $x \ll D$ and the small angle approximation is justified

Compute the slit separation.

$d = \dfrac{\lambda D}{x} = \dfrac{(630\times10^{-9} \text{ m})(1.5 \text{ m})}{0.0135 \text{ m}} = \boxed{7.0\times10^{-5} \text{ m}}$

37. Strategy Apply Eq. (25-10) for each line.

Solution The third-order red line gives
$d\sin\theta_{R3} = m\lambda = 3\lambda_R$.
For the fourth-order blue line, we have
$d\sin\theta_{B4} = m\lambda = 4\lambda_B$.
Since these lines overlap, the angles θ_{R3} and θ_{B4} must be equal, which implies $d\sin\theta_{R3} = d\sin\theta_{B4}$. So, the right-hand sides of the previous equations must also be equal.
$3\lambda_R = 4\lambda_B$, so $\lambda_B = \frac{3}{4}\lambda_R = \frac{3}{4}(630\text{ nm}) = \boxed{470\text{ nm}}$.

39. Strategy Find the slit separation. Then solve Eq. (25-10) for m using $\lambda = 412$ nm and $\theta = 90°$.

Solution Compute the slit separation.
$$d = \frac{1}{5000.0\text{ slits/cm}} = 2.0000\times10^{-6}\text{ m}$$
Find the number of orders.
$$d\sin\theta = m\lambda,\text{ so } m = \frac{d\sin\theta}{\lambda} = \frac{(2.0000\times10^{-6}\text{ m})\sin 90°}{412\times10^{-9}\text{ m}} = 4.85.$$
Since m must be an integer, $\boxed{\text{four}}$ orders can be observed.

41. Strategy Use Eq. (25-10). Draw a diagram. Set $\theta = 90°$ to find the highest-order spectral line.

Solution

(a) In these first-order maxima, the shortest wavelength will appear first and the longest wavelength last, since $\sin\theta \propto \lambda$. Thus, $\boxed{A\text{ is the blue line, }B\text{ is the yellow line, and }C\text{ is the red line}}$.

(b) Solve for the wavelength.
$$d\sin\theta = m\lambda,\text{ so } \lambda = \frac{d\sin\theta}{m}.$$
Here, d = 1870 nm and m = 1. To find θ, use
$$\tan\theta = \frac{\text{Position of line } C}{\text{distance between grating and screen}} = \frac{x}{D}.$$

θ x D

$$\theta = \tan^{-1}\left(\frac{11.5\text{ cm}}{30.0\text{ cm}}\right) = 20.97°$$
Substitute this value for θ and the given values of d and m into the equation for λ.
$$\lambda = \frac{(1870\text{ nm})\sin 20.97°}{1} = \boxed{669\text{ nm}}$$

(c) To find the highest-order of spectral line C, solve for m using $\lambda = 669$ nm, d = 1870 nm, and $\theta = 90°$.
$$d\sin\theta = m\lambda,\text{ so } m = \frac{d\sin\theta}{\lambda} = \frac{(1870\text{ nm})\sin 90°}{669\text{ nm}} = 2.80.$$
Since m must be an integer, the highest order of spectral line C that it is possible to see is $\boxed{2}$.

45. (a) Strategy Use Eq. (25-10). The slit separation is given by $d = 1/N$.

Solution Find the number of slits per cm of the grating.

$d\sin\theta = m\lambda$, so $\frac{\sin\theta}{m\lambda} = \frac{1}{d} = N$.

Since the higher orders are at greater angles than the shorter ones, set $\theta = 90°$; then for the second order, $N = 1/(2\lambda)$. To get the maximum N allowed, the larger wavelength is chosen. An N larger than this maximum value excludes all wavelengths greater than 661.4 nm, but includes smaller wavelengths.

$$N_{max} = \frac{1}{2(661.4\times10^{-7}\text{ cm})} = \boxed{7560\text{ slits/cm}}$$

(b) Strategy Find the average and difference of the wavelengths; then use the given equation for N.

Solution The average of the two wavelengths is

$\lambda = \frac{\lambda_1 + \lambda_2}{2} = \frac{660.0\text{ nm} + 661.4\text{ nm}}{2} = 660.7\text{ nm}.$

The difference between the two wavelengths is

$\Delta\lambda = \lambda_2 - \lambda_1 = 661.4\text{ nm} - 660.0\text{ nm} = 1.4\text{ nm}.$

Substituting these values and $m = 2$ into the given equation yields $N = \frac{\lambda}{m\Delta\lambda} = \frac{660.7\text{ nm}}{2\cdot1.4\text{ nm}} = \boxed{240}$.

47. Strategy Let $x = 1.0$ mm, which is the same as the distance from the center to the first minimum. Since $x = 1.0\text{ mm} \ll D = 1.0$ m, using the small angle approximation is justified, so use the result of Example 25.8 for half the width of the central maximum.

Solution Find the width of the slit.

$$x = \frac{\lambda D}{a},\text{ so } a = \frac{\lambda D}{x} = \frac{(610\times10^{-9})(1.0\text{ m})}{0.0010\text{ m}} = \boxed{0.61\text{ mm}}.$$

49. Strategy Call the width of the central maximum W. Referring to Figure 25.33, we have $W = 2x$, and by trigonometry, $x = D\tan\theta$, so $W = 2x = 2D\tan\theta$. Assume θ is a small angle. Use Eq. (25-12).

Solution Since θ is assumed to be small, we have $W = 2D\tan\theta \approx 2D\sin\theta$.

From Eq. (25-12), we have $\sin\theta = \frac{\lambda}{a}$ since $m = 1$, so W becomes $W = \frac{2L\lambda}{a}$.

If a is replaced with $2a$, the new width is $W_{new} = \frac{2L\lambda}{2a} = \frac{1}{2}\frac{2L\lambda}{a} = \frac{1}{2}W$.

Thus, $\boxed{\text{the new width is half the old width.}}$

53. Strategy The angular width of the central diffraction maximum is two times the angular position of the first minimum. Solve Eq. (25-13) for θ, using $\lambda = 590$ nm and $a = 7.0$ mm.

Solution Find the angular distance to the first minimum.

$$a\sin\theta = 1.22\lambda,\text{ so } \theta = \sin^{-1}\frac{1.22\lambda}{a} = \sin^{-1}\frac{1.22(590\times10^{-9}\text{ m})}{7.0\times10^{-3}\text{ m}} = 0.0059°.$$

The angular width is twice this or $\boxed{0.012°}$.

57. (a) Strategy Use Snell's law to relate the angular separation of the two sources and the angular separation of the two images.

Solution By Snell's law, we have $n_{\text{air}} \sin \Delta\theta = n \sin \beta$ for ray 2.

Using $n_{\text{air}} = 1$ yields $\boxed{\sin \Delta\theta = n \sin \beta}$.

(b) Strategy If $\beta \geq \phi$ then $\sin \beta \geq \sin \phi$. In the fluid, $\lambda = \lambda_0 / n$. Apply Eq. (25-13) to image 1 and solve for $\sin \phi$.

Solution

$a \sin \phi = 1.22\lambda = 1.22 \dfrac{\lambda_0}{n}$, so $\sin \phi = \dfrac{1.22\lambda_0}{an}$.

Taking the result from part (a) and solving for $\sin \beta$ gives $\sin \beta = \sin \Delta\theta / n$. Substitute this expression for $\sin \beta$ and the previous expression for $\sin \phi$ into the inequality $\sin \beta \geq \sin \phi$.

$$\frac{\sin \Delta\theta}{n} \geq \frac{1.22\lambda_0}{an}$$

$$a \sin \Delta\theta \geq 1.22\lambda_0$$

This is equivalent to Eq. (25-14), Rayleigh's criterion.

61. Strategy Find the smallest $\Delta\theta$ the eye can discern using Eq. (25-14) with $a = 7$ mm (diameter of the pupil) and $\lambda_0 = 550$ nm (middle of the visible spectrum). Two objects separated by a distance d, at a distance L from the observer, will have an angular separation $\Delta\theta$ given by $\tan \Delta\theta = d/L$.

Solution Find $\Delta\theta$.

$$a \sin \Delta\theta = 1.22\lambda_0, \text{ so } \Delta\theta = \sin^{-1} \frac{1.22\lambda_0}{a} = \sin^{-1} \frac{1.22(550 \times 10^{-9}\text{ m})}{7 \times 10^{-3}\text{m}} = 0.0055°.$$

Estimate the maximum distance at which the headlights can be resolved. Let the distance between the headlights be $d = 2$ m.

$$L = \frac{d}{\tan \Delta\theta} = \frac{2\text{ m}}{\tan 0.0055°} = \boxed{20\text{ km}}$$

20 km is rather large—clearly diffraction is not the only limitation!

65. Strategy and Solution

(a) All points on the line $\theta = 0$ are equidistant from the two antennas. The difference in the path lengths is $\boxed{0}$.

(b) At $\theta = 90°$, the difference in the path lengths is just the distance between the antennas, $\boxed{0.30\text{ km.}}$

(c) The wavelength is $\lambda = \dfrac{c}{f} = \dfrac{3.00 \times 10^8\text{ m/s}}{3.0 \times 10^6\text{ Hz}} = 100\text{ m} = 0.10\text{ km}$, which is $\dfrac{1}{3}d$. Thus, the double-slit maxima equation will have solutions for $m = 0, \pm 1, \pm 2$, and ± 3, yielding 7 angles between –90° and 90°. By symmetry, there will be 5 maxima in the ranges beyond 90° and –90°. $\boxed{\text{There will be 12 maxima total.}}$

(d) Solve Eq. (25-10) for θ using $d = 0.30$ km, $\lambda = 0.10$ km, and $m = 0, 1, 2,$ and 3.

$d\sin\theta = m\lambda$, so $\theta = \sin^{-1}\frac{m\lambda}{d}$.

$\theta_0 = \sin^{-1}\frac{0(\lambda)}{d} = \boxed{0}$, $\theta_1 = \sin^{-1}\frac{1(0.10\text{ km})}{0.30\text{ km}} = \boxed{19°}$, $\theta_2 = \sin^{-1}\frac{2(0.10\text{ km})}{0.30\text{ km}} = \boxed{42°}$, and

$\theta_3 = \sin^{-1}\frac{3(0.10\text{ km})}{0.30\text{ km}} = \boxed{90°}$.

(e) The answers in parts (a), (b), and (c) would be unchanged. The angles calculated in part (d) for θ_0 and θ_3 would be unchanged, but θ_1 and θ_2 would be different from before.

69. (a) Strategy Light reflects off the top and bottom of the film. Since $n_{\text{film}} > n_{\text{air}}$, the ray that reflects from the top is inverted. Since $n_{\text{lens}} > n_{\text{film}}$, the ray that reflects from the bottom is also inverted. So, destructive interference between the two rays must be due to the path length difference $2t$, which must equal one half the wavelength of the ray in the film for the minimum thickness.

Solution Find the minimum thickness of the coating.

$2t = \frac{1}{2}\lambda$, so $t = \frac{1}{4}\frac{\lambda_0}{n_{\text{film}}} = \frac{1}{4}(560\text{ nm})\frac{1}{1.38} = \boxed{100\text{ nm}}$.

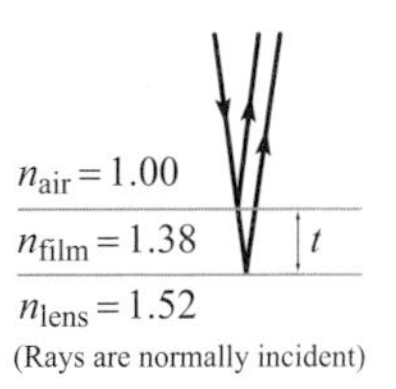

(b) Strategy Constructive interference in this case is given by $2t = m\lambda = m\lambda_0/n_{\text{film}}$; the phase difference must be equal to an integral multiple of the wavelength.

Solution Solve for $\lambda_{0,m}$.

$$\lambda_{0,m} = \frac{2tn_{\text{film}}}{m} = \frac{2(101.4\text{ nm})(1.38)}{m} = \frac{280\text{ nm}}{m}$$

So, $\lambda_1 = \boxed{280\text{ nm}}$ and $\lambda_2 = \boxed{140\text{ nm}}$ are the closest wavelengths to 560 nm giving constructive interference.

(c) Strategy and Solution

Yes; although perfectly constructive interference does not occur for any visible wavelength, some visible light is reflected at all visible wavelengths except 560 nm (the only wavelength with perfectly destructive interference).

73. Strategy Let n be the index of refraction of the mica sheet. Since $n > 1$, rays reflected from the front of the mica will be inverted, while rays reflected from the back of the mica will not be inverted. The phase difference for the reflected rays is 180° plus the phase shift caused by the path length difference $2t$ in the mica. Destructive interference (gaps in the spectrum) occurs when $2t$ is equal to an integral multiple of the wavelength in mica, $\lambda = \lambda_0 / n$.

Solution Calculate the index of refraction of the mica sheet.

$2t = m\lambda = m\frac{\lambda_0}{n}$, so $2tn = m\lambda_0$.

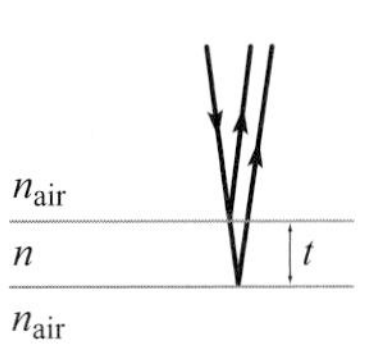

The expression on the left-hand side is constant, so the right-hand side must also be constant; thus, we can write $m_1\lambda_1 = m_2\lambda_2$, where λ_1 and λ_2 are the wavelengths of two gaps in the spectrum and m_1 and m_2 are their respective m-values.

Substitute $\lambda_1 = 630$ nm, $\lambda_2 = 525$ nm, and $m_2 = m_1 + 1$ (adjacent minima; $\lambda_1 > \lambda_2$).

$$m_1 \cdot 630 \text{ nm} = (m_1 + 1) \cdot 525 \text{ nm}$$
$$m_1 \cdot 630 \text{ nm} = m_1 \cdot 525 \text{ nm} + 525 \text{ nm}$$
$$m_1 \cdot 105 \text{ nm} = 525 \text{ nm}$$
$$m_1 = 5.0$$

Solve the equation for destructive interference given above for n, and use $m_1 = 5.0$, $\lambda_1 = 630$ nm, and $t = 1.00\ \mu\text{m}$ to find the value of n.

$$n = \frac{m_1\lambda_1}{2t} = \frac{5.0(630\times10^{-9}\text{ m})}{2(1.00\times10^{-6}\text{ m})} = \boxed{1.6}$$

77. Strategy Use Eq. (25-10). The relationship between the diffraction angle θ, the grating-screen distance D, and the position x on the screen is $\tan\theta = x/D$. The slit separation is given by $d = 1/N$.

Solution Compute the slit separation.

$$d = \frac{1}{10{,}000.0 \text{ slits/cm}} = 1.00000\times10^{-6}\text{ m}$$

Solve for θ.

$$d\sin\theta = m\lambda, \text{ so } \theta = \sin^{-1}\frac{m\lambda}{d}.$$

Solve for x.

$$\tan\theta = \frac{x}{D}, \text{ so } x = D\tan\theta = D\tan\left(\sin^{-1}\frac{m\lambda}{d}\right) = (2.0\text{ m})\tan\left(\sin^{-1}\frac{m\lambda}{1.00000\times10^{-6}\text{ m}}\right).$$

Using $\lambda = 690$ nm and $m = 0, \pm1$ gives the locations of the red lines.

$$x_{R0} = (2.0\text{ m})\tan\left(\sin^{-1}\frac{0\cdot\lambda}{1.00000\times10^{-6}\text{ m}}\right) = 0 \text{ and } x_{R\pm1} = (2.0\text{ m})\tan\left(\sin^{-1}\frac{\pm1\cdot690\times10^{-9}\text{ m}}{1.00000\times10^{-6}\text{ m}}\right) = \pm1.9\text{ m}.$$

Using $\lambda = 460$ nm and $m = 0, \pm1$, and ±2 gives the locations of the blue lines.

$$x_{B0} = 0,\ x_{B\pm1} = (2.0\text{ m})\tan\left(\sin^{-1}\frac{\pm1\cdot460\times10^{-9}\text{ m}}{1.00000\times10^{-6}\text{ m}}\right) = \pm1.0\text{ m, and}$$

$$x_{B\pm2} = (2.0\text{ m})\tan\left(\sin^{-1}\frac{\pm2\cdot460\times10^{-9}\text{ m}}{1.00000\times10^{-6}\text{ m}}\right) = \pm4.7\text{ m}.$$

The only overlap is at $x = 0$, where the lines combine to make purple. The pattern is shown below.

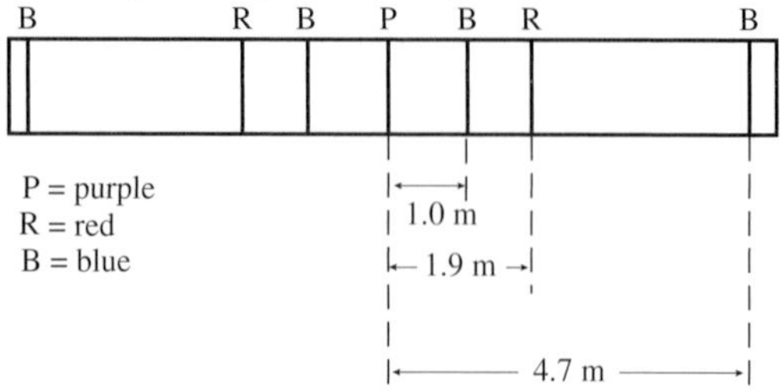

B R B P B R B

P = purple
R = red
B = blue

1.0 m
1.9 m
4.7 m

REVIEW AND SYNTHESIS: CHAPTERS 22–25

Review Exercises

1. Strategy The sound of the bat hitting the ball travels 22 m from the bat to the microphone in a time Δt_1. Then, the EM wave travels 4500 km from the stadium to the television set in a time Δt_2. Finally, the sound travels 2.0 m from the television set to your ears in a time Δt_3. Compute the time it takes for each wave, sound and EM, to travel the stated distances using $\Delta x = v\Delta t$.

Solution Compute the minimum time it takes for you to hear the crack of the bat after the batter hit the ball.

$$\Delta t_1 + \Delta t_2 + \Delta t_3 = \frac{22\text{ m}}{343\text{ m/s}} + \frac{4.50\times10^6\text{ m}}{3.00\times10^8\text{ m/s}} + \frac{2.0\text{ m}}{343\text{ m/s}} = \boxed{85\text{ ms}}.$$

5. Strategy Since the radio waves interfere destructively and the reflection at the plane introduces a 180° phase shift, the minimum path difference must be equal to one wavelength.

Solution

(a) Find the wavelength of the radio waves.

$$\lambda = \frac{c}{f} = \frac{2.998\times10^8\text{ m/s}}{1408\times10^3\text{ Hz}} = 212.93\text{ m}$$

$2\lambda = 425.9$ m and $3\lambda = 638.8$ m. If Juanita is correct that the plane is at least 500 m over her head, $\boxed{\text{the plane must be at least } 3\lambda = 638.8\text{ m over Juanita's head}}$.

(b) The lower estimate is $2\lambda = \boxed{425.9\text{ m}}$ and the higher estimate is $4\lambda = \boxed{851.7\text{ m}}$.

9. Strategy and Solution You don't see an interference pattern on your desk when light from two different lamps are illuminating the surface because $\boxed{\text{the lamps are not emitting coherent light}}$.

13. Strategy The distance between adjacent slits is the reciprocal of the number of slits per unit length. Use Eq. (25-10).

Solution

(a) Compute the distance between adjacent slits.

$$d = \frac{1}{5550\text{ slits/cm}} = \frac{1\text{ cm}}{5550\text{ slits}} = \frac{10^{-2}\text{ m}}{5550\text{ slits}} = \boxed{1.80\times10^{-6}\text{ m}}$$

(b) Find the distance between the central bright spot and the first-order maximum. Find θ.

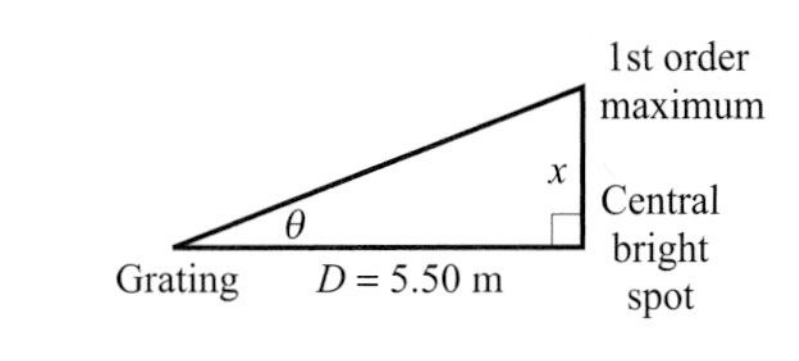

$d\sin\theta = m\lambda = (1)\lambda = \lambda$, so $\theta = \sin^{-1}\dfrac{\lambda}{d}$. Find x.

$\tan\theta = \dfrac{x}{D}$, so

$$x = D\tan\theta = D\tan\left(\sin^{-1}\frac{\lambda}{d}\right) = (5.50\text{ m})\tan\left[\sin^{-1}\frac{0.680\times10^{-6}\text{ m}}{1.8018\times10^{-6}\text{ m}}\right] = \boxed{2.24\text{ m}}.$$

(c) Find the distance between the central bright spot and the second-order maximum. Find θ.

$d\sin\theta = m\lambda = 2\lambda$, so $\theta = \sin^{-1}\frac{2\lambda}{d}$. Find x.

$$\tan\theta = \frac{x}{D}, \text{ so } x = D\tan\theta = D\tan\left(\sin^{-1}\frac{2\lambda}{d}\right) = (5.50 \text{ m})\tan\left[\sin^{-1}\frac{2(0.680\times10^{-6} \text{ m})}{1.8018\times10^{-6} \text{ m}}\right] = \boxed{6.33 \text{ m}}.$$

(d) The assumption that $\sin\theta = \tan\theta$ is not valid because the angles are not small.

17. Strategy Geraldine want to display the zeroth- and first-order interference maxima across the full width of a 20.0-cm screen. To do so, the first-order minima must be at the edge of the screen. Use Eq. (25-11).

Solution Find D, the distance between the double slit and the screen.

$$d\sin\theta = d\frac{x}{\sqrt{x^2+D^2}} = \left(m+\frac{1}{2}\right)\lambda = \left(1+\frac{1}{2}\right)\lambda = \frac{3}{2}\lambda$$

1st order minimum
x
Double slit
θ
θ
D
Screen
x
1st order minimum

Solve for D.

$$\frac{x}{\sqrt{x^2+D^2}} = \frac{3\lambda}{2d}$$

$$\frac{x^2}{x^2+D^2} = \left(\frac{3\lambda}{2d}\right)^2$$

$$x^2+D^2 = \left(\frac{2d}{3\lambda}\right)^2 x^2$$

$$D = x\sqrt{\left(\frac{2d}{3\lambda}\right)^2 - 1} = \frac{0.200 \text{ m}}{2}\sqrt{\left[\frac{2(20.0\times10^{-6} \text{ m})}{3(423\times10^{-9} \text{ m})}\right]^2 - 1} = \boxed{3.15 \text{ m}}$$

21. Strategy Use Eq. (25-10). Let the subscript g stand for the green light and the subscript v stand for the violet light. Assume that the angles are small. Then $\sin\theta \approx \tan\theta = x/D$.

Solution

(a) Since $\sin\theta \propto \lambda$, the green maximum will be farther than the violet maximum from the central maximum. Find the separation $x_g - x_v$.

x_g
θ_g
x_v
Double slit
θ_v
$D = 0.720$ m

$$d\sin\theta_g = d\frac{x_g}{D} = \lambda_g \text{ and } d\sin\theta_v = d\frac{x_v}{D} = \lambda_v, \text{ so}$$

$$x_g - x_v = \frac{D}{d}\lambda_g - \frac{D}{d}\lambda_v = \frac{D}{d}(\lambda_g - \lambda_v) = \frac{0.720 \text{ m}}{0.020\times10^{-3} \text{ m}}(520\times10^{-9} \text{ m} - 412\times10^{-9} \text{ m}) = \boxed{3.9 \text{ mm}}.$$

(b) Find the separation.

$$d\sin\theta_g = d\frac{x_g}{D} = 2\lambda_g \text{ and } d\sin\theta_v = d\frac{x_v}{D} = 2\lambda_v, \text{ so}$$

$$x_g - x_v = \frac{D}{d}(2\lambda_g) - \frac{D}{d}(2\lambda_v) = \frac{2D}{d}(\lambda_g - \lambda_v) = \frac{2(0.720 \text{ m})}{0.020\times10^{-3} \text{ m}}(520\times10^{-9} \text{ m} - 412\times10^{-9} \text{ m}) = \boxed{7.8 \text{ mm}}.$$

MCAT Review

1. **Strategy** Use the transverse magnification equation.

   Solution Find the ratio of the height of the image to the height of the object.

   $$m=\frac{h'}{h}=-\frac{q}{p}=-\frac{\frac{3}{2}f}{3f}=-\frac{1}{2}$$

   The correct answer is $\boxed{\text{A}}$.

2. **Strategy** The object and image distances are the same. Use the mirror equation and the fact that $f = R/2$.

   Solution Find p, the object location.

   $$\frac{1}{p}+\frac{1}{q}=\frac{1}{p}+\frac{1}{p}=\frac{2}{p}=\frac{1}{f}=\frac{2}{R},\text{ so } p=R=50\text{ cm.}$$

   The correct answer is $\boxed{\text{C}}$.

3. **Strategy and Solution** When the telescope is focused on a very distant object, the image is located at the focal point of the mirror. The image is formed in front of the mirror, so it is real. The image and object distances are positive, so according to the magnification equation, $m=-q/p<0,$ therefore, the image is inverted.

   The correct answer is $\boxed{\text{B}}$.

4. **Strategy** The magnification is the equal to the ratio of the focal length of the mirror to that of the eyepiece.

   Solution The magnification of the Hubble is approximately

   $$\frac{f_{\text{mirror}}}{f_{\text{eyepiece}}}=\frac{13\text{ m}}{2.5\times10^{-2}\text{ m}}=520.$$ The correct answer is $\boxed{\text{C}}$.

5. **Strategy** For a very distant object, p is very large, so $1/p$ is approximately zero. Use the mirror equation.

   Solution Find the image location.

   $$\frac{1}{f}=\frac{1}{q}+\frac{1}{p}\approx\frac{1}{q}+0=\frac{1}{q},\text{ so } q\approx f.$$

   The image location is very close to the focal point of the mirror. The correct answer is $\boxed{\text{C}}$.

6. **Strategy and Solution** Visible light has wavelengths greater than ultraviolet light. The wavelengths of visible light are large compared to atoms and molecules, so visible light is not that easily absorbed by the atmosphere. The wavelengths of ultraviolet light are closer to the size of atoms and molecules in the atmosphere, so ultraviolet waves are more readily absorbed by the atmosphere. The correct answer is $\boxed{\text{A}}$.

Chapter 26

RELATIVITY

Problems

1. **Strategy** Compute the times required for the optical signal and the train to reach the station and find the difference.

   Solution The time required for the optical signal to reach the station, as measured by an observer at rest relative to the station (the stationmaster) is
   $$\frac{d}{c}=\frac{1.0\text{ km}}{3.00\times10^{5}\text{ km/s}}=3.3\ \mu\text{s}.$$
   Measured by the same observer, the time needed for the train to arrive is
   $$\frac{d}{v}=\frac{d}{0.60c}=\frac{1.0\text{ km}}{0.60\cdot3.00\times10^{5}\text{ km/s}}=5.6\ \mu\text{s}.$$
   The difference in the arrival times is $5.56\ \mu\text{s}-3.33\ \mu\text{s}=\boxed{2.2\ \mu\text{s}}$.

5. **Strategy** The proper time interval Δt_0 is the time interval measured by the Rolex. The time interval measured at mission control is $\Delta t = 12.0$ h. The time-dilation equation can be used to find Δt_0.

   Solution
   $$\Delta t_0=\frac{\Delta t}{\gamma}=\Delta t\sqrt{1-\frac{v^2}{c^2}}=(12.0\text{ h})\sqrt{1-\frac{(2.0\times10^{8}\text{ m/s})^2}{(3.00\times10^{8}\text{ m/s})^2}}=\boxed{8.9\text{ h}}$$

9. **(a) Strategy** The age of the passenger at the end of the trip is found by adding the elapsed time Δt_0, as measured by the ship's clock, to the passenger's age at the start of the trip. To find the elapsed time Δt_0, use the time dilation equation with $\Delta t=\dfrac{\text{distance relative to Earth}}{\text{speed relative to Earth}}=\dfrac{710\text{ ly}}{0.9999c}=710\text{ yr}.$

   Solution Find Δt.
   $$\Delta t_0=\frac{\Delta t}{\gamma}=\Delta t\sqrt{1-v^2/c^2}=(710\text{ yr})\sqrt{1-0.9999^2}=10\text{ yr}$$
   The passenger's age at the end of the trip is 20 years old + 10 years = $\boxed{\text{30 years old}}$.

   (b) Strategy and Solution The spaceship takes slightly more than 710 years to reach its destination, as measured by an observer on Earth. The radio signal takes 710 years to reach Earth. The total elapsed time between the departure of the ship and the arrival of the signal back on Earth is 710 years + 710 years = 1420 years. The year will be 2000 + 1420 = $\boxed{3420}$.

11. Strategy The proper time interval is 8.0 h. The time interval Δt measured by the clock on the ground is given by the time dilation equation $\Delta t = \gamma\Delta t_0$. Use the binomial approximation.

Solution The time difference is

$$\Delta t - \Delta t_0 = \gamma\Delta t_0 - \Delta t_0 = (\gamma - 1)\Delta t_0 = \left[(1 - v^2/c^2)^{-1/2} - 1\right]\Delta t_0.$$

Since $v << c$, by the binomial approximation, $\gamma = (1 - v^2/c^2)^{-1/2} \approx 1 + \dfrac{v^2}{2c^2}$.

The time difference is $\Delta t - \Delta t_0 \approx \left(1 + \dfrac{v^2}{2c^2} - 1\right)\Delta t_0 = \dfrac{v^2}{2c^2}\Delta t_0 = \dfrac{(220 \text{ m/s})^2}{2(3.00\times10^8 \text{ m/s})^2}(8.0 \text{ h}) = \boxed{7.7 \text{ ns}}$.

13. (a) Strategy The observers on Earth see the contracted heights of the occupants of the spaceship, since the heights are along the direction of motion. Use Eq. (26-4).

Solution Compute the approximate heights of the occupants as viewed by others on the ship.

$$L = \frac{L_0}{\gamma}, \text{ so } L_0 = \gamma L = \frac{L}{\sqrt{1 - v^2/c^2}} = \frac{0.50 \text{ m}}{\sqrt{1 - 0.97^2}} = \boxed{2 \text{ m}}.$$

L_0

L

(b) Strategy The widths of the occupants are perpendicular to their direction of motion, so the widths are not contracted.

Solution The others on the ship see the same widths as the observers on Earth, $\boxed{0.50 \text{ m}}$.

15. (a) Strategy The proper length is $L_0 = 91.5$ m. The length measured in the rest frame of the particle is given by Eq. (26-4).

Solution

$$L = \frac{L_0}{\gamma} = L_0\sqrt{1 - v^2/c^2} = (91.5 \text{ m})\sqrt{1 - 0.50^2} = \boxed{79 \text{ m}}$$

(b) Strategy Since the speed of the particle is constant relative to Earth, the time is found using the equation $\Delta x = v\Delta t$.

Solution

$$\Delta t = \frac{\Delta x}{v} = \frac{91.5 \text{ m}}{0.50(3.00\times10^8 \text{ m/s})} = 6.1\times10^{-7} \text{ s} = \boxed{610 \text{ ns}}$$

(c) Strategy In the rest frame of the particle, the distance traveled is 79 m and the speed is $0.50c$.

Solution The time required is

$$\Delta t = \frac{\Delta x}{v} = \frac{79 \text{ m}}{0.50(3.00\times10^8 \text{ m/s})} = 5.3\times10^{-7} \text{ s} = \boxed{530 \text{ ns}}.$$

17. Strategy Use Eqs. (26-2) and (26-4).

Solution The length measured from the other ship is contracted by a factor $\gamma = (1 - v^2/c^2)^{-1/2} = (1 - 0.90^2)^{-1/2} = 2.3.$

The contracted length is $L = \frac{L_0}{\gamma} = \frac{30.0 \text{ m}}{2.3} = \boxed{13 \text{ m}}.$

19. (a) Strategy and Solution Length contraction occurs only along the direction of motion. Since the rod is held perpendicular to the direction of motion, the Earth observer will measure the same length for the rod as the pilot of the ship did, $\boxed{1.0 \text{ m}}$.

(b) Strategy The pilot still measures the proper length of the rod as $L_0 = 1.0$ m. To the Earth observer, the rod is now contracted. Use Eq. (26-4).

Solution Find the length of the rod according to the Earth observer.

$$L = \frac{L_0}{\gamma} = L_0\sqrt{1 - v^2/c^2} = (1.0 \text{ m})\sqrt{1 - 0.40^2} = \boxed{0.92 \text{ m}}$$

21. Strategy The observer on the ground measures the proper length between the towers as $L_0 = 3.0$ km. The distance is contracted for a passenger on the train. Use Eq. (26-4).

Solution

(a) Relative to the passenger on the train, the distance between the towers is

$$L = \frac{L_0}{\gamma} = L_0\sqrt{1 - v^2/c^2} = (3.0 \text{ km})\sqrt{1 - 0.80^2} = 1.8 \text{ km}.$$

The time interval measured by the passenger is $\Delta t = \frac{L}{v} = \frac{1.8 \text{ km}}{0.80(3.00 \times 10^5 \text{ km/s})} = 7.5 \times 10^{-6} \text{ s} = \boxed{7.5\ \mu\text{s}}.$

(b) The observer on the ground measures the time interval

$$\Delta t = \frac{L_0}{v} = \frac{3.0 \text{ km}}{0.80(3.00 \times 10^5 \text{ km/s})} = 1.3 \times 10^{-5} \text{ s} = \boxed{13\ \mu\text{s}}.$$

25. (a) Strategy and Solution Since the speed of light in vacuum is the same in all inertial reference frames, regardless of the motion of the source or of the observer, the *Galaxa* measures the speed of the laser beam to be $\boxed{c}$.

(b) Strategy Use the relativistic velocity transformation formula, Eq. (26-5). Let the speed of the missile in *Galaxa's* reference frame be v_{mG}, the speed of the missile in *Millenia's* reference frame be v_{mM}, and the speed of the *Millenia* in the *Galaxa's* reference frame be v_{MG}.

Solution Compute the speed of the missile in *Galaxa's* reference frame.

$$v_{\text{mG}} = \frac{v_{\text{mM}} + v_{\text{MG}}}{1 + v_{\text{mM}}v_{\text{MG}}/c^2} = \frac{0.45c + 0.55c}{1 + (0.45)(0.55)} = \boxed{0.802c}$$

27. Strategy Since the ships are approaching the Moon from opposite directions, we use a coordinate system relative to the Moon such that one ship (ship A) approaches from the left at a speed $0.60c$, and the other ship (ship B) approaches from the right at speed $0.80c$. The velocity of ship A relative to ship B is written v_{AB}. If the velocity of ship A relative to the Moon is $v_{\text{AM}} = 0.60c$, then the velocity of ship B relative to the Moon is $v_{\text{BM}} = -0.80c$; so $v_{\text{MB}} = -v_{\text{BM}} = 0.80c$. Use the relativistic velocity transformation equation.

Solution Find the velocity of ship A relative to ship B.

$\vec{\mathbf{v}}_{\text{AM}}$ Moon $\vec{\mathbf{v}}_{\text{BM}}$

$$v_{\text{AB}} = \frac{v_{\text{AM}} + v_{\text{MB}}}{1 + v_{\text{AM}} v_{\text{MB}}/c^2} = \frac{0.60c + 0.80c}{1 + (0.60)(0.80)} = 0.946c$$

$v_{\text{BA}} = -v_{\text{AB}} = -0.946c$, so the relative speed measured by a passenger on either ship is $\boxed{0.946c}$.

29. Strategy The velocity of the electron relative to the lab is v_{EL}. $v_{\text{PL}} = \frac{4}{5}c$ is the velocity of the proton relative to the lab. $v_{\text{EP}} = -\frac{5}{7}c$ is the velocity of the electron relative to the proton. The velocity of the electron relative to the proton is negative since the electron moves to the left and the positive direction is chosen to be to the right. Use Eq. (26-5).

Solution

Lab frame: $\vec{\mathbf{v}}_{\text{PL}}$ p — Proton frame: $\vec{\mathbf{v}}_{\text{EP}}$ e⁻

$$v_{\text{EL}} = \frac{v_{\text{EP}} + v_{\text{PL}}}{1 + v_{\text{EP}} v_{\text{PL}}/c^2} = \frac{-\frac{5}{7}c + \frac{4}{5}c}{1 + \left(-\frac{5}{7}\right)\left(\frac{4}{5}\right)} = \frac{1}{5}c$$

The speed of the electron relative to the lab is $\boxed{\frac{1}{5}c}$.

31. Strategy The speed of electron B in any frame of reference in which electron A is at rest is the same as the speed of electron B relative to electron A. The coordinate system relative to the lab is chosen so that objects traveling west have a positive velocity. Then the velocity of electron B relative to the lab is $v_{\text{BL}} = \frac{4}{5}c$ and the velocity of electron A relative to the lab is $v_{\text{AL}} = \frac{3}{5}c$, so $v_{\text{LA}} = -v_{\text{AL}} = -\frac{3}{5}c$. Use Eq. (26-5).

Solution

Lab frame: $\vec{\mathbf{v}}_{\text{AL}}$, $\vec{\mathbf{v}}_{\text{BL}}$, N — A rest frame: A, B, $\vec{\mathbf{v}}_{\text{BA}} = ?$

$$v_{\text{BA}} = \frac{v_{\text{BL}} + v_{\text{LA}}}{1 + v_{\text{BL}} v_{\text{LA}}/c^2} = \frac{\frac{4}{5}c + \left(-\frac{3}{5}c\right)}{1 + \left(\frac{4}{5}\right)\left(-\frac{3}{5}\right)} = \frac{5}{13}c$$

The speed of electron B in a frame of reference in which electron A is at rest is $\boxed{\frac{5}{13}c}$.

33. Strategy The magnitude of the momentum is given by $p = \gamma mv$.

Solution Find the momentum of the proton.

$$p = \gamma mv = (1 - v^2/c^2)^{-1/2} m_{\text{p}} v$$
$$= (1 - 0.90^2)^{-1/2} (1.673 \times 10^{-27}\ \text{kg})(0.90)(3.00 \times 10^8\ \text{m/s}) = \boxed{1.0 \times 10^{-18}\ \text{kg} \cdot \text{m/s}}$$

35. Strategy Use Eq. (26-6) and $p = mv$. Use Eq. (A-24) to approximate the relativistic momentum. Then find the percent difference.

Solution Find the relativistic momentum of the space shuttle.

$$p = \gamma mv = \frac{mv}{\sqrt{1-v^2/c^2}} \approx mv(1+\tfrac{1}{2}v^2/c^2) = (1\times10^5\text{ kg})(8\times10^3\text{ m/s})[1+\tfrac{1}{2}(8\times10^3/3\times10^8)^2]$$

$$= \boxed{800{,}000{,}000.3\text{ kg}\cdot\text{m/s}}$$

Find the nonrelativistic momentum of the shuttle.

$$p = mv = (1\times10^5\text{ kg})(8\times10^3\text{ m/s}) = \boxed{800{,}000{,}000\text{ kg}\cdot\text{m/s}}$$

Find the percent difference.

$$\frac{0.3}{800{,}000{,}000}\times100\% = \boxed{4\times10^{-8}\ \%}$$

37. (a) Strategy To find the force, use the impulse-momentum equation: $F\Delta t = \Delta p = p_\text{f} - p_\text{i} = p_\text{f} - 0 = p_\text{f}$. The final velocity is a large fraction of c, so the relativistic form of the momentum, $p_\text{f} = \gamma mv$, must be used.

Solution $F\Delta t = \gamma mv$, so

$$F = \frac{\gamma mv}{\Delta t} = \frac{(1-v^2/c^2)^{-1/2}mv}{\Delta t} = \frac{(1-0.70^2)^{-1/2}(2200\text{ kg})(0.70\times3.00\times10^8\text{ m/s})}{3.6\times10^4\text{ s}} = \boxed{1.8\times10^7\text{ N}}.$$

(b) Strategy Use Newton's second law.

Solution Find the initial acceleration.

$$a = \frac{F}{m} = \frac{1.8\times10^7\text{ N}}{2200\text{ kg}} = \boxed{8200\text{ m/s}^2}$$

Since $g \approx 10\text{ m/s}^2$, $a \approx 820g$. $\boxed{\text{This is much larger than any human could survive}}$.

39. Strategy Since no energy is lost to the environment, the change in the kinetic energy of the two lumps can be equated to a change in the rest energy of the system.

Solution $\Delta E_0 = (\Delta m)c^2$, so

$$\Delta m = \frac{\Delta E_0}{c^2} = \frac{\Delta K_1 + \Delta K_2}{c^2} = \frac{\frac{1}{2}m_1v_1^2 + \frac{1}{2}m_2v_2^2}{c^2}.$$

The nonrelativistic form of the kinetic energy is used since the speeds are very small compared to c. Using $m_1 = m_2 = 1.00$ kg and $v_1 = v_2 = 30.0$ m/s gives

$$\Delta m = \frac{\frac{1}{2}(1.00\text{ kg})(30.0\text{ m/s})^2 + \frac{1}{2}(1.00\text{ kg})(30.0\text{ m/s})^2}{(3.00\times10^8\text{ m/s})^2} = 1.00\times10^{-14}\text{ kg}.$$

The mass of the system $\boxed{\text{increased by } 1.00\times10^{-14}\text{ kg}}$.

41. Strategy The mass of the Sun is 1.987×10^{30} kg. 80.0% of the limiting mass of the white dwarf is converted to energy. Use Eq. (26-7).

Solution Compute the energy released by the explosion of the white dwarf.

$$E_0 = mc^2 = 0.800(1.4)(1.987\times10^{30}\text{ kg})(3.00\times10^8\text{ m/s})^2 = \boxed{2.0\times10^{47}\text{ J}}$$

43. Strategy Total energy must be conserved.

Solution Find the energy released in the decay.

rest energy before decay $= m_{\text{Rn}}c^2 =$ rest energy after decay + energy released $= m_{\text{Po}}c^2 + m_\alpha c^2 + E$, so

$E = (m_{\text{Rn}} - m_{\text{Po}} - m_\alpha)c^2 = (221.970\,39\text{ u} - 217.962\,89\text{ u} - 4.001\,51\text{ u})(931.494\text{ MeV/u}) = \boxed{5.58\text{ MeV}}$.

45. Strategy Use Eq. (26-9).

Solution Find the electron's speed.

$$E = \gamma mc^2 = (1 - v^2/c^2)^{-1/2} mc^2$$

$$\sqrt{1 - \frac{v^2}{c^2}} = \frac{mc^2}{E}$$

$$\frac{v^2}{c^2} = 1 - \frac{m^2c^4}{E^2}$$

$$v = c\sqrt{1 - \frac{m^2c^4}{E^2}} = c\sqrt{1 - \frac{(9.109\times10^{-31}\text{ kg})^2(3.00\times10^8\text{ m/s})^4}{(1.02\times10^{-13}\text{ J})^2}} = \boxed{0.595c}$$

47. Strategy The object is moving at $1.80/3.00 = 0.600$ times c. Use Eq. (26-8).

Solution Find the kinetic energy of the object.

$K = (\gamma - 1)mc^2 = [(1 - v^2/c^2)^{-1/2} - 1]mc^2 = [(1 - 0.600^2)^{-1/2} - 1](0.12\text{ kg})(3.00\times10^8\text{ m/s})^2 = \boxed{2.7\times10^{15}\text{ J}}$

49. Strategy Convert 1 MeV to joules. Then divide the result by c.

Solution

$1\text{ MeV} = (1\times10^6\text{ eV})\frac{1.602\times10^{-19}\text{ J}}{1\text{ eV}} = 1.602\times10^{-13}\text{ J}$, so $\frac{1\text{ MeV}}{c} = \frac{1.602\times10^{-13}\text{ J}}{2.998\times10^8\text{ m/s}} = 5.344\times10^{-22}\text{ kg}\cdot\text{m/s}$.

The conversion is $\boxed{1\text{ MeV}/c = 5.344\times10^{-22}\text{ kg}\cdot\text{m/s}}$.

51. Strategy Solve Eq. (26-10) for p using $E = 5.0mc^2$ and $E_0 = mc^2$.

Solution Find the magnitude of the electron's momentum as observed in the laboratory.

$E^2 = (5.0mc^2)^2 = 25m^2c^4 = E_0{}^2 + (pc)^2 = m^2c^4 + p^2c^2$, so $p = \sqrt{24}mc = \boxed{4.9mc}$.

53. Strategy Solve Eq. (26-10) for p using $E_0 = 0.511\text{ MeV}$ and $E = 6.5\text{ MeV}$.

Solution Find the momentum of the electron.

$E^2 = E_0^2 + (pc)^2$, so $p = \frac{1}{c}\sqrt{E^2 - E_0^2} = \frac{1}{c}\sqrt{(6.5\text{ MeV})^2 - (0.511\text{ MeV})^2} = \boxed{6.5\text{ MeV}/c}$.

57. Strategy Use Eq. (26-11) with the condition that $K << E_0$.

Solution Show that $(pc)^2 = K^2 + 2KE_0$ reduces to $K \approx p^2/(2m)$ for nonrelativistic speeds.

$(pc)^2 = p^2c^2 = K^2 + 2KE_0 = K(K + 2E_0)$

If $K << E_0$, then $K + 2E_0 \approx 2E_0$. Solve for K and simplify.

$$p^2c^2 \approx 2KE_0, \text{ so } K \approx \frac{p^2c^2}{2E_0} = \frac{p^2c^2}{2mc^2} = \frac{p^2}{2m}.$$

61. Strategy Use the length contraction equation to find γ. Then, use the time dilation equation to find how long the astronaut would say that she exercised.

Solution Find γ.

$$L = \frac{L_0}{\gamma}, \text{ so } \gamma = \frac{L_0}{L}.$$

Find Δt_0.

$$\Delta t = \gamma \Delta t_0 = \frac{L_0}{L}\Delta t_0, \text{ so } \Delta t_0 = \frac{L}{L_0}\Delta t = \frac{30.5 \text{ m}}{35.2 \text{ m}}(22.2 \text{ min}) = \boxed{19.2 \text{ min}}.$$

65. Strategy Let T be the reading on the Earth clock, as seen by an observer on Earth, when the signal reaches the ship. The distance traveled by the ship according to an Earth observer, is $d = 0.80cT$, so the time reading is $T = d/(0.80c)$.

Solution

(a) The distance traveled by the light signal is the same as the distance traveled by the ship, and can be written $d = c(T - 1.0\times10^4 \text{ s})$. Now substitute for T and solve for d.

$$d = c\left(\frac{d}{0.80c} - 1.0\times10^4 \text{ s}\right) = \frac{d}{0.80} - (1.0\times10^4 \text{ s})c$$

$$0.80d = d - (8.0\times10^3 \text{ s})c$$

$$0.20d = (8.0\times10^3 \text{ s})c$$

$$d = \frac{(8.0\times10^3 \text{ s})(3.00\times10^8 \text{ m/s})}{0.20} = \boxed{1.2\times10^{13} \text{ m}}$$

(b) The clock reading is $T = \frac{d}{0.80c} = \frac{1.2\times10^{13} \text{ m}}{0.80(3.00\times10^8 \text{ m/s})} = \boxed{5.0\times10^4 \text{ s}}$.

69. Strategy Find the total energy of the neutron using Eq. (26-10).

Solution

$$E^2 = E_0^2 + (pc)^2, \text{ so } E = \sqrt{E_0^2 + (pc)^2} = \sqrt{(939.6 \text{ MeV})^2 + (935 \text{ MeV})^2} = \boxed{1.326 \text{ GeV}}.$$

73. Strategy If the collision is elastic, the total kinetic energy before and after the collision should be the same. Use Eq. (26-8).

Solution The total kinetic energy before the collision is

$$K_\text{p}+K_\text{n}=(\gamma_\text{p}-1)m_\text{p}c^2+(\gamma_\text{n}-1)m_\text{n}c^2=[(1-0.70^2)^{-1/2}-1]m_\text{p}c^2+[(1-0^2/c^2)^{-1/2}-1]m_\text{n}c^2=0.4m_\text{p}c^2.$$

After the collision, the total kinetic energy is

$$K_\text{p}+K_\text{n}=(\gamma_\text{p}-1)m_\text{p}c^2+(\gamma_\text{n}-1)m_\text{n}c^2=[(1-0.63^2)^{-1/2}-1]m_\text{p}c^2+[(1-0.128^2)^{-1/2}-1](14m_\text{p})c^2=0.4m_\text{p}c^2.$$

The kinetic energy is conserved, so the collision is elastic.

77. (a) Strategy Use conservation of momentum and Eqs. (26-6) and (26-7).

Solution The magnitude of the momentum for each pion is

$$p=\gamma mv=\frac{\gamma mc^2v}{c^2}=\frac{\gamma E_0v}{c^2}.$$

Since the two pions move at right angles to each other, the magnitude of the vector sum of their momenta is

$$p=\frac{\gamma E_0v/c^2+\gamma E_0v/c^2}{\sqrt{2}}=\sqrt{2}\frac{\gamma E_0v}{c^2}=\sqrt{2}(1-0.900^2)^{-1/2}(140.0\text{ MeV})\frac{0.900c}{c^2}=409\text{ MeV}/c.$$

Since the momentum is conserved, the magnitude of the momentum of the original particle was $\boxed{409\text{ MeV}/c}$.

(b) Strategy The total energy is conserved, so the total energy of the original particle must equal the sum of the total energies of the two pions. Use Eqs. (26-9) and (26-10).

Solution Find the total energy.

$$E=E_1+E_2=\gamma_1E_0+\gamma_2E_0=2\gamma E_0=2(1-0.900^2)^{-1/2}(140.0\text{ MeV})=642\text{ MeV}$$

Find the rest energy of the original particle.

$$E^2=E_0{}^2+(pc)^2,\text{ so }E_0=\sqrt{E^2-(pc)^2}=\sqrt{(642\text{ MeV})^2-(409\text{ MeV})^2}=495\text{ MeV}.$$

Find the kinetic energy.

$$E=K+E_0,\text{ so }K=E-E_0=642\text{ MeV}-495\text{ MeV}=\boxed{147\text{ MeV}}.$$

(c) Strategy The mass is found using $E_0=mc^2$.

Solution

$$m=\frac{E_0}{c^2}=\boxed{495\text{ MeV}/c^2}$$

81. Strategy Use the definition of total energy and Eq. (26-10).

Solution If K is much larger than E_0, then E is also much larger than E_0, since $E=K+E_0>K>>E_0$. Now, $E^2=E_0{}^2+(pc)^2$, so $E^2-E_0{}^2=(pc)^2$. Since $E>>E_0$, $E^2>>E_0^2$, so $E^2-E_0{}^2\approx E^2$. Therefore, $E^2\approx(pc)^2$, or $E\approx pc$.

85. Strategy Use conservation of momentum and total energy, as well as Eq. (26-11).

Solution The relativistic energies are given by $(p_n c)^2 = K_n^2 + 2K_n E_{0n}$ and $(p_\pi c)^2 = K_\pi^2 + 2K_\pi E_{0\pi}$.
By conservation of momentum, $p_n = p_\pi$, so $K_n^2 + 2K_n E_{0n} = K_\pi^2 + 2K_\pi E_{0\pi}$ (1).
The total energy of the system is conserved, so

$$m_\Lambda c^2 = m_n c^2 + m_\pi c^2 + K_n + K_\pi$$

$$(m_\Lambda - m_n - m_\pi)c^2 = K_n + K_\pi = K$$

where K is the total kinetic energy. Substitute $K_n = K - K_\pi$ into (1) and simplify.

$$K^2 - 2KK_\pi + K_\pi^2 + 2KE_{0n} - 2K_\pi E_{0n} = K_\pi^2 + 2K_\pi E_{0\pi}$$

$$2KK_\pi + 2K_\pi E_{0n} + 2K_\pi E_{0\pi} = 2KE_{0n} + K^2$$

$$2K_\pi (K + E_{0n} + E_{0\pi}) = K(K + 2E_{0n})$$

Solve for K_π.

$$K_\pi = \frac{K(K+2E_{0n})}{2(K+E_{0n}+E_{0\pi})} = \frac{(m_\Lambda - m_n - m_\pi)c^2[(m_\Lambda - m_n - m_\pi)c^2 + 2m_n c^2]}{2[(m_\Lambda - m_n - m_\pi)c^2 + m_n c^2 + m_\pi c^2]}$$

$$= \frac{(1115.7\text{ MeV} - 939.6\text{ MeV} - 135.0\text{ MeV})[(1115.7\text{ MeV} - 939.6\text{ MeV} - 135.0\text{ MeV}) + 2(939.6\text{ MeV})]}{2[(1115.7\text{ MeV} - 939.6\text{ MeV} - 135.0\text{ MeV}) + 939.6\text{ MeV} + 135.0\text{ MeV}]}$$

$$= \boxed{35.4\text{ MeV}}$$

Calculate K_n.

$$K_n = (1115.7\text{ MeV} - 939.6\text{ MeV} - 135.0\text{ MeV}) - 35.4\text{ MeV} = \boxed{5.7\text{ MeV}}$$

Chapter 27

EARLY QUANTUM PHYSICS AND THE PHOTON

Problems

1. (a) Strategy and Solution The energy of a photon is inversely proportional to its wavelength, so the photons with the shorter wavelength have the greater energy. Therefore, a single $\boxed{\text{ultraviolet}}$ photon has the greater energy.

(b) Strategy The energy of a photon of EM radiation with frequency f is $E = hf$. The frequency and wavelength are related by $\lambda f = c$.

Solution Compute the energy of a single infrared photon.

$$E_{\text{infrared}} = hf = \frac{hc}{\lambda} = \frac{(6.626\times10^{-34}\ \text{J}\cdot\text{s})(3.00\times10^{8}\ \text{m/s})}{2.0\times10^{-6}\ \text{m}} = \boxed{9.9\times10^{-20}\ \text{J}}$$

Compute the energy of a single ultraviolet photon.

$$E_{\text{ultraviolet}} = hf = \frac{hc}{\lambda} = \frac{(6.626\times10^{-34}\ \text{J}\cdot\text{s})(3.00\times10^{8}\ \text{m/s})}{7.0\times10^{-8}\ \text{m}} = \boxed{2.8\times10^{-18}\ \text{J}}$$

(c) Strategy Use the definition of power. Let N be the number of photons. The total energy is NE, where E is the energy of a single photon.

Solution Find the number of infrared photons emitted per second.

$$P = \frac{NE}{\Delta t},\ \text{so}\ \frac{N_{\text{infrared}}}{\Delta t} = \frac{P}{E} = \frac{200\ \text{W}}{9.9\times10^{-20}\ \text{J/photon}} = \boxed{2.0\times10^{21}\ \text{photons/s}}.$$

Find the number of ultraviolet photons emitted per second.

$$\frac{N_{\text{ultraviolet}}}{\Delta t} = \frac{P}{E} = \frac{200\ \text{W}}{2.84\times10^{-18}\ \text{J/photon}} = \boxed{7.0\times10^{19}\ \text{photons/s}}$$

3. Strategy The energy of a photon of EM radiation with frequency f is $E = hf$. The frequency and wavelength are related by $\lambda f = c$.

Solution

(a) Calculate the wavelength of a photon with energy 3.1 eV.

$$E = hf = \frac{hc}{\lambda},\ \text{so}\ \lambda = \frac{hc}{E} = \frac{1240\ \text{eV}\cdot\text{nm}}{3.1\ \text{eV}} = \boxed{400\ \text{nm}}.$$

(b) Calculate the frequency of a photon with energy 3.1 eV.

$$f = \frac{c}{\lambda} = \frac{3.00\times10^{8}\ \text{m/s}}{400\times10^{-9}\ \text{m}} = \boxed{7.5\times10^{14}\ \text{Hz}}$$

5. (a) Strategy Use Einstein's photoelectric equation.

Solution Calculate the maximum kinetic energy.

$$K_{\text{max}} = hf - \phi = \frac{hc}{\lambda} - \phi = \frac{1240\text{ eV}\cdot\text{nm}}{413\text{ nm}} - 2.16\text{ eV} = \boxed{0.84\text{ eV}}$$

(b) Strategy The threshold wavelength is related to the threshold frequency by $\lambda_0 f_0 = c$. Use Eq. (27-8).

Solution Find the threshold wavelength.

$$hf_0 = \frac{hc}{\lambda_0} = \phi,\text{ so } \lambda_0 = \frac{hc}{\phi} = \frac{1240\text{ eV}\cdot\text{nm}}{2.16\text{ eV}} = \boxed{574\text{ nm}}.$$

7. Strategy The work function for the metal is $\phi = 2.60$ eV. Use Eq. (27-8) to find the maximum wavelength of the photons that will eject electrons from the metal. The frequency and wavelength are related by $\lambda f = c$.

Solution Find the longest wavelength.

$$f_0 = \frac{\phi}{h} = \frac{c}{\lambda},\text{ so } \lambda = \frac{hc}{\phi} = \frac{1240\text{ eV}\cdot\text{nm}}{2.60\text{ eV}} = \boxed{477\text{ nm}}.$$

9. Strategy The stopping potential V_s is related to the maximum kinetic energy of the electrons by $K_{\text{max}} = eV_\text{s}$. Use Einstein's photoelectric equation and $\lambda f = c$.

Solution Find the work function.

$$K_{\text{max}} = eV_\text{s} = hf - \phi = \frac{hc}{\lambda} - \phi,\text{ so } \phi = \frac{hc}{\lambda} - eV_\text{s} = \frac{1240\text{ eV}\cdot\text{nm}}{220\text{ nm}} - 1.1\text{ eV} = \boxed{4.5\text{ eV}}.$$

13. Strategy The frequency intercept gives the threshold frequency, so $f_0 = 43.9\times10^{13}$ Hz. The stopping potential is related to the maximum kinetic energy of the ejected electrons by $K_{\text{max}} = eV_\text{s}$. Einstein's photoelectric equation is $K_{\text{max}} = hf - \phi$, so $V_\text{s} = \frac{hf}{e} - \frac{\phi}{e}$. The work function ϕ is given by the stopping potential intercept $V_0 = -\phi/e$, where $V_\text{s} = mf + V_0$. $V_\text{s} = 0$ when $f = f_0$.

Solution

(a) The slope of the graph is $m = h/e$. Compute the numerical value.

$$m = \frac{1.50\text{ V} - 0\text{ V}}{80.0\times10^{13}\text{ Hz} - 43.9\times10^{13}\text{ Hz}} = 4.155\times10^{-15}\text{ V}\cdot\text{s}$$

Planck's constant is $h = em = (1.602\times10^{-19}\text{ C})(4.155\times10^{-15}\text{ V}\cdot\text{s}) = \boxed{6.66\times10^{-34}\text{ J}\cdot\text{s}}$.

(b) Find the work function.

$$0 = mf_0 - \frac{\phi}{e},\text{ so } \phi = emf_0 = e(4.155\times10^{-15}\text{ V}\cdot\text{s})(43.9\times10^{13}\text{ Hz}) = \boxed{1.82\text{ eV}}.$$

15. Strategy The maximum kinetic energy K of an electron is equal to hf_{max}. $\lambda_{min} = c/f_{max}$ and $K = eV$, where V is the applied voltage.

Solution Find the applied voltage.

$$hf_{max} = \frac{hc}{\lambda_{min}} = K = eV, \text{ so } V = \frac{hc}{\lambda_{min}e} = \frac{1240 \text{ eV}\cdot\text{nm}}{0.46 \text{ nm}\cdot e} = \boxed{2.7 \text{ kV}}.$$

17. Strategy The cutoff (maximum) frequency occurs when the photon's energy and the electron's kinetic energy are equal. The kinetic energy of the electron is equal to the electric potential energy of the electron, eV. Use Eq. (27-9).

Solution Find the cutoff frequency.

$$hf_{max} = K = eV, \text{ so } f_{max} = \frac{eV}{h} = \frac{46\times10^3 \text{ eV}}{4.136\times10^{-15} \text{ eV}\cdot\text{s}} = \boxed{1.1\times10^{19} \text{ Hz}}.$$

21. (a) Strategy Use the Compton shift, Eq. (27-14).

Solution Find the wavelength of the incident photon.

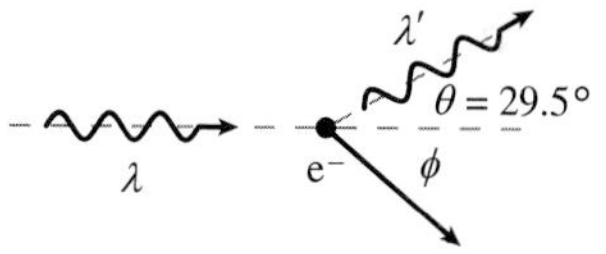

$$\lambda' - \lambda = \frac{h}{m_e c}(1-\cos\theta), \text{ so } \lambda = \lambda' - \frac{h}{m_e c}(1-\cos\theta)$$

$$= 2.81\times10^{-12} \text{ m} - \frac{6.626\times10^{-34} \text{ J}\cdot\text{s}}{(9.109\times10^{-31} \text{ kg})(2.998\times10^8 \text{ m/s})}(1-\cos 29.5°)$$

$$= \boxed{2.50\times10^{-12} \text{ m}}.$$

(b) Strategy The kinetic energy of the electron plus the energy of the scattered photon is equal to the energy of the incident photon. Use Eq. (27-10).

Solution Find the kinetic energy of the electron.

$K_e + E' = E$, so

$$K_e = E - E' = \frac{hc}{\lambda} - \frac{hc}{\lambda'} = (1240\times10^{-9} \text{ eV}\cdot\text{m})\left(\frac{1}{2.4954\times10^{-12} \text{ m}} - \frac{1}{2.81\times10^{-12} \text{ m}}\right) = \boxed{55.6 \text{ keV}}.$$

23. Strategy Use the Compton shift, Eq. (27-14).

Solution

(a) Find the Compton shift in wavelength.

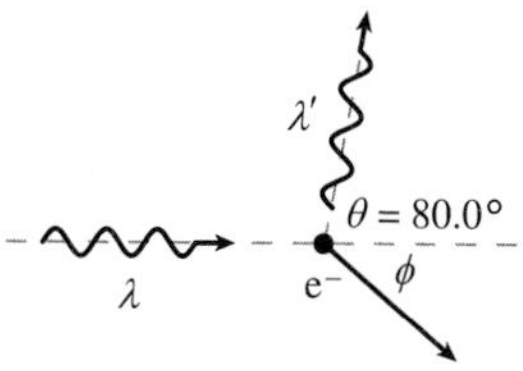

$$\lambda' - \lambda = \frac{h}{m_e c}(1-\cos\theta) = (2.426 \text{ pm})(1-\cos 80.0°) = \boxed{2.00 \text{ pm}}$$

(b) Find the wavelength of the scattered photon.

$$\lambda' - \lambda = \Delta\lambda, \text{ so } \lambda' = \lambda + 2.00 \text{ pm} = 1.50\times10^2 \text{ pm} + 2.00 \text{ pm} = \boxed{152 \text{ pm}}.$$

25. Strategy Use the Compton shift, Eq. (27-14). The frequency and wavelength are related by $\lambda f = c$.

Solution Find the frequency of the scattered photon.

$$\lambda' - \lambda = \frac{c}{f'} - \frac{c}{f} = \frac{h}{m_e c}(1 - \cos\theta), \text{ so}$$

$$f' = \frac{c}{\frac{c}{f} + \frac{h}{m_e c}(1-\cos\theta)} = \frac{3.00\times10^8 \text{ m/s}}{\frac{3.00\times10^8 \text{ m/s}}{3.0\times10^{19} \text{ Hz}} + (2.426\times10^{-12} \text{ m})(1-\cos 90°)} = \boxed{2.4\times10^{19} \text{ Hz}}.$$

$\theta = 90°$, e$^-$, ϕ

29. Strategy The change in kinetic energy of the electron is the difference in the energies of the incident photon and the scattered photon.

Solution Find the change in kinetic energy.

$$\Delta K = E - E' = \frac{hc}{\lambda} - \frac{hc}{\lambda'} = \frac{1240 \text{ eV}\cdot\text{nm}}{0.0100 \text{ nm}} - \frac{1240 \text{ eV}\cdot\text{nm}}{0.0124 \text{ nm}} = \boxed{2.4\times10^4 \text{ eV}}$$

31. Strategy The orbital radius in the *n*th state is given by $r_n = n^2 a_0$.

Solution Compute the orbital radius.

$$r_3 = 3^2 \cdot 52.9 \text{ pm} = 476 \text{ pm} = \boxed{0.476 \text{ nm}}$$

33. Strategy The orbital radius in the *n*th state is given by $r_n = n^2 a_0$.

Solution

(a) Find the difference in the radii of the $n = 1$ and $n = 2$ states for hydrogen.

$$r_n = n^2 a_0, \text{ so } r_2 - r_1 = (2^2 - 1^2)a_0 = 3a_0 = 3(52.9\times10^{-12} \text{ m}) = \boxed{1.59\times10^{-10} \text{ m}}.$$

(b) Find the difference in the radii of the $n = 100$ and $n = 101$ states for hydrogen.

$$r_n = n^2 a_0, \text{ so } r_2 - r_1 = (101^2 - 100^2)a_0 = 201a_0 = 201(52.9\times10^{-12} \text{ m}) = \boxed{1.06\times10^{-8} \text{ m}}.$$

The result for (b) is much larger than that for (a), so the orbital separations are much larger for larger *n* values.

37. Strategy The energy supplied by the photon raises the atom from the ground state to a higher energy level. The energy for a hydrogen atom in the *n*th stationary state is given by $E_n = (-13.6 \text{ eV})/n^2$.

Solution Find the energy level to which the atom is excited.

$$E_n - E_1 = \frac{E_1}{n^2} - E_1 = 12.1 \text{ eV}, \text{ so } n = \sqrt{\frac{E_1}{12.1 \text{ eV} + E_1}} = \sqrt{\frac{-13.6 \text{ eV}}{12.1 \text{ eV} - 13.6 \text{ eV}}} = 3.$$

The atom is excited to the $\boxed{n = 3}$ energy level.

39. Strategy The minimum energy for an ionized atom is $E_{\text{ionized}} = 0$. Use the ground-state energy.

Solution The energy needed to remove a ground-state electron from a hydrogen atom is

$$E = E_{\text{ionized}} - E_1 = 0 - (-13.6 \text{ eV}) = \boxed{13.6 \text{ eV}}.$$

41. Strategy The smallest transition that a ground-state electron can make is to the $n = 2$ state. Use Eq. (27-24).

Solution The energy of the photon is

$$E = E_2 - E_1 = \frac{E_1}{2^2} - E_1 = \frac{-13.6\text{ eV}}{2^2} - (-13.6\text{ eV}) = \boxed{10.2\text{ eV}}.$$

43. (a) Strategy Use Coulomb's law.

Solution Find the magnitude of the force on the electron due to the proton.

$$F = \frac{k|q_1||q_2|}{r^2} = \frac{k|+e||-e|}{r^2} = \boxed{\frac{ke^2}{r^2}}$$

(b) Strategy The electron is undergoing uniform circular motion, so the acceleration is radial. Use Newton's second law.

Solution Find the speed of the electron.

$$\Sigma F_r = \frac{ke^2}{r^2} = ma_r = m_e\frac{v^2}{r}, \text{ so } v = \sqrt{\frac{ke^2}{m_e r}}.$$

(c) Strategy Use the Bohr condition for angular momentum and the result of part (b).

Solution Solve for the speed.

$$m_e v r_n = n\hbar, \text{ so } v = \frac{n\hbar}{m_e r_n}.$$

Find the radius of the nth Bohr orbit.

$$\frac{n\hbar}{m_e r_n} = \sqrt{\frac{ke^2}{m_e r_n}}, \text{ so } \frac{n^2\hbar^2}{m_e r_n} = ke^2 \text{ or } r_n = \frac{n^2\hbar^2}{m_e ke^2}.$$

45. Strategy Use Eq. (27-20) and the values of the fundamental constants.

Solution Compute the numerical value of the Bohr radius.

$$a_0 = \frac{\hbar^2}{m_e ke^2} = \frac{h^2}{4\pi^2 m_e ke^2} = \frac{(6.626\times10^{-34}\text{ J}\cdot\text{s})^2}{4\pi^2(9.109\times10^{-31}\text{ kg})(8.988\times10^9\text{ N}\cdot\text{m}^2/\text{C}^2)(1.602\times10^{-19}\text{ C})^2} = 5.29\times10^{-11}\text{ m}$$

47. Strategy Use the Bohr condition for angular momentum, Eq. (27-18).

Solution Calculate the speed of the electron.

$$m_e v r_n = n\hbar, \text{ so } v = \frac{(1)\hbar}{m_e r_1} = \frac{h}{2\pi m_e a_0} = \frac{6.626\times10^{-34}\text{ J}\cdot\text{s}}{2\pi(9.109\times10^{-31}\text{ kg})(5.29\times10^{-11}\text{ m})} = \boxed{2.19\times10^6\text{ m/s}}.$$

49. Strategy The energy of a photon with the given wavelength is equal to the energy difference between the two helium levels. The energy of a photon of EM radiation with wavelength λ is $E = hc/\lambda$.

Solution Find the energy difference.

$$E = \frac{hc}{\lambda} = \frac{1240\text{ eV}\cdot\text{nm}}{587.6\text{ nm}} = \boxed{2.11\text{ eV}}$$

53. Strategy The energy of a photon of EM radiation with wavelength λ is $E = hc/\lambda$.

Solution The energy of the UV photon is $E_1 = hc/\lambda_1$. The solid dissipates 0.500 eV, so the energy available for the emitted photon is $E_2 = E_1 - 0.500\text{ eV} = \dfrac{hc}{\lambda_1} - 0.500\text{ eV}$.

Calculate the wavelength of the emitted photon.

$$\lambda = \frac{hc}{E_2} = \frac{hc}{\frac{hc}{\lambda_1} - 0.500\text{ eV}} = \frac{1}{\frac{1}{\lambda_1} - \frac{0.500\text{ eV}}{hc}} = \frac{1}{\frac{1}{320\text{ nm}} - \frac{0.500\text{ eV}}{1240\text{ eV}\cdot\text{nm}}} = \boxed{370\text{ nm}}$$

57. Strategy The energy of the photon must equal the sum of the total energy of the electron and the positron. Use $E_0 = mc^2$.

Solution Find the energy of the photon.

$$E_{\text{photon}} = E_\text{e} + E_\text{p} = m_\text{e}c^2 + K_\text{e} + m_\text{p}c^2 + K_\text{p} = 0.511\text{ MeV} + 0.22\text{ MeV} + 0.511\text{ MeV} + 0.22\text{ MeV} = \boxed{1.46\text{ MeV}}$$

61. Strategy and Solution

(a) Since the electron and positron are at rest, their combined momentum is zero. Conservation of momentum requires that the total momentum also be zero after the annihilation. A single photon has momentum $p = E/c$ in the direction that the photon is moving, so a single photon can never have a total momentum of zero. Therefore, the annihilation of an electron and positron can never produce a single photon. Two photons of equal energy moving in opposite directions will have zero total momentum, so the reaction $\text{e}^- + \text{e}^+ \rightarrow 2\gamma$ can occur.

(b) The two photons must have equal energy, so each photon will have the energy of an electron at rest, $E = m_\text{e}c^2 = \boxed{511\text{ keV}}$.

65. Strategy Power is equal to intensity times area. The energy of a photon of EM radiation with wavelength λ is $E = hc/\lambda$. The minimum detected number of photons per second is given by the ratio of the power to the energy per photon.

Solution The minimum amount of power that an owl's eye can detect is

$$P = IA = I\pi r^2 = (5.0\times10^{-13}\text{ W/m}^2)\pi(4.25\times10^{-3}\text{ m})^2 = 2.84\times10^{-17}\text{ W}.$$

The energy of a photon of wavelength 510 nm is

$$E = \frac{hc}{\lambda} = \frac{(6.626\times10^{-34}\text{ J}\cdot\text{s})(3.00\times10^8\text{ m/s})}{510\times10^{-9}\text{ m}} = 3.9\times10^{-19}\text{ J}.$$

Compute the minimum number of photons per second that an owl eye can detect.

$$\frac{2.84\times10^{-17}\text{ W}}{3.9\times10^{-19}\text{ J}} = \boxed{73\text{ photons/s}}$$

69. Strategy Use Eqs. (27-21), (27-24), (27-25), and (27-26).

Solution Equate the energy levels in He^+ and H atoms.

$$E_{n\text{H}} = E_{m\text{He}^+}$$

$$\frac{-13.6\text{ eV}}{n^2} = \frac{-13.6\text{ eV}\times 2^2}{m^2}$$

$$m^2 = 2^2 n^2$$

$$m = 2n$$

The ratio of the orbital radii of H and He^+ is $\frac{r_{n\text{H}}}{r_{m\text{He}^+}} = \frac{n^2 a_0}{\frac{m^2 a_0}{2}} = \frac{2n^2}{m^2}.$

Now substitute $m = 2n.$

$$\frac{r_{n\text{H}}}{r_{m\text{He}^+}} = \frac{2n^2}{(2n)^2} = \frac{2n^2}{4n^2} = \boxed{\frac{1}{2}} = \frac{Z_\text{H}}{Z_{\text{He}^+}}$$

For levels of equal energy, the ratio of orbital radii appears to equal the ratio of atomic numbers.

73. (a) Strategy The photons that can be absorbed have energies equal to the energy differences between energy levels.

Solution Calculate the energies and associated wavelengths needed for a transition from the ground state to the $n = 2$, 3, 4, and 5 states.

$$E = |E_1| - |E_n| = |E_1| - \frac{|E_1|}{n^2}$$

$1 \to 2$: $E = 13.6\text{ eV} - 3.4\text{ eV} = 10.2\text{ eV}$

$$\lambda = \frac{hc}{E} = \frac{1240\text{ eV}\cdot\text{nm}}{10.2\text{ eV}} = 122\text{ nm}$$

$1 \to 3$: $E = 13.6\text{ eV} - 1.51\text{ eV} = 12.09\text{ eV}$

$$\lambda = \frac{1240\text{ eV}\cdot\text{nm}}{12.09\text{ eV}} = 103\text{ nm}$$

$1 \to 4$: $E = 13.6\text{ eV} - 0.85\text{ eV} = 12.75\text{ eV}$

$$\lambda = \frac{1240\text{ eV}\cdot\text{nm}}{12.75\text{ eV}} = 97.3\text{ nm}$$

$1 \to 5$: $E = 13.6\text{ eV} - 0.544\text{ eV} = 13.056\text{ eV}$

$$\lambda = \frac{1240\text{ eV}\cdot\text{nm}}{13.056\text{ eV}} = 95.0\text{ nm}$$

The wavelengths that can be absorbed are 97.3 nm and 103 nm.

(b) Strategy By conservation of momentum, the momentum of the atom must equal the momentum of the photon.

Solution Compute the recoil speeds.

$$p_\text{H} = m_\text{H} v = p_\text{photon} = \frac{E}{c} = \frac{h}{\lambda}, \text{ so}$$

$$v = \frac{h}{m_\text{H}\lambda} = \frac{6.626\times 10^{-34}\text{ J}\cdot\text{s}}{(1.674\times 10^{-27}\text{ kg})(97.3\times 10^{-9}\text{ m})} = \boxed{4.07\text{ m/s for the 97.3-nm photon}}$$

$$\text{and } v = \frac{6.626\times 10^{-34}\text{ J}\cdot\text{s}}{(1.674\times 10^{-27}\text{ kg})(102.6\times 10^{-9}\text{ m})} = \boxed{3.86\text{ m/s for the 103-nm photon}}.$$

(c) Strategy and Solution The transition from the $n = 3$ state to the ground state ($n = 1$) can occur two ways: 3 to 2, then 2 to 1 or 3 to 1. The transition from the $n = 4$ to the ground state can occur four ways: 4 to 3, then 3 to 2, then 2 to 1 or 4 to 3, then 3 to 1 or 4 to 2, then 2 to 1 or 4 to 1. Therefore, there are six ways total for the atom to return to the ground state: two ways when the 103-nm wavelength is absorbed and four ways when the 97.3-nm wavelength is absorbed.

77. Strategy and Solution When the potential difference accelerating electrons in the x-ray tube is doubled, the energies of the characteristic x-rays remain the same because they are characteristic of the target material's energy transitions.

81. (a) Strategy The momentum of the atom must equal the momentum of the incident photon.

Solution Find the recoil speed.

$$p_{\text{H}} = m_{\text{H}}v = p_{\text{photon}} = \frac{h}{\lambda}, \text{ so } v = \frac{h}{m_{\text{H}}\lambda} = \frac{6.626\times10^{-34}\text{ J}\cdot\text{s}}{(1.674\times10^{-27}\text{ kg})(97\times10^{-9}\text{ m})} = \boxed{4.1\text{ m/s}}.$$

(b) Strategy The energy of a photon of EM radiation with wavelength λ is $E = hc/\lambda$. Use Eq. (27-24).

Solution Compute the energy of the incident photon.

$$E = \frac{hc}{\lambda} = \frac{1240\text{ eV}\cdot\text{nm}}{97\text{ nm}} = 12.8\text{ eV}$$

This energy corresponds to a transition from the $n = 1$ state to the $n = 4$ state.

$$E = |E_1| - \frac{|E_1|}{4^2} = 13.6\text{ eV} - 0.850\text{ eV} = 12.8\text{ eV}$$

The transition from the $n = 4$ state back to the $n = 1$ state can occur in 4 ways emitting six different photons: $4 \to 1$, $4 \to 2 \to 1$, $4 \to 3 \to 1$, and $4 \to 3 \to 2 \to 1$.

(c) Strategy There are six photons possible. The classifications are UV for $\lambda < 400$ nm, visible for $400\text{ nm} \le \lambda \le 700\text{ nm}$, and IR for $\lambda > 700$ nm. The wavelengths are given by

$$\lambda = \frac{1}{R\left(\frac{1}{n_\text{f}^2} - \frac{1}{n_\text{i}^2}\right)} = \frac{1}{(1.097\times10^7\text{ m}^{-1})\left(\frac{1}{n_\text{f}^2} - \frac{1}{n_\text{i}^2}\right)}.$$

Solution The results are given in the table.

Transition	λ (nm)	Class
$4 \to 3$	1875	IR
$4 \to 2$	486	visible
$4 \to 1$	97	UV
$3 \to 2$	656	visible
$3 \to 1$	103	UV
$2 \to 1$	122	UV

85. Strategy The minimum wavelength corresponds to the maximum frequency, which occurs when the photon's energy equals the kinetic energy of the electron. Use Eq. (27-9).

Solution Find the minimum wavelength.

$$hf_{\text{max}} = \frac{hc}{\lambda_{\text{min}}} = K_\text{e}, \text{ so } \lambda_{\text{min}} = \frac{hc}{K_\text{e}} = \frac{1240\text{ eV}\cdot\text{nm}}{2.0\times10^3\text{ eV}} = \boxed{0.62\text{ nm}}.$$

89. **(a)** **Strategy** The *Balmer series* ($n_f = 2$) gives visible wavelengths. The wavelengths are given by

$$\lambda = \frac{1}{(1.097\times10^7\ \text{m}^{-1})\left(\frac{1}{2^2}-\frac{1}{n_i^2}\right)}.$$

Solution $n_i = 3$ gives 656.3 nm and $n_i = 4$ gives 486.2 nm, both of which are visible. So, the incident radiation excites the ground-state atoms into the $n = 4$ state. The energy difference between these states is equal to the energy of the incident radiation.

$$\Delta E = E_4 - E_1 = E_1\left(\frac{1}{4^2}-1\right) = (-13.6\ \text{eV})\left(\frac{1}{16}-1\right) = 12.75\ \text{eV}$$

Calculate the wavelength.

$$\lambda = \frac{hc}{E} = \frac{1240\ \text{eV}\cdot\text{nm}}{12.75\ \text{eV}} = \boxed{97.3\ \text{nm}}$$

(b) **Strategy** Since $\lambda \propto 1/E_{\text{photon}}$, the longest wavelength corresponds to the smallest energy for the incident radiation. So, the incident radiation must excite the ground-state atom to its $n = 3$ state for visible light to be emitted.

Solution Calculate the wavelength of the incident radiation.

$$\lambda = \frac{1}{R\left(1-\frac{1}{3^2}\right)} = \frac{1}{(1.097\times10^7\ \text{m}^{-1})\left(1-\frac{1}{9}\right)} = \boxed{102.6\ \text{nm}}$$

As found in part (a), the wavelength of the emitted radiation for the $n = 3$ to the $n = 2$ state is $\boxed{656.3\ \text{nm}}$.

(c) **Strategy** The incident photon must be energetic enough to excite an electron from any state to $n_f = \infty$. The case where the electron is in the ground state, $n_i = 1$, represents the minimum energy required, 13.6 eV.

Solution Find the range of wavelengths.

$$E \geq 13.6\ \text{eV}$$

$$\frac{hc}{\lambda} \geq 13.6\ \text{eV}$$

$$\lambda \leq \frac{1240\ \text{eV}\cdot\text{nm}}{13.6\ \text{eV}}$$

$$\boxed{\lambda \leq 91.2\ \text{nm}}$$

Chapter 28

QUANTUM PHYSICS

Problems

1. Strategy Find the de Broglie wavelength using Eq. (28-1). Compare the result to the diameter of the hoop.

Solution

$$\lambda = \frac{h}{p} = \frac{h}{mv} = \frac{6.626\times10^{-34}\text{ J}\cdot\text{s}}{(0.50\text{ kg})(10\text{ m/s})} = \boxed{1.3\times10^{-34}\text{ m}}$$

The wavelength is much too small compared to the diameter of the hoop for any appreciable diffraction to occur—for a diameter of ~1 m, it's a factor of 10^{-34} smaller!

5. Strategy The electron's kinetic energy is small compared to its rest energy, so the electron is nonrelativistic and we can use $p = mv$ and $K = \frac{1}{2}mv^2$ to find the momentum. Find the de Broglie wavelength using Eq. (28-1) and the wavelength of the photon using $E = hc/\lambda$.

Solution First solve for p in terms of K.

$$K = \frac{1}{2}mv^2, \text{ so } v = \sqrt{\frac{2K}{m}}.$$

The momentum is then $p = mv = m\sqrt{\frac{2K}{m}} = \sqrt{2Km}.$

Find the de Broglie wavelength of the electron.

$$\lambda_e = \frac{h}{p} = \frac{h}{\sqrt{2Km_e}}$$

The wavelength of a photon is $\lambda_p = \frac{hc}{E_p}$. Compute the ratio of the wavelengths.

$$\frac{\lambda_p}{\lambda_e} = \frac{hc/E_p}{h/\sqrt{2Km}} = \frac{c\sqrt{2Km}}{E_p} = \frac{(3.00\times10^8\text{ m/s})\sqrt{2(0.100\times10^3\text{ eV})(1.602\times10^{-19}\text{ J/eV})(9.109\times10^{-31}\text{ kg})}}{(0.100\times10^3\text{ eV})(1.602\times10^{-19}\text{ J/eV})} = \boxed{101}$$

7. Strategy The electron is relativistic, so use $p = \gamma mv$ to find the magnitude of its momentum. Find the de Broglie wavelength using Eq. (28-1).

Solution

$$\lambda = \frac{h}{p} = \frac{h}{\gamma mv} = \frac{h\sqrt{1-v^2/c^2}}{mv} = \frac{(6.626\times10^{-34}\text{ J}\cdot\text{s})\sqrt{1-(3/5)^2}}{(9.109\times10^{-31}\text{ kg})(\frac{3}{5}\times3.00\times10^8\text{ m/s})} = \boxed{3.23\text{ pm}}$$

9. Strategy Assume that the electrons are nonrelativistic. To produce the same diffraction pattern, the electrons must have the same wavelength as the x-rays. Find the de Broglie wavelength using Eq. (28-1) and the wavelength of the x-rays using $E = hc/\lambda$.

Solution Equate the wavelength of an electron and the wavelength of an x-ray photon.

$$\lambda_e = \frac{h}{p_e} \text{ and } \lambda_p = \frac{hc}{E_p}, \text{ so } \frac{h}{p_e} = \frac{hc}{E_p}.$$

Now substitute $p_e = m_e v$ and solve for the speed.

$$\frac{h}{m_e v} = \frac{hc}{E_p}, \text{ so } v = \frac{E_p}{m_e c}.$$

Find the kinetic energy.

$$K = \frac{1}{2}m_e v^2 = \frac{1}{2}m_e \frac{E_p{}^2}{m_e{}^2 c^2} = \frac{E_p{}^2}{2m_e c^2} = \frac{(16\times10^3 \text{ eV})^2}{2(511\times10^3 \text{ eV})} = \boxed{250 \text{ eV}}$$

The kinetic energy is small compared to the rest energy of an election, so the use of the nonrelativistic equations for momentum and kinetic energy was valid.

13. (a) Strategy The energy of a photon is given by $E = hc/\lambda$. The minimum energy corresponds to the maximum wavelength of a photon that is capable of resolving the details of the molecule.

Solution Find the minimum photon energy.

$$E = \frac{hc}{\lambda} = \frac{1240 \text{ eV}\cdot\text{nm}}{0.1000 \text{ nm}} = \boxed{12.4 \text{ keV}}$$

(b) Strategy The minimum kinetic energy corresponds to the maximum wavelength of an electron that is capable of resolving the details of the molecule. Assume that the electrons are nonrelativistic. Use the de Broglie wavelength, $K = \frac{1}{2}mv^2$, and $p = mv$.

Solution Find the kinetic energy of the electrons.

$$K = \frac{1}{2}mv^2 = \frac{(mv)^2}{2m} = \frac{p^2}{2m} = \frac{(h/\lambda)^2}{2m} = \frac{h^2}{2m\lambda^2} = \frac{(6.626\times10^{-34} \text{ J}\cdot\text{s})^2}{2(9.109\times10^{-31} \text{ kg})(1.000\times10^{-10} \text{ m})^2} \times \frac{1}{1.602\times10^{-19} \text{ J/eV}}$$

$$= \boxed{150 \text{ eV}}$$

The kinetic energy is small compared to the rest energy of an electron, so the use of the nonrelativistic equations for momentum and kinetic energy was valid.

(c) Strategy Use conservation of energy.

Solution Find the required potential difference.

$$\Delta U = -e\Delta V = -\Delta K = -K, \text{ so } \Delta V = \frac{K}{e} = \frac{150 \text{ eV}}{e} = \boxed{150 \text{ V}}.$$

17. Strategy Use the position-momentum uncertainty principle, Eq. (28-2).

Solution Find the uncertainty in the position of the basketball.

$$\Delta x \Delta p \geq \frac{1}{2}\hbar, \text{ so } \Delta x \geq \frac{\hbar}{2\Delta p} = \frac{\hbar}{2p\cdot10^{-6}} = \frac{10^6 h}{4\pi mv} = \frac{10^6(6.626\times10^{-34} \text{ J}\cdot\text{s})}{4\pi(0.50 \text{ kg})(10 \text{ m/s})} = \boxed{1\times10^{-29} \text{ m}}.$$

21. Strategy The single-slit diffraction minima are given by $a\sin\theta = m\lambda$. The edge of the central fringe corresponds to $m = 1$. Since the width of the central fringe is small compared to the slit-screen distance, use small angle approximations. According to conservation of energy, the kinetic energy of the electrons is related to the potential difference by $K = eV$. Use the de Broglie wavelength with $p = \sqrt{2Km_e} = \sqrt{2eVm_e}$ for the electrons.

Solution Since $\theta \approx \tan\theta = x/D$ and $\theta \approx \sin\theta = \lambda/a$, we have $\lambda = ax/D$, where x is half the width of the central fringe and D is the distance from the slit to the screen. Find the width of the slit.

θ x D

$$\lambda = \frac{ax}{D} = \frac{h}{p} = \frac{h}{\sqrt{2eVm_e}}, \text{ so}$$

$$a = \frac{hD}{xp} = \frac{hD}{x\sqrt{2eVm_e}} = \frac{(6.626\times10^{-34}\text{ J}\cdot\text{s})(1.00\text{ m})}{\left(\frac{1.13\times10^{-3}\text{ m}}{2}\right)\sqrt{2(38.0\text{ eV})(1.602\times10^{-19}\text{ J/eV})(9.109\times10^{-31}\text{ kg})}} = \boxed{352\text{ nm}}.$$

23. Strategy Use the position-momentum uncertainty principle, Eq. (28-2), and $p = mv$.

Solution

(a) Estimate the minimum uncertainty in the position of the bullet.

$$\Delta x\Delta p_x \geq \frac{1}{2}\hbar, \text{ so } \Delta x \geq \frac{\hbar}{2\Delta p_x} = \frac{\hbar}{2m\Delta v_x}. \text{ Thus,}$$

$$(\Delta x)_{\text{min}} = \frac{\hbar}{2m\Delta v_x} = \frac{\hbar}{2m\left(\frac{\Delta v_x}{v_x}\right)v_x} = \frac{6.626\times10^{-34}\text{ J}\cdot\text{s}}{4\pi(10.000\times10^{-3}\text{ kg})(0.0004)(300.00\text{ m/s})} = \boxed{4\times10^{-32}\text{ m}}.$$

(b) $$(\Delta x)_{\text{min}} = \frac{6.626\times10^{-34}\text{ J}\cdot\text{s}}{4\pi(9.109\times10^{-31}\text{ kg})(0.0004)(300.00\text{ m/s})} = \boxed{0.5\text{ mm}}$$

(c) The uncertainty in the bullet's position is far too small to be observable. The uncertainty in the electron's position is huge on the atomic scale. Thus, the uncertainty principle can be neglected in the macroscopic world, but not on the atomic scale.

25. Strategy Use the energy-time uncertainty principle, Eq. (28-3).

Solution Find the uncertainty in the gamma-ray energies.

$$\Delta E\Delta t \geq \frac{1}{2}\hbar, \text{ so } \Delta E \approx \frac{\hbar}{2\Delta t} = \frac{1\times10^{-34}\text{ J}\cdot\text{s}}{2(1\times10^{-12}\text{ s})}\times\frac{1\text{ eV}}{1.602\times10^{-19}\text{ J}} = \boxed{3\times10^{-4}\text{ eV}}.$$

27. Strategy The kinetic energy of the electron is given by $K = p_n^2/(2m)$, where $p_n = nh/(2L)$. The minimum kinetic energy is the $n = 1$ state.

Solution Find the minimum kinetic energy.

$$K = \frac{p_n^2}{2m} = \frac{n^2h^2}{8mL^2} = \frac{1^2(6.626\times10^{-34}\text{ J}\cdot\text{s})^2}{8(9.109\times10^{-31}\text{ kg})(1.0\times10^{-15}\text{ m})^2}\times\frac{1\text{ eV}}{1.602\times10^{-19}\text{ J}} = \boxed{380\text{ GeV}}$$

29. (a) Strategy The momentum of the marble is given by $p_n = h/\lambda_n = nh/(2L)$.

Solution Find n.

$$p_n = \frac{nh}{2L}, \text{ so } n = \frac{2p_nL}{h} = \frac{2mvL}{h} = \frac{2(10\times10^{-3}\text{ kg})(0.02\text{ m/s})(0.10\text{ m})}{6.626\times10^{-34}\text{ J}\cdot\text{s}} = \boxed{6\times10^{28}}.$$

(b) Strategy Use Eqs. (28-8) and (28-9).

Solution Calculate the energy difference between the n and $n+1$ states.

$$\Delta E = E_{n+1} - E_n = \frac{(n+1)^2h^2}{8mL^2} - \frac{n^2h^2}{8mL^2} = \frac{h^2}{8mL^2}[(n+1)^2 - n^2] = \frac{(2n+1)h^2}{8mL^2} = \frac{(12\times10^{28}+1)(6.626\times10^{-34}\text{ J}\cdot\text{s})^2}{8(0.010\text{ kg})(0.10\text{ m})^2}$$

$$= 7\times10^{-35}\text{ J}$$

We cannot observe the quantization of the marble's energy because the energy difference between levels is too small to observe.

31. Strategy The energy of a photon is related to its wavelength by $E = hc/\lambda$. Use Eqs. (28-8), (28-9), and (28-10).

Solution

(a) Calculate the wavelength.

$$\lambda = \frac{hc}{E} = \frac{hc}{E_2 - E_1} = \frac{hc}{2^2E_1 - E_1} = \frac{hc}{3E_1} = \frac{1240\text{ eV}\cdot\text{nm}}{3(40.0\text{ eV})} = \boxed{10.3\text{ nm}}$$

(b) According to Eq. (28-8), $E \propto L^{-2}$, so doubling the length of the box would reduce the energy to one fourth as much as before.

33. Strategy The graphs are the same as those in Figure 28.10 with $L = 10$ cm. Use Eqs. (28-8), (28-9), and (28-10) to estimate the spacing between the energy levels.

Solution

(a) The wave functions for the four states:

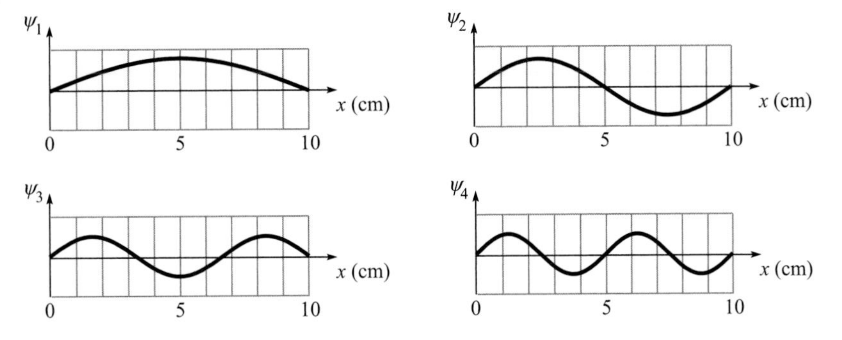

(b) Estimate the energy difference between the ground state and the first excited state.

$$\Delta E = E_2 - E_1 = n^2E_1 - E_1 = (2^2 - 1)E_1 = 3E_1 = \frac{3h^2}{8mL^2}, \text{ so}$$

$$\Delta E = \frac{3h^2}{8mL^2} = \frac{3(6.626\times10^{-34}\text{ J}\cdot\text{s})^2}{8(9.109\times10^{-31}\text{ kg})(0.10\text{ m})^2}\times\frac{1\text{ eV}}{1.602\times10^{-19}\text{ J}} = \boxed{1.1\times10^{-16}\text{ eV}}.$$

37. Strategy Use Table 28.1 and Eq. (28-13).

Solution Since $n = 3$, ℓ can be 0, 1, or 2, and m_ℓ can have values −2, −1, 0, 1, and 2 (for $\ell = 2$). $L_z = m_\ell \hbar$, so it can have values: $\boxed{-2\hbar, -\hbar, 0, \hbar, \text{ and } 2\hbar}$.

41. Strategy and Solution Since $n = 7$, ℓ can have values from 0 to 6. For each value of ℓ, there are $2(2\ell+1)$ states. So, $\boxed{\text{for } \ell = 0, 1, 2, 3, 4, 5, \text{ and } 6, \text{ there are } 2, 6, 10, 14, 18, 22, \text{ and } 26 \text{ electron states, respectively. The total is } 98}$.

43. Strategy Bromine (Br) has atomic number 35, so it has 35 electrons in a neutral atom. Bromine is not an exception to the subshell order, so subshells are filled in the order listed in Eq. (28-16).

Solution The ground-state electron configuration of bromine is $\boxed{1s^2 2s^2 2p^6 3s^2 3p^6 4s^2 3d^{10} 4p^5}$.

45. Strategy Lithium, sodium, and potassium have 3, 11, and 19 electrons in a neutral atom, respectively. None of these appear in the list of exceptions to the subshell order, so subshells are filled in the order listed in Eq. (28-16). The Periodic Table of the elements is arranged in columns by electron configuration.

Solution

(a) The ground-state electron configurations are the following:
$\boxed{\text{Li: } 1s^2 2s^1;\ \text{Na: } 1s^2 2s^2 2p^6 3s^1;\ \text{K: } 1s^2 2s^2 2p^6 3s^2 3p^6 4s^1}$.

(b) All three neutral atoms have valence +1. Their outermost electron is in the $\boxed{s^1 \text{ subshell}}$. This is why they are placed in the same column.

49. Strategy Since the energy of a photon is inversely proportional to its wavelength, 640 nm is the longest wavelength of photon that can supply an electron with enough energy to jump into the conduction band across the band gap; therefore, the band gap equals the energy of a 640-nm photon.

Solution Compute the band gap.

$$E_{\text{gap}} = \frac{hc}{\lambda} = \frac{1240 \text{ eV}\cdot\text{nm}}{640 \text{ nm}} = \boxed{1.9 \text{ eV}}$$

51. Strategy The wavelength of a photon is related to its energy by $E = hc/\lambda$.

Solution The wavelength for this transition is

$$\lambda = \frac{hc}{\Delta E} = \frac{1240 \text{ eV}\cdot\text{nm}}{20.66 \text{ eV} - 18.38 \text{ eV}} = \boxed{544 \text{ nm}}.$$

Light of wavelength 544 nm appears $\boxed{\text{green}}$ in color.

53. (a) Strategy The location of the first diffraction minimum for a beam passing through a circular aperture is given by $a \sin \Delta\theta = 1.22\lambda$. The wavelength is 694.3 nm.

Solution Find the angular spread of the beam.

$$\Delta\theta = \sin^{-1}\frac{1.22\lambda}{a} = \sin^{-1}\frac{1.22(694.3\times10^{-9} \text{ m})}{0.0050 \text{ m}} = \boxed{0.0097^\circ}$$

(b) Strategy If the diameter Δx of the spot on the Moon is approximately equal to the arc length of a circle of radius D subtended by $\Delta\theta$, then $\Delta x \approx D\Delta\theta$, where $\Delta\theta$ is now twice the angle found in part (a) and D is the Earth-Moon distance.

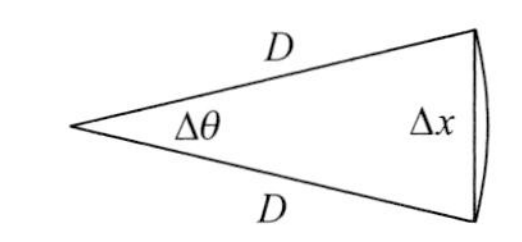

Solution Find the diameter of the spot on the Moon.

$$\Delta x \approx D\Delta\theta = (3.845\times10^8\text{ m})(2)\sin^{-1}\frac{1.22(694.3\times10^{-9}\text{ m})}{0.0050\text{ m}} = \boxed{130\text{ km}}$$

57. Strategy The kinetic energy of the electrons as they reach the slits is equal to $e\Delta V$ due to the 15-kV potential. Use the relationship between momentum and kinetic energy to find the momentum of the electrons at the slits. Then, find the de Broglie wavelength of the electrons and use it and the double-slit interference maxima equation to find the distance between the slits.

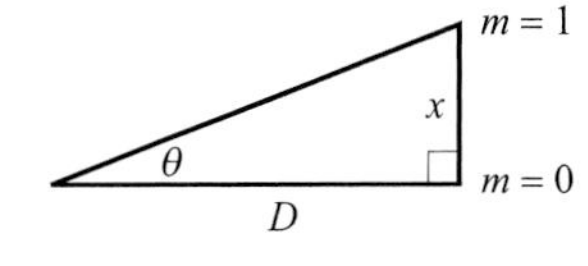

Solution Find the momentum.

$$K = \frac{1}{2}mv^2 = \frac{(mv)^2}{2m} = \frac{p^2}{2m} = e\Delta V, \text{ so } p = \sqrt{2me\Delta V}.$$

Find the de Broglie wavelength.

$$\lambda = \frac{h}{p} = \frac{h}{\sqrt{2me\Delta V}}$$

Find the slit separation.

$$d\sin\theta = m\lambda = (1)\frac{h}{\sqrt{2me\Delta V}}, \text{ so } d = \frac{h}{\sin\theta\sqrt{2me\Delta V}}.$$

If $D = 2.5$ m and $x = 0.0083$ m, then $\sin\theta = \dfrac{x}{\sqrt{x^2+D^2}}$, and

$$d = \frac{h\sqrt{x^2+D^2}}{x\sqrt{2me\Delta V}} = \frac{6.626\times10^{-34}\text{ J}\cdot\text{s}}{0.0083\text{ m}}\sqrt{\frac{(0.0083\text{ m})^2+(2.5\text{ m})^2}{2(9.109\times10^{-31}\text{ kg})(1.602\times10^{-19}\text{ C})(15\times10^3\text{ V})}} = \boxed{3.0\text{ nm}}.$$

61. (a) Strategy The electrons have kinetic energy of 54.0 eV due to the accelerating potential. Since the kinetic energy is so much smaller than the rest energy of the electron, use the nonrelativistic forms of the momentum and kinetic energy. Find the de Broglie wavelength using Eq. (28-1).

Solution Find the momentum.

$$K = \frac{p^2}{2m}, \text{ so } p = \sqrt{2mK}.$$

Calculate the de Broglie wavelength.

$$\lambda = \frac{h}{p} = \frac{h}{\sqrt{2mK}} = \frac{6.626\times10^{-34}\text{ J}\cdot\text{s}}{\sqrt{2(9.109\times10^{-31}\text{ kg})(54.0\text{ eV})(1.602\times10^{-19}\text{ J/eV})}} = \boxed{167\text{ pm}}$$

(b) Strategy Use Bragg's law.

Solution Find the Bragg angle for $m = 1$.

$$2d\sin\theta = m\lambda = (1)\lambda, \text{ so } \theta = \sin^{-1}\frac{\lambda}{2d} = \sin^{-1}\frac{1.669\times10^{-10}\text{ m}}{2(0.091\times10^{-9}\text{ m})} = \boxed{66.5°}.$$

(c) Strategy Make a sketch to show the relationship between the Bragg angle and the scattering angle.

Solution In the figure to the right, $\theta_i = \theta_r$ and $\theta_{Bragg} = \theta_r + \alpha$; but $\theta_i = \alpha$, since they are opposite angles, so $\theta_{Bragg} = \theta_r + \theta_i = 2\theta_i$, or $\theta_{Bragg} = 2(66.5°) = 133°$, which equals 130° to two significant figures. $\boxed{\text{Yes}}$, the results agree.

65. Strategy Use Figure 28.10 as a guide to make a sketch of the wave function for an electron in the $n = 4$ state in a one-dimensional box. Use Eqs. (28-8) and (28-9) to find the energy of the state. Use Figure 28.13 as a guide to make a sketch of the wave function for the electron in the $n = 4$ state of a *finite* box. Compare the energies by comparing the wave functions. Estimate the number of bound states for the *finite* box by using the energy states for an *infinite* box.

Solution

(a) The figure shows the wave function for the third excited state of an electron confined to a one-dimensional box of length L.

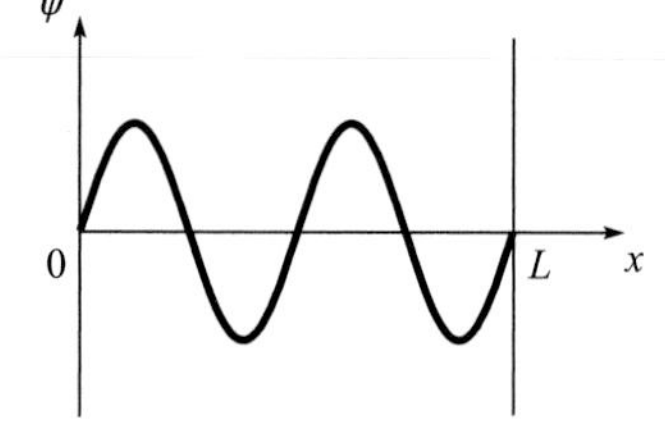

(b) Calculate the energy for the third excited state ($n = 4$).

$$E_4 = \frac{4^2 h^2}{8mL^2} = \boxed{\frac{2h^2}{mL^2}}$$

(c) The figure shows the wave function for the third excited state of the electron in a finite box of length L.

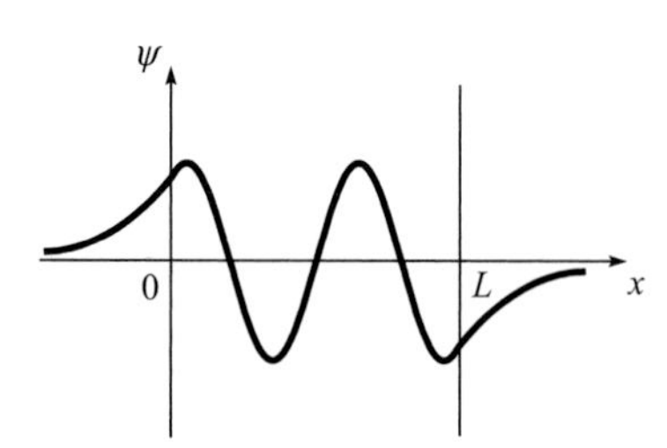

(d) The wavelengths for ψ inside the box are longer for the finite box than for the infinite box, since the wave functions extend past the walls of the finite box. Longer wavelengths correspond to smaller energies, so the energy of the third excited state for the finite box is $\boxed{\text{less than}}$ the energy found in part (b).

(e) Set the energy of the highest bound state equal to U_0 and solve for n.

$$U_0 = \frac{n^2 h^2}{8mL^2}, \text{ so } n^2 = \frac{8U_0 mL^2}{h^2} \text{ or } n = \boxed{\frac{2L}{h}\sqrt{2mU_0}}.$$

69. Strategy The wavelength of the electrons must equal the wavelength of the photons to give the same diffraction pattern. The wavelength of a photon is related to its energy by $E = hc/\lambda$. The wavelength of the electrons is given by the de Broglie wavelength with $p = \sqrt{2Km}$, since the electrons are nonrelativistic $[K = p^2/(2m)]$.

Solution Find the kinetic energy of the electrons.

$$\lambda_{\text{ph}} = \frac{hc}{E} = \lambda_{\text{e}} = \frac{h}{p} = \frac{h}{\sqrt{2Km}}, \text{ so } 2Km = \frac{E^2}{c^2} \text{ or } K = \frac{E^2}{2mc^2} = \frac{(2.0 \text{ eV})^2}{2(511\times10^3 \text{ eV})} = \boxed{3.9\times10^{-6} \text{ eV}}.$$

73. Strategy Use the equation for the ground-state energy, Eq. (28-8).

Solution Calculate the ratio of the ground states.

$$\frac{E_{2L}}{E_L} = \frac{\frac{h^2}{8m(2L)^2}}{\frac{h^2}{8mL^2}} = \boxed{\frac{1}{4}}$$

77. Strategy Use the energy-time uncertainty principle, Eq. (28-3). The rest energy is given by $E_0 = mc^2$.

Solution Calculate the minimum uncertainty in the energy.

$$\Delta E \Delta t \geq \frac{1}{2}\hbar, \text{ so } (\Delta E)_{\text{min}} = \frac{\hbar}{2\Delta t} = \frac{6.626\times10^{-34} \text{ J}\cdot\text{s}}{4\pi(15\times60 \text{ s})} = 5.86\times10^{-38} \text{ J}.$$

Find the inherent uncertainty in the mass of a free neutron.

$$\Delta m = \frac{\Delta E_{\text{min}}}{c^2} = \frac{5.86\times10^{-38} \text{ J}}{(3.00\times10^8 \text{ m/s})^2} = \boxed{6.5\times10^{-55} \text{ kg}}$$

Compare to the average neutron mass.

$$\frac{6.5\times10^{-55} \text{ kg}}{1.67\times10^{-27} \text{ kg}} = 3.9\times10^{-28}, \text{ so } \boxed{\frac{\Delta m}{m} = 3.9\times10^{-28}}.$$

Chapter 29

NUCLEAR PHYSICS

Problems

1. **Strategy** Nucleons have masses of approximately 1 u, so dividing the mass of the person by 1 u will give an estimate of the total number of nucleons in the person's body.

   Solution Estimate the number of nucleons.

   $$\frac{75\text{ kg}}{1\text{ u}} = \frac{75\text{ kg}}{1.660\,539\times10^{-27}\text{ kg}} = \boxed{4.5\times10^{28}}$$

3. **Strategy** For a spherical object, the density equation is $\rho = M/V = M/(\frac{4}{3}\pi r^3)$.

   Solution Solving for r and using $M = M_{\text{Sun}} = 1.99\times10^{30}$ kg and $\rho = 2.3\times10^{17}\text{ kg/m}^3$ (the density of nuclear matter found in Problem 2) yields the result.

   $$r^3 = \frac{M}{\frac{4}{3}\pi\rho}, \text{ so } r = \left(\frac{M}{\frac{4}{3}\pi\rho}\right)^{1/3} = \left[\frac{1.99\times10^{30}\text{ kg}}{\frac{4}{3}\pi(2.3\times10^{17}\text{ kg/m}^3)}\right]^{1/3} = \boxed{13\text{ km}}.$$

5. **Strategy** The nucleon number A is the sum of the total number of protons Z and neutrons N. Use the Periodic Table of the elements to find the number of protons.

   Solution Find the number of protons.
   Potassium has atomic number 19, so $Z = 19$.
   $N = \#$ of neutrons $= 21$
   Find the nucleon number.
   $A = Z + N = 19 + 21 = 40$
   So, the symbol is $\boxed{{}^{40}_{19}\text{K}}$.

7. **Strategy** Use the Periodic Table of the elements to find the number of protons.

   Solution Xe has atomic number 54, so the number of protons is $\boxed{54}$.

9. **Strategy** $A = 107$ for technetium-107. Find the radius of the nucleus using Eq. (29-4). Use the volume of a sphere to find the volume of the nucleus.

   Solution Find the radius.
   $r = r_0 A^{1/3} = (1.2\text{ fm})(107)^{1/3} = \boxed{5.7\text{ fm}}$
   Find the volume.
   $$V = \frac{4}{3}\pi r^3 = \frac{4}{3}\pi r_0^3 A = \frac{4}{3}\pi(1.2\times10^{-15}\text{ m})^3(107) = \boxed{7.7\times10^{-43}\text{ m}^3}$$

11. Strategy A deuteron has 1 proton and 1 neutron. Use Eqs. (29-7) and (29-8) to find the mass defect and binding energy.

Solution Find the mass defect.
$\Delta m = 1\times 1.007\ 276\ 5\text{ u} + 1\times 1.008\ 664\ 9\text{ u} - 2.013\ 553\text{ u} = 0.002\ 388\text{ u}$

The binding energy is $E_\text{B} = (\Delta m)c^2 = 0.002\ 3884\text{ u}\times 931.494\text{ MeV/u} = \boxed{2.225\text{ MeV}}$.

13. Strategy The binding energy of a nucleus is equal to the total energy of *Z* protons and *N* neutrons less the total energy of the nucleus. The binding energy per nucleon is the binding energy divided by the total number of nucleons in the nucleus. The mass of the atom is 39.962 383 1 u. The mass of 18 electrons must be subtracted from the mass of the atom to obtain the mass of the nucleus. Use Eq. (29-8).

Solution There are $Z = 18$ protons and $N = A - Z = 40 - 18 = 22$ neutrons in the nucleus of argon-40. Find the binding energy per nucleon.

$$\frac{E_\text{B}}{A} = \frac{(\Delta m)c^2}{A} = \frac{[18\times 1.007\ 276\ 5\text{ u} + 22\times 1.008\ 664\ 9\text{ u} - (39.962\ 383\ 1\text{ u} - 18\times 0.000\ 548\ 6\text{ u})]\times 931.494\text{ MeV/u}}{40}$$
$$= \boxed{8.595\ 28\text{ MeV/nucleon}}$$

15. Strategy The nucleon number *A* is the sum of the total number of protons *Z* and neutrons *N*. Use Eqs. (29-7) and (29-8) to find the mass defect and binding energy. The binding energy per nucleon is the binding energy divided by the total number of nucleons in the nucleus.

Solution The $^{31}_{15}\text{P}$ atom has 15 protons and 16 neutrons. Its mass is 30.973 761 5 u. Find the mass defect.

$$\Delta m = (\text{mass of 15 } ^1\text{H atoms and 16 neutrons}) - (\text{mass of } ^{31}_{15}\text{P atom})$$
$$= 15\times 1.007\ 825\ 0\text{ u} + 16\times 1.008\ 664\ 9\text{ u} - 30.973\ 761\ 5\text{ u} = 0.282\ 251\ 9\text{ u}$$

Calculate the average binding energy per nucleon.

$$\frac{E_\text{B}}{A} = \frac{(\Delta m)c^2}{A} = \frac{0.282\ 251\ 9\text{ u}\times 931.494\text{ MeV/u}}{31\text{ nucleons}} = \boxed{8.48116\text{ MeV/nucleon}}$$

17. Strategy The nucleon number *A* is the sum of the total number of protons *Z* and neutrons *N*. Use Eq. (29-7) to find the mass defect.

Solution The ^{14}N atom has 7 protons, 7 neutrons, 7 electrons, and a mass of 14.003 074 0 u. Find the mass defect.

$$\Delta m = (\text{mass of 7 } ^1\text{H atoms} + \text{mass of 7 neutrons}) - (\text{mass of } ^{14}\text{N atom})$$
$$= 7\times 1.007\ 825\ 0\text{ u} + 7\times 1.008\ 664\ 9\text{ u} - 14.003\ 074\ 0\text{ u} = \boxed{0.112\ 355\ 3\text{ u}}$$

21. Strategy In beta-minus decay, the atomic number *Z* increases by 1 while the mass number *A* remains constant. Use Eq. (29-11).

Solution

For the parent $\left(^{40}_{19}\text{K}\right)$ $Z = 19$, so the daughter nuclide will have $Z = 19 + 1 = 20$, which is the element Ca. The symbol for the daughter is $\boxed{^{40}_{20}\text{Ca}}$.

23. Strategy In electron-capture decay, the atomic number Z is decreased by 1 while the mass number A stays the same.

Solution In this case, the parent nuclide has $Z = 11$ and $A = 22$, so the daughter nuclide will have $Z = 10$ and $A = 22$, which is the element neon. Write out the reaction.

${}^{A}_{Z}P + {}^{0}_{-1}\text{e} \to {}_{Z-1}^{A}D + {}^{0}_{0}\nu$, so $\boxed{{}^{22}_{11}\text{Na} + {}^{0}_{-1}\text{e} \to {}^{22}_{10}\text{Ne} + {}^{0}_{0}\nu}$. The daughter nuclide is $\boxed{{}^{22}_{10}\text{Ne}}$.

25. Strategy The kinetic energy of the decay products equals the energy associated with the change in mass.

Solution Find the change in mass during the decay using atomic masses.

$\Delta m = (\text{mass of } {}^{222}_{86}\text{Rn} + \text{mass of } {}^{4}_{2}\text{He}) - (\text{mass of } {}^{226}_{88}\text{Ra}) = (222.017\,570\,5\text{ u} + 4.002\,603\,2\text{ u}) - 226.025\,402\,6\text{ u}$
$= -0.005\,228\,9\text{ u}$

Find the kinetic energy.

$K = E = |\Delta m|c^2 = 0.005\,228\,9\text{ u} \times 931.494\text{ MeV/u} = 4.8707\text{ MeV}$

Assuming the ${}^{222}_{86}\text{Rn}$ nucleus takes away an insignificant fraction of the kinetic energy, the alpha particle's kinetic energy will be $\boxed{4.8707\text{ MeV}}$.

29. Strategy In Problem 21, the daughter nuclide in this decay was found to be ${}^{40}_{20}\text{Ca}$. The maximum kinetic energy of the beta particle is equal to the disintegration energy.

Solution The reaction is

${}^{40}_{19}\text{K} \to {}^{40}_{20}\text{Ca} + {}^{0}_{-1}\text{e} + {}^{0}_{0}\bar{\nu}$

The atomic masses of ${}^{40}_{19}\text{K}$ and ${}^{40}_{20}\text{Ca}$ are 39.963 998 7 u and 39.962 591 2 u, respectively. To get the masses of the nuclei, we subtract Zm_e from each. The mass of the electron is 0.000 548 6 u, and the neutrino's mass is negligible. The mass difference is

$\Delta m = [(M_{\text{Ca}} - 20m_e) + m_e] - (M_{\text{K}} - 19m_e) = M_{\text{Ca}} - M_{\text{K}} = 39.962\,591\,2\text{ u} - 39.963\,998\,7\text{ u} = -0.001\,407\,5\text{ u}.$

The disintegration energy is $E = |\Delta m|c^2 = 0.001\,407\,5\text{ u} \times 931.494\text{ MeV/u} = 1.3111\text{ MeV}.$

The maximum kinetic energy of the β^- particle is $\boxed{1.3111\text{ MeV}}$.

33. Strategy The activity is reduced by a factor of two for each half-life. Use Eqs. (29-18), (29-20), and (29-22) to find the initial number of nuclei and the probability of decay per second.

Solution

(a) Find the number of half-lives.

$$600.0\text{ s} = \frac{600.0\text{ s}}{200.0\text{ s/half-life}} = 3.000\text{ half-lives}$$

The activity after 3.000 half-lives will be $R = \left(\frac{1}{2}\right)^{3.000} \times R_0 = \frac{1}{8.000} \times 80{,}000.0\text{ s}^{-1} = \boxed{10{,}000\text{ s}^{-1}}$.

(b) Find the initial number of nuclei.

$$N_0 = \frac{1}{\lambda}R_0 = \tau R_0 = \frac{R_0 T_{1/2}}{\ln 2} = 80{,}000.0\text{ s}^{-1} \times \frac{200.0\text{ s}}{\ln 2} = \boxed{2.308\times 10^7}$$

(c) The probability per second is $\lambda = \frac{1}{\tau} = \frac{\ln 2}{T_{1/2}} = \frac{\ln 2}{200.0\text{ s}} = \boxed{3.466\times 10^{-3}\text{ s}^{-1}}$.

35. Strategy Use Eq. (29-22) to find the time constant. The number of nuclei is related to the mass by $N = mN_A/M$, where N_A is Avogadro's number and M is the molar mass. The activity is equal to the number of nuclei divided by the time constant.

Solution Convert the half-life of uranium-238 to seconds.

$$4.468\times10^9 \text{ yr}\times3.156\times10^7 \text{ s/yr} = 1.410\times10^{17} \text{ s}$$

The time constant is $\tau = \frac{T_{1/2}}{\ln 2} = \frac{1.410\times10^{17} \text{ s}}{\ln 2} = 2.034\times10^{17}$ s.

Find the number of nuclei in 1.0 kg of ^{238}U.

$$N = \frac{mN_A}{M} = \frac{1.0\times10^3 \text{ g}}{238.051 \text{ g/mol}}\times6.022\times10^{23} \text{ nuclei/mol} = 2.5\times10^{24} \text{ nuclei}$$

The activity is $R = \frac{N}{\tau} = \frac{2.5\times10^{24} \text{ nuclei}}{2.034\times10^{17} \text{ s}} = \boxed{1.2\times10^7 \text{ Bq}}$.

37. Strategy The activity as a function of time is given by $R = R_0e^{-t/\tau}$. Use Eq. (29-22) to find the time constant. Assume that the original activity was 0.25 Bq per g of carbon.

Solution Find the age of the bones.

$$e^{-t/\tau} = \frac{R}{R_0}, \text{ so } -\frac{t}{\tau} = \ln\frac{R}{R_0} \text{ or } t = -\tau\ln\frac{R}{R_0} = -\frac{5730 \text{ yr}}{\ln 2}\times\ln\frac{0.242}{0.25} = \boxed{270 \text{ yr}}.$$

39. Strategy The ratio of C-14 to C-12 in the bone is 1/4 as much as in a living sample. The ratio is reduced by a factor of 1/2 for each half-life.

Solution Since $2^{-2} = 1/4$, we conclude that the age of the bone is 2 half-lives, or $2\times5730 \text{ yr} = \boxed{11{,}500 \text{ yr}}$.

41. Strategy The activity is given by $R = R_0e^{-t/\tau} = R_0e^{-t\ln 2/T_{1/2}}$.

Solution Solve for the half-life.

$$e^{-t\ln 2/T_{1/2}} = \frac{R}{R_0}$$

$$-\frac{t\ln 2}{T_{1/2}} = \ln\frac{R}{R_0}$$

$$T_{1/2} = \frac{-t\ln 2}{\ln\frac{R}{R_0}} = -\frac{12 \text{ min}\times\ln 2}{\ln\frac{2.0\times10^3 \text{ Bq}}{6.4\times10^4 \text{ Bq}}} = \boxed{2.4 \text{ min}}$$

45. Strategy The number of ionized molecules is equal to the kinetic energy of the alpha particle divided by the ionization energy per molecule.

Solution

$$\text{\# of ionized molecules} = \frac{6\times10^6 \text{ eV}}{20 \text{ eV/molecule}} = \boxed{3\times10^5 \text{ molecules}}$$

49. Strategy Write out the reaction using variables for the unknown quantities. Balance the reaction to find the unknowns. The total charge and total number of nucleons must remain the same. The emission of an electron is beta-minus decay.

Solution

(a) The reaction is

$${}_0^1\text{n} + {}_b^a(?_1) \to {}_d^c(?_2) \to {}_f^e(?_3) + {}_{-1}^{0}\text{e}$$

$${}_f^e(?_3) \to {}_2^4\alpha + {}_2^4\alpha$$

Working backward, $e = 4+4 = 8$ and $f = 2+2 = 4$, so $(?_3) = \boxed{{}_4^8\text{Be}}$. Next, $c = e = 8$ and $d = f - 1 = 4 - 1 = 3$, so $(?_2) = \boxed{{}_3^8\text{Li}}$. Finally, $a + 1 = c = 8$, so $a = 7$ and $b = d = 3$, which means $(?_1) = \boxed{{}_3^7\text{Li}}$.

(b) $\boxed{\text{Yes; the emission of an electron (beta-minus decay) is accompanied by the emission of one antineutrino.}}$

53. (a) Strategy The mass numbers on the two sides of the reaction must be equal. Let x be the number of neutrons.

Solution Find the number of neutrons.

$235 + 1 = 141 + 93 + x$, so $x = \boxed{2}$.

(b) Strategy From Figure 29.2, the binding energies per nucleon of ${}^{235}\text{U}$, ${}^{141}\text{Cs}$, and ${}^{93}\text{Rb}$ are approximately 7.6 MeV, 8.35 MeV, and 8.7 MeV, respectively. The energy released is equal to the increase in the binding energy.

Solution The binding energies of the three nuclides are as follows:

${}^{235}\text{U}$: $235 \times 7.6\ \text{MeV} = 1786\ \text{MeV}$, ${}^{141}\text{Cs}$: $141 \times 8.35\ \text{MeV} = 1177\ \text{MeV}$, ${}^{93}\text{Rb}$: $93 \times 8.7\ \text{MeV} = 809\ \text{MeV}$.

Find the binding energy.

$1177\ \text{MeV} + 809\ \text{MeV} - 1786\ \text{MeV} = 200\ \text{MeV}$

The energy released is approximately $\boxed{200\ \text{MeV}}$.

(c) Strategy The atomic masses of ${}_{92}^{235}\text{U}$, ${}_{55}^{141}\text{Cs}$, and ${}_{37}^{93}\text{Rb}$ are 235.043 923 1 u, 140.920 044 0 u, and 92.922 032 8 u, respectively. Atomic masses can be used, since both sides include the same number of electrons (92).

Solution Find the change in mass.

$\Delta m = 140.920\,044\,0\ \text{u} + 92.922\,032\,8\ \text{u} + 2 \times 1.008\,664\,9\ \text{u} - 235.043\,923\,1\ \text{u} - 1.008\,664\,9\ \text{u} = -0.193\,181\,4\ \text{u}$

The energy released is $E = |\Delta m| c^2 = 0.193\,181\,4\ \text{u} \times 931.494\ \text{MeV/u} = \boxed{179.947\ \text{MeV}}$.

(d) Strategy Divide the energy released by the rest energy.

Solution

$$\frac{|\Delta E|}{E_0} = \frac{|\Delta m|}{m} = \frac{0.193\,181\,4\ \text{u}}{235.043\,923\,1\ \text{u}} \boxed{\approx 0.000\,822}$$

57. Strategy The total charge and total number of nucleons must remain the same. The energy released is equal to the difference between the binding energy of the reaction product and that of the deuteron. Compare the thermal energy to the coulomb repulsion.

Solution

(a) The atomic number of X must be 1 + 1 = 2, so X is helium. The mass number of X must be 1 + 2 = 3. The reaction product is $\boxed{{}^{3}_{2}\text{He}}$.

(b) The binding energies are as follows:
${}^{1}_{1}\text{H}$: 0, ${}^{2}_{1}\text{H}$: $2\times 1.1\text{ MeV} = 2.2\text{ MeV}$, ${}^{3}_{2}\text{He}$: $3\times 2.6\text{ MeV} = 7.8\text{ MeV}$.
Find the energy released.
$7.8\text{ MeV} - 2.2\text{ MeV} = \boxed{5.6\text{ MeV}}$

(c) At room temperature, the kinetic energies of the proton and the deuteron are much too small to overcome their Coulomb repulsion.

61. (a) Strategy Find the mass of a water molecule. Then, divide the mass of 1.00 L of water by the mass of one molecule.

Solution Find the mass of a water molecule.
$$m_{\text{H}_2\text{O}} = 2M_\text{H} + M_\text{O} = [2(1.007\,94\text{ u}) + 15.9994\text{ u}](1.6605\times 10^{-27}\text{ kg}) = 2.9914\times 10^{-26}\text{ kg}$$
The mass of 1.00 liter of water is 1.00 kg. Compute the approximate number of water molecules per liter of water.
$$\frac{1.00\text{ kg}}{2.9914\times 10^{-26}\text{ kg/molecule}} = \boxed{3.34\times 10^{25}\text{ molecules}}$$

(b) Strategy Use Eqs. (29-4) and (29-5) and the volume of a sphere to compute the volume of a water *nucleus*; that is, the volume of one oxygen nucleus and two hydrogen nuclei. Then, find the total volume of the nuclei by using the result from part (a). The volume of 1.00 liter of water is equal to $1.00\times 10^{-3}\text{ m}^3$.

Solution Compute the volume of one water *nucleus*.
$$V = \frac{4}{3}\pi r_\text{O}^3 + 2\times\frac{4}{3}\pi r_\text{H}^3 = \frac{4}{3}\pi(r_\text{O}^3 + 2r_\text{H}^3) = \frac{4}{3}\pi(r_0^3 A_\text{O} + 2r_0^3 A_\text{H}) = \frac{4}{3}\pi r_0^3(A_\text{O} + 2A_\text{H}) = \frac{4}{3}\pi r_0^3(16+2)$$
$$= 24\pi r_0^3$$
Find the fraction of the liter's volume occupied by the water *nuclei.*
$$\frac{24\pi r_0^3(3.34\times 10^{25})}{1.00\times 10^{-3}\text{ m}^3} = \frac{24\pi(1.2\times 10^{-15}\text{ m})^3(3.34\times 10^{25})}{1.00\times 10^{-3}\text{ m}^3} = \boxed{4.4\times 10^{-15}}$$

65. Strategy The activity is related to the number of nuclei N and the time constant τ by $R = N/\tau$. The number of nuclei is related to the mass by $N = mN_\text{A}/M$, where N_A is Avogadro's number and M is the molar mass. The time constant is related to the half-life by $T_{1/2} = \tau\ln 2$.

Solution Compute the mass of the sample of radon.
$$m = \frac{NM}{N_\text{A}} = \frac{R\tau M}{N_\text{A}} = \frac{RT_{1/2}M}{N_\text{A}\ln 2} = \frac{(2050\text{ Bq})(3.8235\text{ d})(86{,}400\text{ s/d})(222\times 10^{-3}\text{ kg/mol})}{(6.022\times 10^{23}\text{ mol}^{-1})\ln 2} = \boxed{3.60\times 10^{-16}\text{ kg}}$$

69. Strategy The original number of samarium-147 nuclei in the rock is equal to the current number plus the number of neodymium-143. The number of nuclei is related to the mass by $N = mN_A/M$, where N_A is Avogadro's number and M is the molar mass. Use Eq. (29-23).

Solution Find the age of the rock.

$$N = N_0 \times 2^{-t/T_{1/2}}$$

$$N_{Sm} = (N_{Sm} + N_{Nd}) \times 2^{-t/T_{1/2}}$$

$$\frac{N_{Sm}}{N_{Sm} + N_{Nd}} = 2^{-t/T_{1/2}}$$

$$\log_2 \frac{N_{Sm}}{N_{Sm} + N_{Nd}} = \log_2 2^{-t/T_{1/2}} = -\frac{t}{T_{1/2}}$$

$$t = T_{1/2} \log_2 \frac{N_{Sm} + N_{Nd}}{N_{Sm}}$$

$$t = T_{1/2} \log_2 \frac{m_{Sm}N_A/M_{Sm} + m_{Nd}N_A/M_{Nd}}{m_{Sm}N_A/M_{Sm}}$$

$$t = T_{1/2} \log_2 \left(1 + \frac{m_{Nd}M_{Sm}}{m_{Sm}M_{Nd}}\right) = (1.06 \times 10^{11}\text{ yr}) \log_2 \left[1 + \frac{(0.150)(146.914\,897\,9)}{(3.00)(142.909\,814\,3)}\right] = \boxed{7.67 \times 10^9\text{ yr}}$$

73. Strategy Compare the relative decay rates.

Solution The amount of tellurium-106 decreased by a factor of $3.00/4.00 = 0.750$ in the time $\Delta t = 25\ \mu\text{s}$. So, it will decrease again be this factor in the next $\Delta t = 25\ \mu\text{s}$. Therefore, at $t = 50\ \mu\text{s}$, there will be 0.750(3.00 mol) = 2.25 mol of tellurium-106 remaining. Notice that in the first time period, the entire amount of tellurium-106 that decayed into tin-102 remained. This is due to the fact that the half-life of tin-102 is much longer than $\Delta t = 25\ \mu\text{s}$. Essentially, not enough time has passed for an appreciable amount of tin-102 to decay. In the next time period, it is likely that all of the tellurium-106 that decays into tin-102 will still be tin-102. Since an additional 0.75 mol of tellurium-106 has decayed, there will be $2.50\text{ mol} + 0.75\text{ mol} = \boxed{3.25\text{ mol}}$ of tin-102 at $t = 50\ \mu\text{s}$.

77. (a) Strategy The total charge and total number of nucleons must remain the same. Use the Periodic Table of the elements to identify the element.

Solution The reaction is

$${}^4_2\alpha + {}^{14}_7\text{N} \rightarrow {}^1_1\text{p} + {}^a_b\text{X}.$$

$4 + 14 = 1 + a \Rightarrow a = 17$ and $2 + 7 = 1 + b \Rightarrow b = 8$.

Therefore, X must be ${}^{17}_8\text{O}$.

(b) Strategy Use Eq. (29-4) to find the radii of the nuclei. The sum of the radii is equal to the distance between the centers when the nuclei touch.

Solution The radii of ${}^4_2\alpha$ and ${}^{14}_7\text{N}$ are $1.2\text{ fm} \times 4^{1/3} = 1.9\text{ fm}$ and $1.2\text{ fm} \times 14^{1/3} = 2.9\text{ fm}$, respectively. The distance between their centers when they touch is $1.9\text{ fm} + 2.9\text{ fm} = \boxed{4.8\text{ fm}}$.

(c) Strategy The minimum kinetic energy required is equal to the electric potential energy when the two particles touch.

Solution

$$K_{\text{min}} = U_{\text{E}} = \frac{kq_\alpha q_{\text{N}}}{r} = \frac{k \times 2e \times 7e}{d} = \boxed{\frac{14ke^2}{d}}$$

(d) Strategy If the mass of the products is less than the mass of the reactants, then the kinetic energy will increase; if the reverse is true, it will decrease. Use atomic masses since the electrons cancel.

Solution Find the change in mass.
$\Delta m = 16.999\,131\,5 \text{ u} + 1.007\,825\,0 \text{ u} - 14.003\,074\,0 \text{ u} - 4.002\,603\,2 \text{ u} = 0.001\,279\,3 \text{ u}$

The mass of the products is greater than the mass of the reactants. The total kinetic energy of the products will be $\boxed{\text{less than}}$ the initial kinetic energy of the reactants by an amount

$E = (\Delta m)c^2 = 0.001\,279\,3 \text{ u} \times 931.494 \text{ MeV/u} = \boxed{1.1917 \text{ MeV}}.$

81. (a) Strategy Assume the particles are nonrelativistic and use the classical relationship between kinetic energy and speed.

Solution Find the speed of the alpha particles.

$$K = \frac{1}{2}mv^2, \text{ so } v = \sqrt{2K/m} = \sqrt{\frac{2 \times 4.17 \times 10^6 \text{ eV} \times 1.602 \times 10^{-19} \text{ J/eV}}{4.00 \text{ u} \times 1.66 \times 10^{-27} \text{ kg/u}}} = \boxed{1.42 \times 10^7 \text{ m/s}}.$$

$\gamma \approx 1.001,$ so using the nonrelativistic form for the kinetic energy was reasonable.

(b) Strategy To pass through undeflected, the net force on the alpha particles must be zero. The magnetic and electric forces must be equal in magnitude.

Solution Find the required magnitude of the electric field.

$F_{\text{M}} = qvB = F_{\text{E}} = qE, \text{ so } E = Bv = 0.30 \text{ T} \times 1.42 \times 10^7 \text{ m/s} = \boxed{4.3 \times 10^6 \text{ V/m}}.$

(c) Strategy Use Newton's second law and $F = qvB.$

Solution Find the radius.

$$\Sigma F = qvB = m\frac{v^2}{r}, \text{ so } r = \frac{mv}{Bq} = \frac{4.00 \text{ u} \times 1.66 \times 10^{-27} \text{ kg/u} \times 1.42 \times 10^7 \text{ m/s}}{0.30 \text{ T} \times 2 \times 1.602 \times 10^{-19} \text{ C}} = \boxed{98 \text{ cm}}.$$

(d) Strategy Use the result for the radius found in part (c).

Solution Find the charge-to-mass ratio.

$$r = \frac{mv}{Bq}, \text{ so } \frac{q}{m} = \frac{v}{Br}.$$

B, v, and r are the only quantities that can be measured in this experiment, so only the ratio q/m can be determined—$\boxed{\text{both } m \text{ and } q \text{ affect the radius of the trajectory}}$.

Chapter 30

PARTICLE PHYSICS

Problems

1. Strategy The difference in the rest energy of the particles before and after the decay is related to the kinetic energy of the particles by Einstein's mass-energy relation. Neglect the relatively small rest energy of the neutrino.

Solution Compute the change in mass.

$\Delta m = (\text{mass of muon}) - (\text{mass of pion}) = 0.106\ \text{GeV}/c^2 - 0.140\ \text{GeV}/c^2 = -0.034\ \text{GeV}/c^2$

The total kinetic energy of the two particles is $K = E = |\Delta m| c^2 = \boxed{34\ \text{MeV}}$.

5. Strategy The difference in the rest energy of the particles before and after the decay is related to the kinetic energy of the particles by Einstein's mass-energy relation. Conservation of momentum requires that the magnitude of the momenta of the two pions be the same. Since the pions have the same mass, their velocities must be equal and opposite; thus, their kinetic energies are the same.

Solution Compute the change in mass.

$\Delta m = 2(0.14\ \text{GeV}/c^2) - 2(0.938\ \text{GeV}/c^2) = -1.6\ \text{GeV}/c^2$

The energy released in the decay is $E = |\Delta m| c^2 = 1.6\ \text{GeV}$.

Find the kinetic energies of the two pions.

$$K_{\pi^-} = K_{\pi^+} = \frac{E}{2} = \frac{1.6\ \text{GeV}}{2} = \boxed{0.80\ \text{GeV}}$$

7. Strategy Replace each particle in the decay reaction with its corresponding antiparticle to write the two decay modes.

Solution Since π^+ is the antiparticle of π^-, the decay products of π^+ must be antiparticles of the decay products of π^-. The decay modes of π^+ are then $\boxed{\pi^+ \to \mu^+ + \nu_\mu \text{ and } \pi^+ \to e^+ + \nu_e}$.

9. Strategy The energy-time uncertainty principle is $\Delta E \Delta t \geq \hbar/2$. Use $\Delta E \Delta t = \hbar/2$ to estimate the range of the weak force carried by the Z particle.

Solution The range is approximately $c\Delta t$.

$\Delta E \Delta t = \frac{\hbar}{2}$, so $\Delta t = \frac{\hbar}{2\Delta E}$. The range is $\frac{\hbar c}{2\Delta E} = \frac{(1.055\times10^{-34}\ \text{J}\cdot\text{s})(3.00\times10^8\ \text{m/s})}{2(92\times10^9\ \text{eV})(1.602\times10^{-19}\ \text{J/eV})} = \boxed{1.1\times10^{-18}\ \text{m}}$, $\boxed{\text{which is approximately the same as } 1\times10^{-17}\ \text{m in Table 30.3}}$.

13. Strategy Choose the correct fundamental force for each decay. Use Tables 30.2 and 30.3.

Solution

(a) The decay products are a muon and a muon neutrino, which are leptons. Leptons are associated with the $\boxed{\text{weak}}$ force.

(b) The decay products are photons, which are associated with the $\boxed{\text{electromagnetic}}$ force.

(c) Two of the decay products are an electron and an electron antineutrino, which are leptons. Leptons are associated with the $\boxed{\text{weak}}$ force.

17. Strategy The rest energy of a proton is 938 MeV, so a proton with energy 1.0 TeV is extremely relativistic.

Solution Calculate the de Broglie wavelength.

$$E \approx pc \text{ and } p = \frac{h}{\lambda}, \text{ so } \lambda = \frac{hc}{E} = \frac{1240\times10^{-9}\text{ eV}\cdot\text{m}}{1.0\times10^{12}\text{ eV}} = \boxed{1.2\times10^{-18}\text{ m}}.$$

21. Strategy The proton has a charge of $+e$. The kinetic energy gained in each revolution is $K = eV = 2.5$ MeV. The number of revolutions required is equal to 1 TeV divided by the energy gained per revolution. The distance traveled is equal to the number of revolutions times the circumference of the ring.

Solution Find the number of revolutions.

$$N = \frac{1\text{ TeV}}{2.5\text{ MeV}} = \frac{1\times10^{12}}{2.5\times10^{6}} = \boxed{4\times10^{5}\text{ revolutions}}$$

Find the distance traveled.

$$(\text{circumference})\times(\#\text{ of revs}) = 2\pi rN = 2\pi\times1.0\text{ km}\times4\times10^{5} = \boxed{2.5\times10^{6}\text{ km}}$$

REVIEW AND SYNTHESIS: CHAPTERS 26–30

Review Exercises

1. Strategy Use Eq. (26-5) to find the velocity v_{pE} of the escape pod relative to Earth. Then, use Eq. (26-4) to find how long the pod appears to the people on Earth. Let the positive direction be toward the Earth.

Solution The velocity of the starship relative to Earth is $v_{\text{sE}} = 0.78c.$ The velocity of the pod relative to the starship is $v_{\text{ps}} = 0.63c.$ Find the velocity of the pod relative to Earth.

$$v_{\text{pE}} = \frac{v_{\text{ps}} + v_{\text{sE}}}{1 + v_{\text{ps}} v_{\text{sE}} / c^2} = \frac{0.63c + 0.78c}{1 + (0.63)(0.78)} = 0.945c$$

Find the length of the pod as it appears to the people on Earth.

$$L = \frac{L_0}{\gamma} = L_0 \sqrt{1 - v^2/c^2} = (12.0\text{ m})\sqrt{1 - 0.945^2} = \boxed{3.91\text{ m}}$$

5. Strategy Use Eq. (26-8).

Solution Find the speed of the electron.

$$K = (\gamma - 1)mc^2$$

$$\frac{K}{mc^2} = \gamma - 1$$

$$1 + \frac{K}{mc^2} = \frac{1}{\sqrt{1 - \frac{v^2}{c^2}}}$$

$$1 - \frac{v^2}{c^2} = \left(1 + \frac{K}{mc^2}\right)^{-2}$$

$$v = c\sqrt{1 - \left(1 + \frac{K}{mc^2}\right)^{-2}} = c\sqrt{1 - \left[1 + \frac{1.02 \times 10^{-13}\text{ J}}{(9.109 \times 10^{-31}\text{ kg})(2.998 \times 10^8\text{ m/s})^2}\right]^{-2}} = \boxed{0.895c = 2.68 \times 10^8\text{ m/s}}$$

9. Strategy The width of the central maximum is the distance between the first minimum on either side of the central maximum. Use conservation of energy to find the speed of the electrons. Use the de Broglie wavelength for the wavelength of the electrons.

Solution Find the wavelength of the electrons.

$$p = mv = \frac{h}{\lambda} \text{ and } K = \frac{1}{2}mv^2 = \frac{p^2}{2m} = e\Delta V, \text{ so } \lambda = \frac{h}{\sqrt{2me\Delta V}}.$$

The first minimum on either side of the central maximum is given by $a \sin\theta = \lambda.$ Solve for $\theta.$

$$\theta = \sin^{-1}\frac{\lambda}{a} = \sin^{-1}\frac{h}{a\sqrt{2me\Delta V}} = \sin^{-1}\frac{6.626 \times 10^{-34}\text{ J}\cdot\text{s}}{(6.6 \times 10^{-10}\text{ m})\sqrt{2(9.109 \times 10^{-31}\text{ kg})(1.602 \times 10^{-19}\text{ C})(8950\text{ V})}} = 1.12555°$$

If x is the distance to the first minimum on either side of the central maximum and D is the distance from the slit to the screen, then $\tan\theta = x/D.$ Find $2x$, the width of the central maximum.

$$2x = 2D\tan\theta = 2(2.50\text{ m})\tan 1.12555° = \boxed{9.8\text{ cm}}$$

13. Strategy The rest energy of the lambda particle not used to create the proton and the pion will become kinetic energy of the proton and the pion. Use conservation of momentum and energy.

Solution Find the energy not used to create the proton and the pion.
$K = 1116\text{ MeV} - (938\text{ MeV} + 139.6\text{ MeV}) = 38.4\text{ MeV}$

Let the magnitudes of the momenta of the proton and pion be p_p and p_π, respectively. Then, according to conservation of momentum, $p_\text{p} = p_\pi$. Let the kinetic energies of the proton and pion be K_p and K_π, respectively. Then, according to conservation of energy, $K = K_\text{p} + K_\pi$. According to Eq. (26-11) and conservation of momentum, $(p_\pi c)^2 = K_\pi{}^2 + 2K_\pi E_{\pi 0} = (p_\text{p} c)^2 = K_\text{p}{}^2 + 2K_\text{p} E_{\text{p}0}$, where $E_{\text{p}0}$ and $E_{\pi 0}$ are the rest energies of the proton and pion, respectively. Substituting for K_π in $(p_\text{p} c)^2 = K_\pi{}^2 + 2K_\pi E_{\pi 0}$ gives $(p_\text{p} c)^2 = (K - K_\text{p})^2 + 2(K - K_\text{p})E_{\pi 0}$.

Subtracting $(p_\text{p} c)^2 = K_\text{p}{}^2 + 2K_\text{p} E_{\text{p}0}$ from $(p_\text{p} c)^2 = (K - K_\text{p})^2 + 2(K - K_\text{p})E_{\pi 0}$ gives
$0 = K^2 - 2KK_\text{p} + 2KE_{\pi 0} - 2K_\text{p} E_{\pi 0} - 2K_\text{p} E_{\text{p}0}$.
Solving this equation for the kinetic energy of the proton K_p gives

$$K_\text{p} = \frac{K^2 + 2KE_{\pi 0}}{2(K + E_{\pi 0} + E_{\text{p}0})} = \frac{(38.4\text{ MeV})^2 + 2(38.4\text{ MeV})(139.6\text{ MeV})}{2(38.4\text{ MeV} + 139.6\text{ MeV} + 938\text{ MeV})} = \boxed{5.5\text{ MeV}}.$$

Therefore, the kinetic energy of the pion is $K_\pi = K - K_\text{p} = 38.4\text{ MeV} - 5.46\text{ MeV} = \boxed{33\text{ MeV}}$.

17. (a) Strategy Use Eqs. (29-21) and (29-22).

Solution Find the activity of the sample of gold-198 after 8.10 days.
$R = R_0 e^{-t/\tau} = R_0 e^{-t \ln 2/T_{1/2}} = (1.00\times 10^{10}\text{ Bq})e^{-8.10 \ln 2/2.70} = \boxed{1.25\times 10^9\text{ Bq}}$

(b) Strategy The atomic number *Z* increases by 1, while the mass number *A* stays the same, so the isotope undergoes beta-minus decay.

Solution Since the isotope undergoes beta-minus decay, the particles emitted in this process are an electron and an antineutrino.

21. Strategy Use Eq. (28-2).

Solution Find the order of magnitude of the minimum uncertainty in the momentum of the electron.

$$\Delta x \Delta p_x \ge \frac{\hbar}{2}, \text{ so } (\Delta p_x)_\text{min} = \frac{\hbar}{2\Delta x} \sim \frac{10^{-34}\text{ J}\cdot\text{s}}{(10^{-11}\text{ m})(10^{-19}\text{ J/eV})} = \boxed{10^{-4}\text{ eV}\cdot\text{s/m}}.$$

25. Strategy According to Eq. (27-9), The cutoff frequency f_{max} of x-rays produced by bremsstrahlung (braking radiation) is directly proportional to the kinetic energy of the incident electrons. Since $\lambda_{\text{min}} \propto 1/f_{\text{max}}$, the minimum wavelength of the x-rays is inversely proportional to the kinetic energy of the electrons. Now, by conservation of energy, an electron accelerated through a potential difference is given kinetic energy equal to $e\Delta V$. Therefore, the minimum wavelength of the x-rays is inversely proportional to the potential difference through which the electrons are accelerated.

Solution Find the ratio of the minimum wavelength of x-rays in tube A to the minimum wavelength in tube B.

$$\frac{\lambda_{\text{A}}}{\lambda_{\text{B}}} = \frac{\Delta V_{\text{B}}}{\Delta V_{\text{A}}} = \frac{40 \text{ kV}}{10 \text{ kV}} = \frac{4}{1}$$

The ratio is $\boxed{4:1}$.

MCAT Review

1. Strategy The rest energy of an alpha particle is approximately 3.7 GeV. This is much greater than the kinetic energy of the alpha particle, so $K = \frac{1}{2}mv^2$ is a reasonable approximation of the kinetic energy of the particle.

Solution Compute the approximate speed of the alpha particle.

$$K = \frac{1}{2}mv^2, \text{ so } v = \sqrt{\frac{2K}{m}} = \sqrt{\frac{2(4.8\times10^6 \text{ eV})(1.602\times10^{-19} \text{ J/eV})}{(4 \text{ u})(1.66\times10^{-27} \text{ kg/u})}} = 1.5\times10^7 \text{ m/s}.$$

The correct answer is $\boxed{\text{C}}$.

2. Strategy For each alpha emitted, the nucleus loses two protons and two neutrons. For each beta emitted, the nucleus gains a proton and loses a neutron. Count the number of protons, neutrons, and betas emitted.

Solution A total of $1+4+1=6$ alphas are emitted, so the nucleus loses 12 protons and 12 neutrons. A total of $2+1+1=4$ betas are emitted, so the nucleus gains 4 protons and loses 4 neutrons. The new atomic number is $Z = 90-12+4 = 82$ and the new nucleon number is $A = 232-12-12+4-4 = 208.$ The correct answer is $\boxed{\text{A}}$.

3. Strategy There are three protons and $7-3=4$ neutrons in the nucleus of lithium-7. Use Eq. (29-8) to find the binding energy.

Solution Compute the approximate binding energy.

$$E_{\text{B}} = (\Delta m)c^2 = [(3\times1.0073 \text{ u} + 4\times1.0087 \text{ u}) - 7.014 \text{ u}]\times931 \text{ MeV/u} = 40.0 \text{ MeV}$$

The correct answer is $\boxed{\text{D}}$.

4. Strategy The external magnetic field only exerts forces on moving charged particles.

Solution Alphas are positively charged, betas are negatively charged, and gammas have no charge. Therefore, gamma rays travel straight and alpha and beta rays are bent in opposite directions. The correct answer is $\boxed{\text{D}}$.

5. **Strategy** The nucleus decreases by $|\Delta m| = E/c^2$, where E is the energy of the gamma ray.

   Solution Compute the decrease in mass of the nucleus.

$$|\Delta m| = \frac{E}{c^2} = \frac{(2.5\times10^6\ \text{eV})(1.602\times10^{-19}\ \text{J/eV})}{(2.998\times10^8\ \text{m/s})^2} = 4.5\times10^{-30}\ \text{kg}$$

   The correct answer is $\boxed{\text{C}}$.

6. **Strategy** Use Eqs. (29-21) and (29-22).

   Solution Find the original activity R_0 of the sodium-24 sample.

$$R = R_0e^{-t/\tau} = R_0e^{-t\ln 2/T_{1/2}},\ \text{so}\ R_0 = Re^{t\ln 2/T_{1/2}} = (100\ \text{mCi})e^{24\ln 2/15} = 300\ \text{mCi}.$$

   The correct answer is $\boxed{\text{B}}$.

7. **Strategy** Subtract the mass equivalent of the energy required to break the nucleus of the neon-20 atom into its constituent parts from the masses of $Z = 10$ protons and $N = A - Z = 20 - 10 = 10$ neutrons to find the atomic mass of the atom.

   Solution Find the atomic mass of the atom.

$$10\times1.0073\ \text{u} + 10\times1.0087\ \text{u} - 0.173\ \text{u} = 19.987\ \text{u}$$

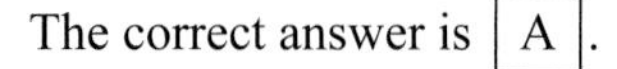

   The correct answer is $\boxed{\text{A}}$.

8. **Strategy** The intermediate product uranium-234 has $238 - 234 = 4$ fewer neutrons than does uranium-238. Count the number of neutrons lost in each sequence.

   Solution Beta emission does not effect the nucleon number A, but does increase the atomic number by one, so choice A is not the correct answer. Alpha emission reduces the nucleon number by four and the atomic number by two. Evaluate the remaining three sequences.

   B: $A = 238 - 4 = 234$ and $Z = 92 - 2 + 3 = 93$.

   C: $A = 238 - 4 - 4 = 230$ and $Z = 92 - 2 - 2 + 1 + 1 = 90$.

   D: $A = 238 - 4 = 234$ and $Z = 92 - 2 + 1 + 1 = 92$.

   The correct answer is $\boxed{\text{D}}$.

9. **Strategy** Every 8 months, the sample has been reduced by half.

   Solution Find the fraction of the sample still remaining after 2 years.

$$2\ \text{yr} = 24\ \text{mo and}\ \frac{24}{8} = 3,\ \text{so the sample has been reduced to}\ \left(\frac{1}{2}\right)^3 = \frac{1}{8}\ \text{of its original amount.}$$

   The correct answer is $\boxed{\text{C}}$.

10. **Strategy and Solution** In gamma decay, the atomic number Z and the mass number A stay the same. The correct answer is $\boxed{\text{B}}$.

11. Strategy The nucleon number A is the sum of the total number of protons Z and neutrons N.

Solution The nucleon number of thallium-201 is $A = 201$, which is equal to the total number of protons and neutrons in the nucleus. The atomic number is $Z = 81$, which is equal to the total number of protons in the nucleus. In a neutral atom, the number of electrons is equal to the number of protons; in this case, there are 81 electrons. There are $N = A - Z = 201 - 81 = 120$ neutrons. The correct answer is $\boxed{\text{D}}$.

12. Strategy The activity R is given by $R = R_0 e^{-t/\tau}$, where R_0 is the initial activity, t is the time elapsed, and τ is the time constant.

Solution The activity decreases exponentially with time. The correct answer is $\boxed{\text{C}}$.

13. Strategy The energy of a photon is related to its wavelength by $E = hc/\lambda$.

Solution Compute the wavelength of the gamma ray photon.

$$\lambda = \frac{hc}{E} = \frac{(4.15\times10^{-15}\ \text{eV}\cdot\text{s})(3.0\times10^{8}\ \text{m/s})}{1.35\times10^{5}\ \text{eV}} = \frac{(4.15\times10^{-15})(3.0\times10^{8})}{1.35\times10^{5}}\ \text{m}$$

The correct answer is $\boxed{\text{A}}$.

Notes